HIGHER POWER

An American Town's Story of
Faith, Hope, and Nuclear Energy

CASEY BUKRO

A MIDWAY BOOK

AGATE

CHICAGO

First printed in June 2023

Library of Congress Cataloging-in-Publication Data

Names: Bukro, Casey, author.
Title: Higher power : an American town's story of faith, hope, and nuclear energy / Casey Bukro.
Description: Chicago : Midway, an Agate imprint, [2023] | Includes index. | Summary: "An in-depth, timely examination of one town's nuclear power plant, the scandal that plagued it, and the reporter who was allowed inside"-- Provided by publisher.
Identifiers: LCCN 2022057949 (print) | LCCN 2022057950 (ebook) | ISBN 9781572843233 (hardcover) | ISBN 9781572848740 (ebook)
Subjects: LCSH: Zion Nuclear Power Station (Ill.) | Nuclear power plants--Illinois--Zion--History. | Nuclear energy--United States--History. | Dowie, John Alexander, 1847-1907. | Bukro, Casey. | Zion (Ill.)--History.
Classification: LCC TK1345.Z56 B85 2023 (print) | LCC TK1345.Z56 (ebook) | DDC 621.48/30977321--dc23/eng/20230111
LC record available at https://lccn.loc.gov/2022057949
LC ebook record available at https://lccn.loc.gov/2022057950

Printed in the United States of America

10 9 8 7 6 5 4 3 2 1   23 24 25 26 27

Author photo by Stuart-Rodgers Photography
Jacket photo by the Nuclear Regulatory Commission

Midway is an imprint of Agate Publishing. Agate books are available in bulk at discount prices. For more information, visit agatepublishing.com.

To my Muses, my guiding spirits: My two daughters, Bevin and Molly, and my best friend and companion, Diana Darnell.

Table of Contents

Atoms for Peace

Determined to solve "the fearful atomic dilemma," President Dwight D. Eisenhower made his "Atoms for Peace" speech before the United Nations General Assembly on December 8, 1953.

"The United States knows that peaceful power from atomic energy is no dream of the future. That capability, already proved, is here—now—today. Who can doubt, if the entire body of the world's scientists and engineers had adequate amounts of fissionable material with which to test and develop their ideas, that this capability would rapidly be transformed into universal, efficient, and economic usage?"

Introduction

IMAGINE A PLACE CALLED HEAVEN ON EARTH, CREATED BY A WORLD-FAMOUS faith-healer and theocrat who demanded that residents follow a holy lifestyle of shunning tobacco, alcohol, newspapers, dancing, gambling, and medical doctors. Police carried Bibles in their holsters and whistling on Sunday was punishable by jail time.

And imagine a place that was famous for sinful wickedness, gangsters, and corrupt politicians. A brawling, hell-raising, muscular, and hard-driving town known as a "City on the Make" where painted women under street gas lamps lured farm boys.

You'd think they were worlds apart, on opposite sides of the planet.

Actually, they are 40 miles apart.

Chicago is that gritty, morally loose metropolis, exactly the opposite of what its heavenly neighbor, Zion, Illinois, was intended to be.

Zion began as a holy city in 1900, the creation of John Alexander Dowie, a flamboyant Scottish Australian evangelical minister dressed in flowing robes with a thick white beard covering his chest, appearing like a biblical prophet only five feet, four inches tall and weighing a portly 200 pounds.

Dowie was not America's first faith-healer, but he was the first to get rich doing it.

Yet Chicago and the City of Zion had something in common: nuclear power.

Chicago invented the atomic age with the world's first sustained nuclear chain reaction at the University of Chicago on June 2, 1942, as part of the Manhattan Project, which desperately was trying in wartime secrecy to be the first to make an atomic bomb. They called it harnessing the power of the atom, a basic unit of matter.

Nuclear power branched into two parts: the first is that devastating bomb, famous for its terrifying mass destruction and signature mushroom-shaped clouds.

The other branch is nuclear electric power generating stations dating back to an "Atoms for Peace" speech in 1953 by President Dwight David Eisenhower, who wanted nuclear energy to be more than a technology for death and destruction.

Both branches are among the most controversial and emotional issues in American life. They came with what we call the atomic age, a monumental technological shift that affects all of us.

This book began as a look at the peaceful, commercial side of nuclear energy and the way it transformed a small town in northern Illinois, the United States, and the world in the 20th century and beyond as scientists work to develop safer nuclear technology at a time of climate change.

Calling it the peaceful side might be wrong, though, since nuclear safety is a contentious topic, but commercial nuclear power touches all of us every time we flip a light switch. Twenty percent of America's electricity comes from nuclear power plants, and one of the big producers in that statistic has long been the state of Illinois. Commonwealth Edison's Dresden Nuclear Power Station, the first privately owned commercial nuclear generating station, sprang to atomic life in Morris, Illinois, in 1960, another reason for the Midwest to stake a claim as the birthplace of commercial nuclear energy.

Though Zion's early city leaders scoffed at science and believed the world was flat, the city north of Chicago eventually became the home of a monument to science: a Commonwealth Edison Co. nuclear power generating station. It was the only nuclear station ever investigated by the U.S. Nuclear Regulatory Commission for allegations of sexual misconduct, drug use, and alcohol use inside that station.

Sex, drugs, and alcohol in the holy city? It's too much of an irony, and one I could not ignore as a staff reporter for the *Chicago Tribune* covering nuclear power in the 1980s. And there were nuclear safety implications if drug-addled operators had their hands on the controls of a nuclear power reactor. I was drawn to this story. I also was drawn to the Zion power station because its superintendent called and wanted to talk to me. I was the *Chicago Tribune*'s environment and energy writer.

The superintendent was distressed by all the media attention his station, on the shore of Lake Michigan, was getting over the vice scandal. He thought that was unfair. "I wish you could see what it's *really* like here," he said. That led to an opportunity most reporters don't get—a 28-month-long inside look at what goes on in a nuclear power plant, including the radioactive zones. Radiation protection training was required.

In this way, I met what I call my nuclear family—the men and women who worked in the Zion nuclear power plant located halfway between Chicago and Milwaukee. It was an uneasy relationship. Many of them hated reporters. They blamed the media for turning the public against them. Some of that hostility intensified when I wrote about the station's worst radioactive contamination accident that occurred when I was in the plant one day. They resented that.

A nuclear power plant is an amazing technological achievement, even more amazing when you get a rare chance to see how it works. But it did not take long to recognize that the power station was in an equally amazing city, where the streets are named for people and places in the Bible and it's still difficult to buy an alcoholic beverage, a lingering vestige of its beginnings as a holy city. This, too, deserved some scrutiny because of its colorful history as a theocracy created through the vision of one very controversial man—John Alexander Dowie. A forerunner of Pentecostalism, his fast-growing Protestant Christian movement emphasized direct personal experience of God. Dowie's innovative faith-healing ministry connected people with God; even his critics agreed that he cured people. His ministry still claims about 15 million followers. They call themselves "Children of Zion," referring to that little town in northern Illinois.

I like to think that beginnings are important to understanding what is happening today and appreciating what came before us as a backdrop. Zion's history deserves to be recognized because of its influence, then and now. Besides, it's a fascinating story that touches on theology and life in an earlier America.

For me, the Zion Nuclear Power Station was part of a lifetime of eyewitness reporting on the 60-year evolution of commercial nuclear energy, including the nation's worst commercial nuclear power plant accident at Three Mile Island near Middletown, Pennsylvania, in 1979, which stalled nuclear power development in the U.S. for 30 years. The Pennsylvania accident is the most consequential event in the nation's nuclear power history, changing governmental policies and regulations and public attitudes toward nuclear energy. That first sustained chain reaction in Chicago in 1942 happened in wartime secrecy. The Three Mile Island accident unfolded for the entire world to see, and it was frightening.

As a reporter, I was among the 150,000 people who fled Harrisburg, Pennsylvania, when federal officials warned that the crippled TMI reactor might explode. Three Mile Island was the nation's closest brush with nuclear disaster, and its specter still haunts America.

No longer the cheapest form of energy as predicted, the nuclear power industry is struggling to survive, despite early predictions that it would make electricity too cheap to meter and change every aspect of American life. Today, critics say commercial nuclear power is in the decommissioning era, when aging nuclear power plants are being demolished and decontaminated so the station's property can be returned to public use. Nuclear advocates say a new class of safer, cheaper atomic reactors now under development will restore the nuclear power industry at a time when climate change requires technology, like nuclear power, that does not add greenhouse gases to the atmosphere.

This book at first fell into three sections, beginning with the life and times of John Alexander Dowie, then describing my time spent reporting from inside the Zion nuclear power plant, and, I thought, ending with the Biden administration's $700 billion Inflation Reduction Act that includes billions of dollars for energy and nuclear power, which critics say is a dead end. That nuclear power plant in Zion, Illinois? It was demolished 15 years earlier than expected after a rocky history unlike any other in the United States.

After a distinguished history as a pioneer in nuclear energy, Commonwealth Edison seemed to lose its nuclear Midas touch, condemned by industry peers as the worst in the nation. Edison transferred ownership of all its nuclear stations to Exelon Corp., which later handed them off to a spinoff company, Constellation Energy Corp.—a double hand-off as though they were radioactive hot potatoes.

Then a new, and unexpected, element of this book suddenly opened like an explosion.

Russia's 2022 invasion of Ukraine was a thunderbolt moment, changing the world outlook on energy supplies, nuclear safety, and food security while nations still grappled with the ravages of the Covid-19 pandemic, climate change, and economic uncertainty.

It marked a new nuclear era, when Russian president Vladimir Putin threatened to do what was thought unthinkable—unleashing nuclear weapons, the evil sister of nuclear technology. Exploding missiles landed dangerously close to two Ukraine nuclear power generating stations in the conflict, creating the potential hazard of blowing open power reactors and spreading radioactive debris across the countryside, similar to the 1986 Chernobyl disaster in northern Ukraine.

The war in Europe, the worst since World War II, caused millions of Ukrainians to flee their homeland, and hundreds of thousands of Russians to flee their country when Putin sought new recruits for the battle, which he had expected to last a few days. War refugees added to the world's immigration problems. As of this writing, the dead were counted in the tens of thousands.

Another example of how world events can change in an instant, altering the landscape of unexpected issues—like nuclear power.

In a way I never imagined, the life and times of John Alexander Dowie hinted at the calamity that eventually swept Zion too, a company town that first was defined by religion, then 70 years later by nuclear power. Both enterprises failed. Dowie fell from grace in scandal, and so did the nuclear power plant. Zion was a place where high ideals came to die.

Part I: John Alexander Dowie and His Holy City

Chapter 1: Despair

FAILURE AND MISFORTUNE DOGGED JOHN ALEXANDER DOWIE EVERY STEP OF the way in his first 10 years as a Congregational church minister, first in Scotland and then in South Australia.

Impatient, uncompromising, and iconoclastic, Dowie wanted to preach to the masses. Instead, after 10 years, he had served as pastor at three small churches and found his flocks unmoved and largely unchanged from their sinful ways by his thundering sermons against tobacco and alcohol, "the dark pit of intemperance." Despite his impassioned sermons, they still smoked and drank the devil's concoction.

Disappointed and disillusioned, the young preacher moved three times, hoping to find somewhere, something, more satisfying and receptive, despite pleas from his congregation to stay.

Dowie revealed the depths of his despair in a letter to his "beloved wife," Jane, whom he called Jeanie, dated March 28, 1882, from Sydney, Australia:

"It is hard and bitter for me to have to write to you today. . . . Once more, I have to write you the discouraging word 'failed.'" Even though he worked hard, the preacher considered himself a wretched failure.

The liquor traffic in Sydney grew stronger, and Dowie grew weaker, physically exhausted, poverty-stricken, and shabby. Alone "in this great, cold city."

Then he added words that would prove to be prophetic: "I will try again in another direction—indeed, I am already at it. . . ."

Clearly, Dowie had concluded that the ways of a conventional preacher were not getting him where he wanted to go, and he needed "another direction" that was not conventional. It was time for a drastic change. He became an independent, big-show healing ministry evangelist.

That was a big leap from the humble beginning where he started, and where the traits that formed him began at an early age.

Dowie was born May 25, 1847, in Edinburgh, Scotland, a town known at the time as "Auld Reeky," Scots for "Old Smoky," because of the smoke and reek spread over it from residential coal fires, industry, and railways. A character in Walter Scott's *The Abbot* says, "yonder stands Auld Reeky—you may see the smoke hover over her at twenty miles distance." Polluted skies marked Edinburgh's location.

Recognized as the capital of Scotland since at least the fifteenth century, Edinburgh was home to about 190,000 people when Dowie was born, the son of John Murray Dowie, a tailor and part-time preacher, and his wife, Ann, an illiterate widow at the time of her marriage. Many of Edinburgh's residents lived in squalor in filthy, overcrowded tenements, often with no water supply and with little or no sanitation. Disease was rampant. Two recent cholera epidemics had swept the area, while typhoid, diphtheria, and smallpox were endemic. To make conditions worse, the winter of 1846–47 was remembered for severe frosts and heavy rains.

Young Dowie was a sickly child whose parents feared more than once that he might die. Birth was a perilous time for mother and child in Scotland at the time, when 120 of every 1,000 infants died in their first year. It was a bad time for babies. Childbirth was a predominantly female event, with neighbors and midwives lending support. As was the custom, the father was present or near the birthing chamber to admit his paternity.

The lad's school attendance was irregular, partly because he was sick so often and partly because his clothes were shabby. But he had a keen mind, often borrowing books from friends, and was considered precocious.

Young John Alexander was a child prodigy, if measured by Wolfgang Amadeus Mozart, who composed music by the age of five. Dowie is credited with reading the entire Bible at the age of six, and he signed a pledge at that age against the use of intoxicating liquors as the temperance movement rose in Scotland. No doubt he was guided by his preacher father, who also was involved in temperance activity. In all the literature describing Dowie's life, nothing much is said about the influence of his parents, except that they were not known "for any unusual accomplishments."

Years later, in a sermon, Dowie thanked God for his lifelong aversion to tobacco and for teaching him a lesson by making him sick as "a wee, wee chap" at six years old.

Playing with friends within sight of Edinburgh Castle, young Dowie had boasted that he knew all about tobacco from watching his father smoke a pipe and declared, "We'll be men." He filled a pipe with Cavendish tobacco, lit it, and took deep puffs.

"By the time I got my third draw, I began to feel—oh my!" He thought something "from the depths of hell had got me now." The castle appeared to be spinning, and he vomited, learning the misery of becoming what he called a "stinkpot," his term for smokers. Upon returning home that evening, the boy's kindly mother recognized he was sick and comforted him. She did not know the reason for his illness, and he did not explain, but the boy secretly blamed his father for setting a poor example.

When John Alexander turned 13, his parents emigrated with him to Adelaide, Australia, where a paternal uncle, Alexander Dowie, owned a shoe shop and was in the import business.

This was a dangerous journey of 12,169 nautical miles of treacherous waters. The intrepid family boarded a sailing ship, the kind that plied the oceans with stacks of wind-filled, billowing canvas sails, at the Port of Leith, adjacent to Edinburgh.

Young John Alexander tutored children on board to make some money. That was all records say about the journey, except that it took an astonishing six months to reach Australia. According to archives, sailing between Great Britain and Australia typically took about 100 days on a fast clipper ship.

Safely, the Dowie family reached the welcoming shores of Adelaide, known as the "City of Churches" for its diversity of faiths, religious freedom, and progressive political reforms. It never was a colony for convicts, a history shared by many Australian communities.

Uncle Alexander Dowie hired the teenaged Dowie, who quit after a few months to become a clerk in a wholesale dry goods firm. The boy got business training and rose to a junior partnership. Many young men might be satisfied with such advancement, but not young Dowie.

A restlessness took hold of him and did not abate for many years. It drove him, as though he had a mission, although he didn't know yet what that mission should be. Business did not suit him. His yearnings lay in another direction. Clergymen like to think he was getting a "call to the ministry."

Oddly, this is another gaping hole in the Dowie saga that otherwise is so full of description. Even Dowie's biographer marvels at the lack of information about Dowie's thoughts at this critical point in his life.

"The writings of Dr. Dowie do not particularly elaborate on God's dealings with him during this time, but it is known that even from early years, he felt a distinct call to God's service," wrote Gordon Lindsay, in biographical accounts that could be described as adoring hagiography.

Maybe John Alexander's call was a whisper, rather than a full-throated summon. He seemed to creep up on it by hiring a private tutor with the money he earned to study for the ministry. He might have been testing the idea.

After 15 months of tutelage, Dowie boarded another ship in 1869 and sailed back to Scotland, where he enrolled at the University of Edinburgh as an arts student, not as a divinity student. The diary he kept on the voyage to Scotland reveals an interest in scripture and evangelism, a fierce dislike for liquor, and an eagerness to discuss its harmful effects. He helped to plan and lead worship services aboard the ship, which would not be unusual for a preacher's son. As a boy, John Alexander followed his father on his occasional preaching outings.

Only sketchy accounts tell of his experience at the university, where he was not regarded as a model student because he quarreled with professors over the dogmatic theology of the day. His classes included Latin, Greek, logic, and moral philosophy. Like other students, Dowie had not entirely made up his mind about his future.

Two learning experiences during his three years at the university stand out, suggesting they made lasting impressions. Dowie took voluntary lessons in the Free Church School, which would have introduced him to a tumultuous time in Scottish religious history known as the Great Disruption of 1843 and the Ten Years' Conflict.

On May 18, 1843, 121 ministers and 73 elders left the national Church of Scotland to form a splinter church, the Free Church of Scotland. This ecclesiastic rebellion came after bitter conflict within the established church and caused havoc in the church and in Scottish civic life. One major issue was the right of patronage, meaning the patron of a parish could install a minister of his choice in a church. The Church of Scotland regarded this a matter of property under the state's jurisdiction. The Free Church held the right of patronage infringed on the spiritual independence of the church, including clerical appointments and benefits. Dowie would have been steeped in this conflict at an impressionable age.

Another lesson Dowie learned at Edinburgh was a lifelong, deep distrust of medicine and medical doctors. As an unofficial chaplain in the Edinburgh infirmary, Dowie attended clinics, lectures, and surgeries by professors and surgeons. While patients were chloroformed, he heard professors "admit that they were only guessing in the dark" about what they were doing, and Dowie saw what surgery could do to a person. He developed a lasting skepticism toward the medical profession and came to offer his healing services as an alternative. But that part of his life was yet to come.

Unexpectedly, a cablegram from his father called Dowie home to Australia, without explanation, cutting short his time at the University of Edinburgh after three years. Upon arrival at home, the youth learned that the firm for which his father was a senior member was going bankrupt. Dowie's student days were over.

Chapter 2: Peripatetic Pastor

UPROOTED FROM HIS UNIVERSITY STUDIES, JOHN ALEXANDER DOWIE DID NOT experience a sudden "call to Jesus" moment in Australia.

Dowie's admirers have said on his behalf that the forced hiatus meant he was ready now to begin work in the ministry, to start preaching the Gospel. But that's not how he acted. He was thinking about returning to Scotland.

Before fully making up his mind, Dowie visited Alma, a small community in South Australia, to look the place over. While there, apparently making a good impression, he was invited to be the Congregational Church's pastor. He declined. If the ministry was supposed to be his dedicated life's work, he was showing a lot of hesitation toward the idea.

But after thinking about it further, Dowie decided that he was being prodded by Divine Providence to take the job. On April 21, 1872, Dowie was ordained into the ministry and became pastor of the Alma Congregational Church. Apparently, he had studied enough theology at the University of Edinburgh to qualify.

When ordained, Dowie was 25 years old. A photo of him about that time shows a stern gaunt man clad in black clerical robes, with a heavy dark mustache and a beard down to his chest, his broad head in the advanced stages of balding. He was short, standing five feet, four inches tall. When photographed with others, he vainly sat or stood in a way that did not call attention to his height.

It's easy to trace Dowie's history and thoughts. He was a prolific letter-writer and preserved his sermons throughout his life.

Work at Alma was divided between several congregations. The central church was located two miles from Alma, which was about 60 miles north of Adelaide.

His ministry included appointments at preaching stations several miles apart.

Upon taking the pulpit, Dowie did what he did best, vigorously denouncing the popular evils of the day, especially the use of intoxicating liquors. It was not an especially popular message among imbibing members of the parish. Dowie was bright enough to detect open resentment toward him. He lasted less than eight months.

In his letter of resignation dated December 5, 1872, Dowie said prayer and divine guidance prompted him to relinquish his office as pastor because "my hopes in accepting your call have not been realized; but I can only view this result as God's appointment." In other words, the congregation failed to live up to Dowie's expectations, and maybe God agreed. The church accepted his resignation "with profound sorrow," and maybe a snicker.

Now that Dowie was committed to the ministry, he tried again in 1873, and "accepted a call" to be pastor of the Manly Congregational Church near Sydney, Australia. He got a warm welcome with a crowded church auditorium. Dowie saw a general impenitence of the population and remarked in a December 3, 1873, letter on "the possibilities of judgment being visited upon the people because of their sins." He saw sin everywhere he looked.

While scolding his flock, the young minister discerned desires of his own. He wanted a wife, a life companion.

In a letter to his parents, Dowie said the good folks of his congregation were trying to find a mate for the lonesome bachelor and introduced him "at least six times to widows and maidens of all sorts." He compared the experience to boys throwing stones at frogs for fun, which was no fun for the frog.

"Seriously though, I am feeling that if I am to settle in New South Wales or elsewhere, I ought to marry, and if I do, I mean to," he wrote. But who? "How can I tell?" He decided to let God decide.

The Bible says, "a good wife is from the Lord," wrote Dowie. "And since I want a good one at all risks, I will ask the Lord to send her to me."

He was not as detached as that might sound. Always a nonconformist, Dowie already had decided he was in love with Jane, his first cousin, whom he called Jeanie. Since it was a highly unconventional choice even at that time, some might argue that God had nothing to do with it.

The young lady learned of the preacher's ardor when he discovered that Jeanie was planning on attending a ball. He wrote a letter warning her against the moral hazards of such a worldly affair and went further, admitting "a very deep and special care" for her welfare.

Jeanie's exact words are not known, but they were curt and essentially told the brash, lovelorn swain to butt out. She was not interested in his thoughts on the matter. Wounded, Dowie began making plans for his next move.

Six weeks later, Dowie became pastor of the Newtown Congregational Church, in a suburb of Sydney. It was 1874 now, and this was the restless preacher's third church in two years.

Sydney would be a life-changing proving ground for Dowie, offering some happiness, some torment, and an inkling of the healing ministry for which he became famous.

Suffering from the cold shoulder he got from his cousin, Dowie buried himself in temperance and social reform work to the point of exhaustion. After recovering, he thought he was in love with another young woman, but decided it was an illusion.

"I cheated myself with a vain illusion of another love at the end of the year, but that soon vanished, a good deal to my pain for awhile, but now I see it was for the best, for it was only a beautiful, transient, desert mirage," he wrote.

Jeanie still was on his mind, and he was convinced that he would be a better minister if he had a wife. He thought and prayed on it. Then his parents told Dowie that his uncle and Jeanie were coming to visit him in Newtown. The young man didn't know how to react to that at first, but decided to play it cool, be agreeable, and no more.

Toward the end of the visit, the weary uncle retired early, leaving Dowie and Jeanie alone to chat in the glow of the hearth in the cozy, humble parsonage. The intimate closeness led to thoughts about what good friends they had been, and how that changed abruptly.

The passage of two years gave Jeanie time to think of Dowie's letter as less of an intrusion and more of a kindness, and a step toward wooing her. She was the first to bring it up. The letter contained good advice, she admitted, opening the conversation to long-suppressed thoughts about how they felt for each other.

Jeanie said she cared very much for the young minister, so much that she would be willing to be his wife, except for one problem: they were cousins.

Now free to express his love for Jeanie, Dowie said the relationship barrier was a mere superstition that could be ignored.

The next day, after waking, Jeanie told her father about the conversation she'd had the night before with young Dowie. He was livid, although he had sensed something was going on between the youngsters, and he did not approve. Just before boarding a steamer back to Scotland, Jeanie told her cousin that her father strongly opposed their marriage.

But the young minister was aflame with a reawakened love for Jeanie and no doubt believed God had answered his prayers.

Even with God as an ally, Dowie realized he needed some earthbound allies who could soften the resistance from his proposed father-in-law—his father's brother. He decided to write to his parents to recruit them.

Convinced now of Jeanie's love, and the "strange intensity" of his love for her, he described that evening in the Newtown parsonage and urged his parents to show his uncle the letter he was writing as a "permanent statement of my feeling regarding Jeanie," and ask him to consider Jeanie's future happiness.

"I know that he is a reasonable man who loves his child greatly," Dowie wrote on, and that the uncle would reconsider the matter if it was "properly laid out before him." The uncle had always been friendly toward young Dowie, who pointed out that the only obstacle was the family relationship.

Dowie turned to biblical scholarship in an attempt to brush that aside. Throughout Jewish law and history, he argued, cousins married cousins, and nobody was more strict and correct than the Jews. The practice was not only permitted, but approved.

"To take an instance," Dowie wrote, "Jacob married Rachel and Leah, his full cousins—and from these were descended the founders of the Jewish nation."

Next, Dowie gave instructions to his parents on how to soften his uncle. Persuasion is among a preacher's tools, and he applied them gently, but with determination.

"Now, father, I constitute you my ambassador to uncle; mother will do her part in a loving way, I know, should opportunity offer, and I beg you as early as you can, have a long chat with uncle about it, presenting this letter as your credentials, and as my plea."

It worked, along with Jeanie's cooperation. The uncle gave his reluctant permission to allow the marriage, which took place May 26, 1876, one day after the groom's 29th birthday, in a quiet religious ceremony. Dowie immediately went back to work, while Jeanie worked to find her place in the community as a preacher's wife.

Chapter 3: Tested by Plague and In-Laws

Now that he was a properly married minister, John Alexander Dowie settled down with his bride, Jeanie, as newlyweds in Newtown.

Jeanie, described as "a young woman of excellent character" by Dowie's biographer, worked to find her place in the row-house community that was rapidly taking shape in what had been a farming area outside Sydney.

It also was Sydney's burial ground. From 1849 to 1868, the Camperdown Cemetery saw 15,000 burials of Sydney's dead. About half of them were paupers buried in unmarked or communal graves, sometimes as many as 12 a day during a measles epidemic.

The cemetery became Newtown's main green space, and a rare example of mid-19th century cemetery landscaping dotted with huge fig and oak trees. The Dowies could not have known that the idyllic greensward was a warning of life-changing events to come.

In a letter to Jeanie before they were married, Dowie already had plotted their lives together as a married couple: "We shall ask God every day to chase all self-love, and self-will, away from our hearts and lives." He claimed it would be "a joyous thing to live the life God's will appoints."

Convinced that a God-driven life would be peaceful, Dowie was in for a surprise. What they got was not joyous. What they got was a scarlet fever epidemic, one of the worst disasters in Australia's history.

Scarlet fever. The words struck terror when the epidemic ripped through Australia between 1875 and 1876, leaving an estimated 8,000 dead, many of them children. This was before antibiotics and public health improvements that eventually snuffed out the disease in much of the world. From 1840 to 1883, scarlet fever was one of the most common infectious childhood diseases causing death in most of the major metropolitan centers of Europe and the United States. Fatality rates reached 30 percent or more in some areas, worse than measles, diphtheria, and whooping cough.

Scarlet fever takes its name from a crimson rash that spreads over a victim's body. The first sign of the dreaded disease was a severe sore throat, known later as a "strep throat," caused by streptococcus bacteria spread via airborne droplets from coughing and sneezing.

Children aged five to 15 years were most vulnerable. With the sore throat came the sudden onset of a headache, chills, a high fever, weakness, and sometimes severe abdominal pain. The scarlet rash appeared 12 to 24 hours later. The rash was described like a "sunburn with goose pimples," with a rough, sandpaper-like texture. After a week, the victim's skin would begin peeling or flaking off.

In the worst cases, all of a family's children died within a week or two. Those who did not die sometimes suffered serious heart, kidney, and ear infections. The disease was so common, it had a macabre nickname, Scarlatina, which appeared in a children's tale as a warning against the rapacious, child-killing disease. Adults, too, fell victim. England and Wales saw major scarlet fever outbreaks between 1825 to 1885.

Contrary to his expectations of a tranquil life as a small-town pastor, Dowie was hit full force by the epidemic in 1876. He tended to disease-tormented members of his church, appalled by what he saw as God's failure to respond to the devastation.

"My heart was very heavy," he wrote of his life at the time, "for I had been visiting the sick and dying beds of more than thirty of my flock, and I had cast the dust to its kindred dust into more than forty graves within a few weeks. Where, oh where, was he who used to heal his suffering children?"

Dowie witnessed strong men "sickened with a putrid fever" who "suffered nameless agonies, passed into delirium, sometimes with convulsions, and then died," leaving families without a father or husband.

Then, "one by one, the little children, the youths and the maidens were stricken, and after hard struggling with the foul disease, they too, lay cold and dead."

During this time, Dowie was in the parsonage study, meditating in sorrow, tearful and praying for help against "the defiler," when he heard the stamping of feet, a loud ring and knocking at the outer door. He found two panting messengers who told him, "Mary is dying; come and pray." Dowie ran with them to the stricken girl's house, and found her "groaning, grinding her clenched teeth in the agony of the conflict with the destroyer, the white froth, mingled with her blood, oozing from her pain-distorted mouth."

Watching the girl suffer at her bedside, Dowie became angry. He called it divinely imparted anger and wished "for some sharp sword of heavenly temper" to attack this cruel disease.

In the room, he encountered a medical doctor, whom Dowie identified as "Dr. K—," who was sympathizing with the anguished mother. According to Dowie, the doctor turned to the minister, saying, "Sir, are not God's ways mysterious?" suggesting God was responsible for the child's illness.

Furious, Dowie responded hotly, "How dare you...call that God's way of bringing his children home from earth to heaven? No, sir, that is the devil's work, and it is time we called on him who came to destroy the work of the devil, to slay that deadly foul destroyer, and to save the child."

Dowie asked the doctor to pray for the child. Offended by the minister's outburst, the doctor encouraged Dowie to calm himself and said, "You are too much excited, sir. 'Tis best to say, 'God's will be done,'" and he left the room.

Turning to the girl's mother, Dowie asked why she sent for him. "Do pray, oh pray for her that God may raise her up," she said. And they prayed together.

In the faith-healing ethos, this scene is mentioned often by Dowie and his followers in the years to come as a turning point in Dowie's career as a healing minister, although even he did not fully recognize it then. Blaming the devil for illness was key, the linchpin, to his healing ministry.

What came next is seen as his first healing miracle. Approaching the unconscious Mary, Dowie "lay hands in Jesus's name on her," following the scripture's advice to "lay hands on the sick, and they shall recover."

According to Dowie, Mary lay still in sleep, so deeply that her mother asked, "Is she dead?" No, said Dowie, "Mary will live, the fever is gone." Soon after, the story goes, Mary woke and drank some cocoa and hungrily ate two slices of buttered bread.

Then Dowie went into another room where Mary's brother and sister laid sick with the same fever. Dowie claimed he prayed for them and they also recovered.

"And this is the story of how I came to preach the gospel of healing through faith in Jesus," Dowie wrote about the Newtown incident. After he healed Mary, the rest of Dowie's congregation had no more epidemic illnesses or death from that day on.

Healing ministries were largely unknown at that time, and even Dowie did not fully understand how to interpret what happened in Newtown until later. In a strange way, he said, "I found the sword I needed [to attack illness] was in my hands, and in my hand I hold it still and never will I lay it down."

But once the idea of divine healing took hold, Dowie explained what he meant. And that included his contempt for anyone who said "God's will be done" in answer to any misfortune. He was outraged by that expression.

"It cannot be for God's glory that any of his children should be unhealed, since God is never glorified in our sickness any more than in our sin, for both sickness and sin are clearly Satan's work," he wrote. "He is glorified in delivering us from sickness, and nowhere is it written he is glorified in sickness."

Dowie blamed St. John for giving a "false impression" that God is glorified by sickness, where the Bible mentions Jesus raising Lazarus from the dead, after he was sick, "for the glory of God." St. John got it wrong, insisted Dowie.

Because he had little formal education in theology, Dowie's beliefs were based mainly on his own interpretation of scripture, though he spent long hours studying the Bible on his own. He was a scriptural literalist: whatever the Bible said was "gospel" or unquestionably true.

The Bible describes 72 accounts of exorcisms and healings performed by Jesus, 41 of which were distinct episodes of healing, many of them before crowds of people who saw it happen. His apostles also are credited with such acts. The early church later sanctioned faith-healing by anointing and by the placing of hands on those being healed. It also is associated with miracles by saints.

The young minister fought the plague to a standstill in Newtown, his small corner of the world. Dowie's thoughts on illness evolved into a mantra: the devil is the father and sin is the mother of illness, physical and mental. It was a good slogan for a religious crusade. Dowie was catching his stride.

But slogans are useless against the agonies of in-law problems. He had no miracle for his father-in-law dilemma.

Pregnant with her first child, Jeanie Dowie went to live with her parents in Adelaide for their tender care while she waited to give birth, leaving the prospective father to fend for himself in Newtown. He made some missteps, including trusting some people on financial matters that left him short of funds. Dowie wrote to his uncle, explaining his difficulties, which turned out to be a mistake.

A three-way exchange of letters between the uncle, Dowie, and his wife resulted in an absolute kerfuffle. Beside telling his uncle of his financial difficulties, Dowie freely admitted he was planning on leaving the Congregational Church to start a Free Christian Church in Sydney.

Uncle Dowie scolded his young nephew and was beginning to suspect his daughter made a mistake in marrying an erratic, ne'er-do-well young preacher who was unable to sink his roots anywhere and apparently did not have a proper appreciation for money. Resigning his present pastorate, where his income was fairly substantial, to start a new church where his income was uncertain, seemed like a bad idea. The uncle also was irked at learning that his son-in-law and daughter had sold household furniture to augment their finances.

In the midst of this domestic upheaval, Jeanie gave birth to a son, Alexander John Gladstone, named for a British prime minister, in the fall of 1877.

In the first of a volley of letters, Dowie told his wife that he did not ask her father for money, and that he would rather go back into business than to ask his in-laws for their help. He claimed he was just explaining his situation to the uncle. Dowie also told Jeanie that he rejected an offer to be pastor at a Congregational church in Waterloo.

Prompted by her father, Jeanie wrote a letter to her husband saying she agreed with her father's criticisms.

Young Dowie turned bitter at what he saw as a betrayal. "I dare say that you thought you were doing a smart thing in writing it, and imparting some very necessary chastisement to a foolish and weak-minded fellow who was too fond of you to resent it; but you missed your aim completely and have only fallen in my esteem as a consequence of your ill-timed and ungenerous smartness. You are not the same wife now as when you left me alone…"

Jeanie and their infant son could stay with her parents, if that was her choice, Dowie wrote, while he set his heart "supremely upon God," ending the letter with: "O Jeanie, you don't know how deeply you have wounded my heart."

The young wife responded with "two long and loving and satisfactory letters," noted Dowie, mending the domestic rift. Each of them apologized to the other. Someone with modern sensibilities might argue that Dowie bullied his wife into agreeing with him.

"Let your heart be perfectly at rest concerning our future, for it is in the best of hands, come what may, I can see the future far more clearly than I can solve the mysteries of the immediate present," Dowie wrote in the aftermath.

As for the immediate future, Dowie saw clearly a rift with the Congregational church and prepared to make another leap to another church. Practice makes perfect, even in the church world.

Chapter 4: Evangelism

THIS TIME, THE LEAP WAS NOT MERELY TO ANOTHER CHURCH BUILDING IN another city.

This time, the fiercely independent clergyman was setting up shop for himself as a solo evangelist, but not before taking a few scornful swipes at the Congregational church he was leaving, an organized church.

Spelling out his disdain in a letter to his wife, Dowie said the church "really killed individual energy, made denominational tools of many ministers, or worse, made them rich and worldly minded men's flunkies, and which separated the churches more than it united them, and then tying them in a heartless union together, left them high and dry and useless for the most part—good ships, but badly steered, and terribly over laden with worldliness and apathy. . . ."

The man liked run-on sentences, and he was developing a habit of bashing anyone who disagreed with him, including other denominations, even those that believed in divine healing. He was a take-no-prisoners kind of minister.

Sheep should follow their shepherd, not the other way around, he demanded, and he thought it was wrong for a minister "to sell and for the church to buy any man's spiritual power or services."

Dowie hungered to preach to the masses, and recognized that included the ignorant, uncared for, and dying people in teeming big cities, where the big money was too.

Moving to Sydney in 1878, Dowie set to work creating a Free Christian Church in a city with a history of cruelty and hardship that would deal him some harsh setbacks.

Sydney began in 1788 as a British penal colony for hardened criminals, only 16 years after the territory's discovery. The first 850 convicts, men and women, arrived in a fleet of 11 ships at Sydney Cove, after another location was rejected because of poor soil and no fresh water.

The 8,254-nautical-mile journey from Great Britain to Australia itself was considered extreme punishment, since up to a quarter of the "passengers" died of sickness en route. Those who survived often were sick and lacked the skills to start a new settlement in a place that lacked housing, agriculture, and planning. Supplies from overseas were scarce.

Early Sydney, named for a British home secretary who authorized the new colony, was molded by the suffering of its early settlers, mostly convicts and their guardians. Together, they fought starvation, drought, and disease. Convicts were forced laborers. Those who rebelled were flogged or hanged. And the European new arrivals added to the local miseries by bringing a plague with them. It's estimated that half of the indigenous native Aboriginal population in Sydney died of a smallpox epidemic from contact with the infected Europeans.

Convict transport to Sydney continued until 1840, when the city's population reached 35,000, only 52 years after the penal colony began.

Then gold was discovered in 1851 only 125 miles from Sydney, triggering one of the biggest gold rushes in world history. Within a year, more than 500,000 people nicknamed "diggers" stampeded into Australia's gold fields. Prospectors came from Britain, the United States, Germany, Poland, China, and other parts of Australia.

Almost overnight, this onslaught of new arrivals changed the character of Sydney's population, which reached 200,000 by 1871, just 20 years after the gold rush started, and included many former convicts who became free citizens by government proclamation.

Prosperity reigned in Sydney, reflected by elaborate temperance coffee palaces, alcohol-free alternatives to corner pubs, and residential hotels for the working man. Coffee was a respectable alternative to the "demon drink"—alcohol. With the coffee palaces came libraries, museums, and transportation to support a growing population while boasting of newfound wealth in the country. Sydney's waterfront consisted of a series of natural bays, making it one of the best harbors in the world.

Into this swirl of hucksters and temperance movement promoters came Dowie, eager to make his mark in 1878.

Money was his first problem. He had none. With his wife's approval, Dowie auctioned off household belongings, including furniture and a treasured collection of pictures, one in particular of a bird by a renowned Australian artist. And they moved into more modest housing.

"My beautiful furniture and pictures were gone," Dowie said later of that episode, "but there came in place of them men and women that were brought to the feet of Jesus by the sale of my earthly goods."

With money from the auction, Dowie rented an auditorium in Sydney's Royal Theater and began preaching to small groups that grew to a thousand in a month. He could not afford to keep paying rent for the Royal Theater, so he moved to another hall, but went into debt. Promises of financial assistance collapsed, but the young minister forged on. His work in Sydney gathered strength and financial aid from donations by new converts grew along with attendance at his services.

Dowie hit upon a new strategy: advertising. He circulated 100,000 printed sheets promoting his healing ministry across Sydney, which also reached the homes of members of various churches. Some pastors objected to this intrusion on their religious territory, one of them calling Dowie's advertising "obnoxious papers."

Dowie responded at length, calling the minister rude. "I consider your judgment to be as feeble and incapable as your ministry," he wrote. "I do not reckon it to be the slightest value, and it would be foolish to be angry or vexed about it much less to be 'filled with indignation,' as you say you were with my 'obnoxious paper.'"

Despite his best efforts, "moral wickedness" and the liquor traffic continued in Sydney, and Dowie was not making as much headway as he'd hoped, although he was a rousing speaker and his talents in the pulpit were recognized.

Impressed with Dowie, members of temperance groups asked him in 1880 to run for a seat in Australia's parliament. At first, he turned them down. His goal was to start a Free Christian Church in Sydney, and he wondered how running for political office might advance that goal.

The temperance groups persisted, and Dowie decided politics might be a way to gain recognition by preaching his brand of evangelism in parliament. That idea appealed to him, so he agreed to run, and lost by a wide margin. He blamed the loss on "liquor interests" he had attacked in his election campaign. "Mammon and Bacchus are the supreme rulers in the political arena here," he complained in a letter to his parents, predicting that those forces would "enchain and drag down fair Australia into the depths of an awful political hell" unless God intervened.

Dowie learned the hard way that politics was not his game. Friends of other candidates offered him money to withdraw from the race, he said, and newspapers printed false rumors that he had withdrawn, affirming his hatred for newspapers, especially those printing unflattering stories about him.

The election effort left him more impoverished and had distracted him from his ministry. He trudged on, into the arms of a confidence man, revealing what his biographer describes as "a strange capacity at times to be deceived."

In short, Dowie, desperate for funds, encountered George Holding, who professed to be wealthy and offered the clergyman $100,000 to build a tabernacle. In a letter, Dowie called the offer like "cold water to a thirsty soul." But it was a scheme to get money from Dowie's relatives, including his father. Dowie never got the $100,000, though he was accused of getting the money and never accounting for it.

Deceived by a swindler, considered a swindler himself, losing money and friends through his failed election attempt, and struggling to provide for his wife and family, Dowie fell into a despair he described in a letter to his wife dated March 28, 1882, four years after he landed in Sydney.

"Beloved wife: It is hard and bitter for me to have to write to you today.... Once more, I have to write you the discouraging word 'failed.' But I live and God lives, and it cannot be that the night will long endure, and that one who strives to do his will shall always fail. I will try in another direction—indeed, I am already at it...."

Dowie does not explain exactly what he meant by that new direction, but goes on to describe his poverty, weakness, "my growing shabbiness," and hunger. Going days without eating, the minister sponged off friends who invited him to dinner, identifying them as "Dr. T" and "Mr. C—, a Christian bookseller."

"I am a good deal thinner, and a little paler, and there are a few more grey hairs in my head, but this is no doubt due to my fasting, added to my sad thoughts and disappointments." Dowie asks for her prayers and for her faith in him.

Dowie sounds like a very desperate man.

Chapter 5: New Direction

If necessity is the mother of invention, so is desperation.

In 1882, Dowie and his family moved to dazzling Melbourne, where he began to perform astonishing miracles that tend to explain what Dowie meant when he said he was moving in a new direction in his divine healing ministry.

Dowie told his wife of receiving deeper spiritual experiences, including a gift of "discerning of spirits" that enabled him to "penetrate into the deepest, most secret thoughts of men."

The minister, now 35 years old, believed he had finished the preparatory phase of his ministry and was on the fringes of something new and wonderful, only 10 years after taking his first pastorate at Alma in 1872.

In Melbourne, 443 miles from Sydney, Dowie's career appears to shift into high gear, with no traces of the frustration and disappointment that dogged him before as a conventional Congregational church minister.

Like Sydney, Melbourne, too, was transformed by the gold fever that swept the region in the 1850s. Its population grew from practically zero to 123,000 during the gold rush, then to 280,000 by 1880 and to 490,000 by 1890. The city on Port Phillip Bay also had ornate coffee palaces and other ostentatious embellishments of newfound wealth. It was named by Queen Victoria after the second viscount Melbourne, British prime minister and political mentor to the young queen.

The Melbourne area had been home to indigenous Australians for at least 40,000 years, but they were ousted and their lands seized by Europeans.

In these rapacious times, the independent evangelist thrived. He dove headlong into a swirling human maelstrom in high motion. In February 1883, he organized the Free Christian Church, achieving a longtime goal. In 1884, construction on a new tabernacle was completed, and Dowie established the International Divine Healing Association. Each year in February, he conducted a convention celebrating the anniversary of his healing ministry. He also published a magazine about healing, called *Jehovah Rophi*.

Neither Dowie nor his followers explain how the penniless minister could afford to build a new tabernacle, except to say that he overcame the devil's fierce resistance and "many wonderful deliverances" were taking place in the Melbourne Tabernacle.

One of those "deliverances," as described by Dowie's biographer, involved Mrs. Lucy Parker, pregnant and blind in her left eye from cancer. Some of Melbourne's best doctors had been treating Mrs. Parker's eye cancer for two years and nine months, and she was in agony. Her doctor feared she would die when she gave birth, or before. Hearing about Dowie, the woman went to the Melbourne Tabernacle to see the healing minister.

"He laid hands upon her and prayed," goes the story. "The miracle happened at once. The cancer burst and discharged into two handkerchiefs. The swelling disappeared and the opening closed. When she opened that eye, she was immediately able to see, and that perfectly." The cancer disappeared and she gave birth to a healthy child a few months later.

"This case of healing was published far and wide in many newspapers, and was never challenged," his biographer notes. "It was miracles of this nature that caused Dr. Dowie's work to achieve rapid prominence in Australia."

But look closely at the Parker "miracle." There is method and stagecraft at work. Expelling a bloody object into a handkerchief and holding it up for an audience to see would be awe-inspiring.

Though Dowie detested newspapers, they were instrumental in spreading the word about his activities. And Dowie was either feeding information to the press or inviting reporters to his tabernacle to see for themselves. That's often how stories get into newspapers. Reporters are suckers for an interesting story, true or not. Let the reader decide, they say. The public did not hear about newspapers that ignored the Dowie story, only about those that bit for whatever reason, including a slow news day.

Consider a second story, about a 16-year-old boy who could not walk and had to be carried everywhere because of "tuberculosis of the bones." When he learned of Dowie, the boy "expected to be healed" that night.

Talking to the boy and "leading him to Christ," Dowie told him to rise and stamp his feet against the floor. The boy did, crying "Oh, praise God, I'm healed, I am healed."

The mother said, "Oh, Arthur, are you healed? My boy, are you healed?"

Arthur answered, "Yes, mother, I am healed." He began walking and said, "I believe I can run," and did, running around the aisles of the tabernacle, likely full of people, although the account does not mention an audience. But that's why tabernacles exist, as a place where people congregate to see and listen to a performance.

The dialogue seems scripted and stilted, but useful for an audience—particularly one that wouldn't wonder how likely it is that muscles atrophied from long lack of use could suddenly spring into a full run.

Clearly, Dowie was operating under a new business model now. Where did that come from?

Dowie had been in the preaching business long enough to recognize that ordinary preaching did not turn people on. But jaded city folks were "greatly taken by big shows of any kind," he wrote in his personal letters, and realized he needed to provide that kind of excitement.

The charismatic minister had loads of charisma and transformed himself into a strident temperance advocate, leading large street demonstrations and singing against the evil "liquor interests." This got him jailed. He was convicted of violating an ordinance against street meetings in Melbourne in 1885.

"The day came for trial and I stood before the court; I gave my reasons for my course. The law of God and the law of England were in my favor, but there was a corrupt petty court judiciary; and the infidels had gotten hold of the Supreme Court." That's Dowie's version.

Since he held street services for two years before the arrest, Dowie believed his attacks on social issues, especially liquor, provoked the authorities. Refusing to pay a fine, Dowie and seven of his followers were jailed for 30 days.

"My people followed me to prison," Dowie claimed, and more than 500 men and women stood up for him in the tabernacle.

Released after serving the sentence, Dowie continued demonstrating in the streets and was arrested again and jailed. But this time, according to the minister, the governor of Victoria intervened because of public pressure after Dowie had served two days.

"Many thousands flocked to hear our preaching when we came out of that prison, and many were saved," he said.

Released from prison, Dowie continued preaching in Melbourne and discovered that because of the publicity he got while in prison, bigger crowds were coming to his tabernacle.

Just when the 38-year-old minister's luck seemed to be turning in his favor, heartbreaking misfortune struck in 1885.

The Dowies had a six-year-old daughter, Jeanie. Dowie recalled her toddling about the house, stronger than he'd ever seen the frail girl he called his "dear little angel." She appeared "happy and bright," but developed one or two spots that looked like measles.

At dinner that evening, the girl was sitting on a maid's lap when Mrs. Dowie called, "Come here, John, and look at Jeanie's eyes." The girl appeared to be suffering from a seizure. Dowie called a neighbor doctor, but the girl died of what the minister describes as "an effusion of the brain."

"Daylight saw only a beautiful, white, marble-like form lying with closed eyes," Dowie wrote to a friend days after the death. Her hands "gently folded on her breast, and a look of holy peace upon her little face, which looked so calm, with the dark hair parted from her placid broad brow."

The minister who had officiated at many funerals now performed one for his little daughter.

"Again, I have stood over the open grave, and laid aside the earthly garments of my little 'angel,' whose spirit quietly stole away just as the day was dawning on Lord's Day morning last. I can scarcely realize it yet, for it was so sudden and unexpected, but I bow, with my dear wife, in resignation, though in grief."

As Dowie saw it, heaven had one more angel.

Grieving, Dowie threw himself into his supercharged temperance work. Only months after his daughter's death, another serious blow awaited him.

In September 1886, Dowie said he was having dreams of himself dead. But he went about his temperance work, preaching and getting signatures to the Christian Temperance pledge. One night, he went to his office in the tabernacle and was dictating to the church secretary when he began hearing voices saying, "Rise! Go!" The secretary denied hearing anything, but the voices continued. Dowie decided to leave the tabernacle office and continue working with the secretary in his stone house, a six- or seven-minute walk away.

While dictating, Dowie said he heard a "strange thud," like an explosion. "That whole part of town had awakened, and people ran out to see what had happened," but saw nothing in the dark night.

The next morning, upon arriving at the tabernacle, Dowie found that a wall of his office was blown out and the wreckage of the wall and his office furniture scattered around. Dynamite had been placed under the spot where he normally sat, and Dowie claimed it "would have been the place of my murder."

He was convinced that the warning voices he heard the night before "were the words of an angel."

Chapter 6: The Man Behind the Curtain

Don't believe everything John Alexander Dowie and his followers said about him. Or about how God acted through him.

In Melbourne, Dowie was performing what amounted to a high-wire healing act that turned him into one of the most famous Australians and Christians of his age.

The minister knew he needed a "big show" style to be successful, and temperance alone was not going to do it. He needed pizzazz, and he studied hard to find it.

From 1880 to 1882, Dowie took extreme interest in séances and Spiritualism, a 19th century form of quackery that convinced paying customers they could communicate with the spirits of the dead through a medium. He attended séances in various cities and "began a thorough examination of all the literature upon the subject that I could find, until I had acquired a very large Spiritualistic library."

Dowie's library most likely included works of early and contemporary faith-healers and other "divine healing" advocates, some of whom might be described as shady. Séances typically required associates using lights and sound effects to create a convincing mystique.

Through close examination and participation with mediums conducting those séances, and conducting some himself, Dowie learned to master those techniques and proclaimed himself an expert, boasting in a sermon: "Now, friends, I made many such tests. I tested these mediums on many points, and I got to know them, and to know their ways so well that, at last, they all held I was the greatest medium of them all."

That explains what Dowie meant in his 1882 letter to his wife, Jeanie, of embarking "in another direction" after earlier failures.

Then he took a giant leap and did what no one else had ever done.

Dowie "invented a new form of faith healing spectacle in the 1880s that was substantively different to all previous forms of divine healing," according to Barry Morton, a research fellow at the University of South Australia, in the state of South Australia. "Where Dowie was original was in conducting healings in front of mass audiences . . . in devising elaborate public performances." His public healing spectacles began in Melbourne in late 1883.

Morton's in-depth research on Dowie and the remarkable history of his ministry provides a more dispassionate look at the complicated figure than those by fawning sycophants, one that is more suspicious and delves into the clergyman's devious methods learned from spiritualists. Morton is no fan of Dowie's and describes him as a con man, a conniving swindler, but one who also managed to do a lot of good. Dowie presents a very complicated picture that is not easily dismissed.

Dowie is credited with being a forerunner of Pentecostalism, a Protestant Christian movement that emphasizes direct personal experience with God. For Dowie, that meant divine healing, with him acting as God's instrument.

Healing is a persistent theme in the history of Christianity, as well as almost all world religions. But Dowie took his cues from Jesus, a healer and an exorcist who cast out evil spirits.

Christian healing is more about a sense of relationship with a divine person than about religious doctrine, and Dowie was one of those religious figures who made it happen. Seeing is believing. If many were healed because they believed, others believed because they were healed. Even his critics agree that Dowie caused healing, perhaps because he came to understand the human mind and the power of suggestion. Other faith-healers were operating at the time, but they received less notoriety than Dowie.

Dowie's genius also lay in combining two beliefs, Spiritualism and the late 19th century "mind cure" movement, which emphasized the healing power of positive emotions, beliefs, and prayer. It worked by manipulating the mind, rather than relying on medicine, allowing the mind to triumph over bodily ailments.

The charismatic minister hit upon what is known as the "placebo effect," improving symptoms through psychological factors like expectations or conditioning and suggestions. A placebo can be some harmless substance, like a sugar pill, given on the pretext that it will cause a cure. Research shows the placebo effect can ease pain, fatigue, depression, and other symptoms because a person expects it will.

"Dowie's primary method of healing relied on placebo cures for psychosomatic illness,"insists Morton,"with most of his success achieved in treating adult women."

The unfortunate side effect of the placebo effect is that it cannot be relied upon for long, says Morton. As the power of "suggestion" wore off, many illnesses the faith-healer cured returned. Dowie would blame this failure on the followers themselves for their immorality, insufficient faith, lack of holiness, or their failure to tithe.

If Dowie healed someone, he publicized their cases and encouraged them to join his church, settle nearby, and pay him a 10th of their monthly income in tithes, which accounted for much of his income. The Divine Healing Association that Dowie started was a one-man operation that strongly encouraged devotees to tithe to continue receiving good health.

Though Dowie said his ministry was modeled after Jesus's, that was not true in one major way: Jesus sternly ordered many who received healings from him, "Do not tell anyone." He did not approve of anyone asking for a cure just for the spectacle of it.

John Alexander Dowie was all about spectacles—the bigger, the better. His healing ministry depended on them. He was a master showman and used his talents to set the pattern for the rest of his life's work. Barnum & Bailey would have admired Dowie's spectacles.

At these performances, Dowie dressed in flowing robes that made him look like a bearded Biblical prophet. Dowie's healings were carefully constructed, highly organized events, writes Morton. To gain a private audience with Dowie, a person had to pass a screening by Dowie's minions. Only Christians who believed in the possibility of divine healing were allowed to see him or given tickets to appear in front of him during public worship ceremonies. Screening allowed Dowie to focus on individuals highly predisposed to a placebo cure. The process singled out those most likely to mentally submit to the process, as well as those most suggestible to it. Belief in the authority of the healer or the healing process is basic to the psychosomatic cure.

These religious ceremonies were models of stagecraft and lively drama. "Dowie's altars were always decorated with large numbers of crutches and canes that were allegedly discarded by the formerly lame and crippled," describes Morton. "Large, emotional crowds of believers worked up by Dowie's fierce oratory made up most of the audience, and guards instantly ejected any visitors who actively questioned or dissented from the process."

In the midst of this religious frenzy, Dowie used deceptions, including one called the "fake cripple." This might be somebody crawling feebly on hands and knees to the stage, or approaching haltingly on crutches. Dowie laid hands on them, and they were instantly cured. Dowie said Jesus cured immediately, and like Jesus, so did he. People should not have to wait for a miracle, he insisted.

Once "the healing" happened, the actor jumped, danced, or ran around the stage praising the lord and shouting, "I am cured!" And another set of crutches was tossed upon the altar as the audience watched in amazement, transfixed and rapturous.

Blind and deaf people also were cured and presented to the audience as proof of another miracle. This could backfire if someone in the audience recognized one of the actors as someone they knew who was perfectly healthy.

Other ploys were false testimony about feigned maladies, demon-possessed epileptics, tumor extraction by sleight of hand (then presenting a jar containing a bloody object for the audience to see), and the "distant cure" for audience members asking Dowie to cure faraway friends and relatives who were not in the audience.

While Dowie engaged in deception, it's not easy to pin him down as a total fake.

"That Dowie was able to cure or alleviate the suffering of large numbers of people is not in doubt—even his greatest detractors admitted as much," writes Morton. Also, without doubt, Dowie was "an original, someone completely different from all preceding religious healers."

But even a holy man is human. Dowie's success was driven in part by desperation and by the shame of in-laws harping about his poor financial situation, making his life miserable. He had to do something about it.

But where did his ideas come from? Who influenced him? Yes, Dowie was a religion innovator. But that's not to say he invented faith-healing single-handedly. There were influences in his life and other clergy men and women practicing what they called faith-healing, or something like it, while Dowie was active in that arena or before.

To be fair, some of them should be named to show that Dowie rose to fame in an era when faith-healing was practiced in the United States and in Europe. Some researchers find it interesting that American religious historians have paid little attention to the healing ministers. Their names are seldom found in leading texts on American religion.

Most likely, Dowie's first influencer was his father, John Murray Dowie, who in 1867 was president of the South Adelaide chapter of the Total Abstinence Society, where his son was an active member.

Dowie worked with the Salvation Army in Adelaide and in Melbourne, but little about that is said by him or his followers.

Edward Irving (1792–1834), a Scotsman and pastor at the National Scottish Church in London, believed pastors should preach healing and said all healing came from God. While at the University of Edinburgh, Dowie was exposed to Irving's teachings and was greatly influenced by him. Irving had attended that university, too.

Johann Christoph Blumhardt (1805–1880) and his son Christoph Friedrich Blumhardt (1842–1919) were German Lutheran theologians credited with faith-healing. The son was especially known as a mass evangelist and faith-healer.

James William Wood (1830–1916) worked with Dowie for several years around 1884 on healing campaigns in various parts of Victoria.

Charles Haddon Spurgeon (1834–1892) prayed for the sick and preached on divine healing and is credited with healing thousands.

Dorothea Trudel (1813–1862) lived in the remote village of Mamiendorf, Switzerland, where hundreds of travelers went to be cured by her prayers. She started several faith-healing centers.

Ethan Otis Allen (1813–1902) is called the father of the divine healing movement and was the first American to have a full-time healing ministry. He prayed for the sick for 50 years around New England and in 1881 published his book, *Faith Healing*.

R. Kelso Carter (1849–1928) wrote *Miracles of Healing* in 1880.

A.B. Simpson (1849–1919) wrote *The Gospel of Healing* in 1888.

Maria Woodworth-Etter (1844–1924) was a Pentecostal evangelist whose healing meetings drew huge crowds.

George Fox (1624–1691) founded the Society of Friends (Quakers) and prayed for the sick by laying on hands.

Carrie Frances Judd Montgomery (1858–1946) was a woman preacher and faith-healer who promoted faith-healing and Pentecostalism through her writings.

Before, during, and after Dowie's ministry, faith-healing was preached in America and Europe. It is likely that Dowie noticed.

Not only did he notice them, he attacked some of them. Dowie's early healing career began with lectures against "diabolical forces" and "Spiritual Unmasked" harangues, which evolved into "Supernatural Showdowns" that became part of his spectacular healing shows. You could say he jumped to the gray side, embracing some of the activities that he had criticized earlier.

One of his favorite targets was George Milner Stephen (1812–1894), a faith-healer who operated in Sydney and Melbourne by "laying on of hands" and got hundreds of letters testifying to the benefits. In 1883, Dowie repeatedly attacked Stephen, saying his healings came through the devil while Dowie's came through Christ.

Dowie accused competing faith-healers, infidels, spiritualists, and mind healers of being frauds and imposters and staged dramatic onstage battles showing him defeating these enemies of God in supernatural showdowns, cheered by the rapt audience.

He was winning battles, but losing the war against penury. Mass advertising and his healing extravaganzas were expensive and he was deep in debt. Despite his rapid rise to world fame, his career in Australia came to a swift end.

Dowie scammed a businessman into building a church for his congregation and then signing it over to Dowie as his personal property. The insured building soon was destroyed in a suspicious fire that Dowie blamed on "liquor interests." Using insurance money, Dowie paid off his debts.

"One step ahead of the law," writes Morton, "Dowie then decamped to the United States." Which is a fancy way of saying Dowie skipped town, in keeping with his peripatetic past.

A further advantage was that the minister could leave those pesky in-laws far behind him and sail away.

Chapter 7: America

THE "GILDED AGE" WAS GLEAMING IN AMERICA TOWARD THE END OF THE 19TH century, a time of gross materialism, ruthless robber barons who grew rich through monopolies, and blatant political corruption obscured by the golden shine of a nation rolling in prosperity.

Magnates John D. Rockefeller, Andrew Carnegie, Cornelius Vanderbilt, and J.P. Morgan were household names and objects of envy for their enormous wealth. One of the presidents of the era was Rutherford B. Hayes, known for keeping an alcohol-free White House by serving lemonade and restoring popular faith in the presidency.

If following the money was part of John Alexander Dowie's plan, he was coming to the right place. The healing minister and his family steamed into the Golden Gate, the strait that connects San Francisco Bay to the Pacific Ocean, on June 7, 1888, fully 49 years before the Golden Gate Bridge existed.

Victorian San Francisco in 1880 was a city of 233,959 residents, the ninth largest city in the United States, a city of hills, sand dunes, and fog. It also was a city with a gold rush and a population explosion in its history, which was becoming a theme wherever Dowie landed.

By now, Dowie was 41 years old and had grown into the full-blown persona shown in the many photos of him: portly at around 200 pounds, short, bald and with a dense thicket of white hair covering his face down to his chest. He looked older than his actual age. Black brows topped dark, commanding eyes. In appearance and speech, he resembled an Old Testament patriarch. He was a powerful, charismatic speaker with a tenor's voice, a rising and falling cadence, and a slight

Scottish burr revealing the land of his birth.

In public, Dowie wore what fashionable men of good taste wore: black knee-length frock coats over a fitted vest and a white shirt, with loose-cut trousers matching the coat or vest. Mrs. Dowie wore what women thought fashionable at the time, which included heavy and ornate fabrics (similar to cloth used for drapery and furnishings) cut in the vertical "princess line."

With only $75 in his pockets, Dowie and family checked into the Palace Hotel, also known as the "Bonanza Inn," 755 rooms of luxury, each with a private bathroom and an electric call button to summon hotel staff. It was San Francisco's tallest building and the largest hotel in the Western United States.

Perhaps it was here that Dowie developed a taste for opulence, which continued until his dying day.

Reporters looking for a good story featured the healing preacher in their local newspapers, causing a burst of interest. Soon, horse-drawn carriages of the upper crust grand dames from Oakland and Berkeley clopped their way up Market Street to the Palace Hotel.

Large numbers of these fine ladies came to see the miracle man, but it did not go well. Dowie bluntly told them to give up their sins and follies before he would talk to them about healing. They went away, saying the preacher "was one of the most attractive men up to a certain point and then he was a terror."

Still others filled the Palace Hotel corridors, asking to see Dowie, but he refused to pray for any of them, calling them "godless Christian Scientists and church members, and fine-feathered birds with polluted hearts."

Tired and hungry, Dowie ordered his secretary to clear the corridors so he and Mrs. Dowie could go to lunch. Walking away, Dowie passed an old woman in the corridor with a long white crutch made of common pine and a painful, diseased foot.

"There was something in the eye of that old woman that went to my heart," Dowie said in recounting the episode later. "It was a spirit looking out of the windows of a house of suffering. I could not go to lunch." He led the wizened woman and her daughter into a room so they could talk.

Engaging the woman in conversation, Dowie learned that she had to borrow money to pay the $1 fare from Sacramento, about 90 miles away, to San Francisco. She was tall and gaunt, and her non-Christian husband ordered her to go to Dowie for a cure when he read about the minister in a newspaper. "Go!" he said, and "So I have come," she explained.

"Doctor, I am a hard case; my husband is a much harder case; we are very poor and I am very ignorant," describing herself as illiterate "poor white trash" who grew up in the South as a "white-skinned slave" who "was beaten, half starved and cruelly treated by a drunken step-father."

Dowie asked if she was a Christian, and she replied, "I don't know," refusing to believe in anything unless she was sure, but then exclaimed, "Oh, doctor, I want to be sure of salvation," and she asked how to do that, with tears on her face. "Oh, tell me how I can be sure of salvation."

Deeply moved, the preacher said, "She was speaking with a natural, or per-haps I should say, a supernatural eloquence that was irresistible." He was as much impressed by her as she was with the white-haired clergyman.

If Jesus entered the room, said Dowie, would she ask him to heal her, believing that he would?

"Oh, yes, doctor," she replied.

"He is present," said the minister, adding that he was "invisibly present" when she looked around the room for him. Jesus was always present, Dowie explained, and "he is here now in spirit and in power."

"Doctor," the woman said, "I believe he is."

At that point, Dowie kneeled and took the woman's diseased foot in his hand and prayed for healing in Jesus's name. She was crying when he finished the prayer, then Dowie rose and said, "In Jesus's name, rise and walk!" She hesitated, looking for her crutch, which Dowie put out of reach. He commanded her again to rise and walk. She stood up and walked several times across the room.

Dowie described a tearful and emotional farewell. As the woman was leaving, Dowie reminded her she was leaving behind something that belonged to her.

"What?" she asked.

"Your crutch," said the minister, to which she answered, "I don't need it any-more; I am healed." She invited Dowie to keep it, and walked eight blocks to her daughter's home.

The unnamed woman returned in two days to tell Dowie she was walking com-fortably. In the previous two and a half years, she said, she could not walk without a crutch. She vowed to tell everyone.

Believe it or not, this is the sort of story that propelled Dowie to fame and fortune in the United States. It was his first recorded healing on American soil, and featured all the elements needed for a Dowie healing: a strong belief that a cure was possible through faith in Jesus and Dowie's forceful prayers leading the way. The discarded crutch was a leitmotif in the Dowie story, another souvenir from his battle with Satan.

For the next two years, Dowie conducted a series of healing campaigns up and down the Pacific coast: Oakland, San Jose, Los Angeles, San Diego, Seattle, and Portland. These campaigns went so well, the minister decided to stay in America.

About this time, the preacher assumed the title of "doctor." It's what people were calling him, and Dowie apparently decided he deserved the title.

Abruptly, this stage of his ministry ended, as it did in Melbourne, with dark overtones.

Although Dowie funded his lifestyle largely through tithes, he also liked to buy securities in bankrupt companies and sell them to his faithful followers. Two wealthy women whose afflictions were eased by the minister bought such securities and then sued Dowie for fraud. They won their cases in court, and Dowie was fined and disgraced.

These legal and public relations setbacks forced Dowie to consider his options. So he resorted to a familiar strategy: he skipped town.

Chapter 8: An Annoying World's Fair

IT'S EASY TO SEE WHY JOHN ALEXANDER DOWIE LANDED NEXT IN EVANSTON, Illinois, home of the Women's Christian Temperance Union since 1874, and a dry, teetotaling town since 1858.

Dowie might have even been inspired by WCTU president Frances Willard, a fierce advocate of temperance, prohibition, and women's rights. Evanston had no bars or saloons and was nicknamed "Heavenston," just as its stern Methodist founders liked it.

Located on Chicago's northern border, Evanston was one part of a side-by-side municipal odd couple. Chicago was sin city, fueled by alcohol even during prohibition. Its suburb Evanston was the quiet home of Methodist-founded Northwestern University, and a safe harbor for Dowie.

It's possible Dowie wanted a perch from which to examine Chicago, a bigger challenge than anything he had seen in San Francisco or Melbourne. And if he had larceny in mind, Chicago could be the place for that too.

Years later, author Nelson Algren satirizes 120 years of Chicago history as "a tangle of hustlers, gangsters and corrupt politicians." Dowie might have been deciding whether to join the tangle as one of the hustlers.

Dowie made his home in Evanston in the summer of 1890. From 1890 to 1893, he preached in various churches in the Chicago area and made evangelistic tours to cities in the Midwest and Eastern United States and in Canada.

Leaves of Healing, a weekly periodical that reported healing testimonials and Dowie's teachings, boosted the minister's visibility. The idea for such a periodical first occurred to the preacher in 1886, and eventually gained worldwide circulation.

Dowie had long recognized the power of advertising, and *Leaves of Healing* proved to be a powerful instrument.

Timing is everything, goes the aphorism, as well as seizing opportunities when they come along.

Dowie knew plans were taking shape for the 1893 World's Columbian Exposition, also known as the Chicago World's Fair, to celebrate the 400th anniversary of Christopher Columbus's arrival in the New World in 1492. This was only 22 years after the Great Chicago Fire of 1871 had reduced much of the city to ashes, killing 300 residents, destroying 18,000 buildings, and leaving 100,000 of the city's 300,000 inhabitants homeless. Municipal leaders wanted to show that the city was rising from those ashes, like a phoenix, and well on its way to becoming the toddling town that Frank Sinatra would sing about.

Covering 690 acres of former swampland, the exhibition featured nearly 200 new buildings of predominately neoclassical architecture, canals and lagoons, a large pool symbolizing the ocean that Columbus crossed, and people and cultures from 46 countries. It was an influential social and cultural event with profound impact on architecture, sanitation, the arts, Chicago's self-image, and American industrial optimism.

Since the 1850s, Chicago has been one of the dominant cities in the midwestern United States, becoming a major transportation hub with the construction of railroads. It became a trans-shipment and warehousing center, followed by factories such as Cyrus Hall McCormick's famous harvester factory.

The festival would attract millions of visitors from around the world, and the cagey minister wanted to be part of the action. Fair organizers would not recognize Dowie's Divine Healing Association as a church, and denied him a place on the fair's Avenue of Churches.

Undeterred, Dowie built a one-story frame structure near the fair's front gates. A sign on its side read "Zion Tabernacle" in large letters. Overhead, a flag fluttered in Chicago's brisk breezes with the words "Christ is All." On the front was a towering two-story sign covering the entire front of the building, reading "International Divine Healing Association" in a radiant semicircular pattern on the top. In the lower right-hand corner were the letters "Rev. John Alex. Dowie." The man knew how to advertise.

The opening service in the new building was held May 7, 1893, days after the fair itself opened to the public on May 1.

In a sermon years later, Dowie said fairgoers streamed past "our despised 'little wooden hut.'" It's odd that he calls it little or a hut, since photos of the place show a sizeable long, rectangular building. "If anybody noticed it, they simply noticed it with supreme contempt and passed on through the gates" to the fair, as Dowie told it, maybe trying to describe himself as an underdog.

"But there were some that looked, and there were some that stepped in, and there were some that listened," he went on. "They were very few, however."

Judging from what Dowie wrote later about the fair, poor attendance was not his main concern. The Zion Tabernacle was across 62nd Street from the amphitheater where Colonel W.F. (Wild Bill) Cody and an assortment of "rough riders" and American Indians performed a world-famous Wild West show featuring loud gun battle enactments.

"Sunday and week day alike it was our misery to be compelled to hear the yells of the Indians and shouts of tens of thousands of spectators in the great amphitheater constructed for that show, throughout the whole period of the World's Fair," wrote Dowie. "Oh what agonies we suffered all these long months. In defiance of law, the Sunday was the maddest, wildest day of all the week: for the mayor and the police authorities protected Cody in his disobedience to the laws of God and man." It was sacrilege, Dowie believed, to be performing entertainment on Sunday, the Lord's Day.

"There was no rest for us or anyone near the howling hideous cries of Indians who 'massacred Custer and his cavalry' or 'attacked the stage coach.' Whilst reading or praying, showers of small shot would fall on the tabernacle, or the strains of the Wild West band playing 'Marseillaise' or 'Yankee Doodle' would break forth, in on our hymns."

For six months, for the duration of the fair, Dowie held almost daily meetings in the tabernacle, all in "this diabolical din." But Dowie would have his revenge, in a very odd way. One of the visitors to the fair was Sadie Cody, Buffalo Bill's niece. She fell seriously sick while attending the fair. But more about Sadie later.

Covering 690 acres, the fair attracted 27 million people. Expected to be a celebration of revival and discovery, the fair instead coincided with the panic of 1893, a serious economic depression that lasted until 1897. People rushed to banks to withdraw their money, expecting banks to fail. In such rushes, banks ran out of cash and went bankrupt. These were unsettling times.

The fair closed on October 30, 1893, Dowie commented, "amidst horror, and blood, and ruin, financial and moral on every side," a terrible crash that turned millions of people into beggars.

Through the winter of 1893–94, Dowie's tabernacle drew small audiences, 20 to 40, so small that prayer services were held in a small back room of the tabernacle. The tabernacle was too big for such a small group of people, said the minister, belying accounts that the tabernacle was merely a hut.

For Dowie, it was a "dark, terrible winter." The frigid blasts off Lake Michigan added to his misery, since winter is regarded as one of the most pleasant times in Australia, where temperatures in Sydney range from the 40s to the 60s (in degrees Fahrenheit) that time of year.

During that winter Dowie reminisced over how far he had fallen; he had sometimes preached to 20,000 people in Australia. He got large audiences when he preached in Canada and in Eastern United States, each larger than the one before. Not in Chicago. But his faith did not waver. He prayed for a break in this streak of being largely ignored—and then it came.

"The mighty power of God descended upon us," Dowie recalled in a sermon. "One after another people were brought from long distances and were wonderfully healed. We moved back into the main part of the tabernacle again." Long before meetings began, people crowded into the aisles of the tabernacle.

"For months people stood in snow or sat on improvised seats and stood where they could hear if they could not see. God blessed and the revival of his work has been going on from that hour to this."

Chapter 9: Chicago

Growing attendance at his healing events convinced Dowie that Chicago was his kind of town.

By word of mouth and through the *Leaves of Healing* publication, stories spread across Chicago and the world that Dowie was causing amazing miracles of healing. Hundreds of people were coming to the city to be healed, and encountering difficulty finding lodging.

Dowie leased and furnished several large rooming houses that were converted into healing houses, where people paid to stay and get meals, while receiving spiritual encouragement and to be close to evening healing services.

The diminutive minister struggled for a great revival of primitive Christianity through the ministry of healing.

As usual, Dowie attacked alcohol, tobacco, doctors, newspapers, brothels, gambling, dancing, and other "filthy pleasures of sin." But this was Chicago! Those were lucrative business activities. It was not long before the law and city officials came after him.

Judging from the backlash, you'd think Dowie was Public Enemy Number One. Ministers, medical doctors, and newspapers called him a quack, a fraud, a charlatan, an imposter, and a swindler. This was long before Al Capone became Chicago's most notorious gangster, so Dowie was the menace of the day.

Near the end of 1894, the *Chicago Tribune* reported that the Illinois State Board of Health was preparing to investigate Dowie because he was "practicing medicine without a license."

The *Chicago Dispatch*, which began publication in 1876, attacked Dowie, saying he was operating "a private lunatic asylum where gibbering idiots are confined and from whose keeping Dowie receives a handsome revenue." When their money runs out, according to the *Dispatch*, "the unfortunates are thrown into the streets."

The *Dispatch* also reported that "Dowie's homes are a haven for low prostitutes from the avenues of sin," and that these "women of lost reputation" were "the most objectionable feature of Dr. Dowie's aggregation of freaks."

The same newspaper, on December 13, 1895, carried another story calling Dowie's healing houses "death houses" and falsely accused Dowie of playing a role in a woman's death. Such attacks continued, but the preacher considered them an honor.

If you knew Dowie, you knew he enjoyed a good fight and attacks did not discourage him. He believed he was God's instrument, healing through divine intervention. Those who attacked him, he said, were Satan's henchmen. The more they maligned him, the better he liked it.

"We expected stormy times and they have come," Dowie wrote in *Leaves of Healing* on January 18, 1895. "We have no right to complain nor to be surprised. We sought the conflict with the powers of hell and we have found it. The hellish forces in Chicago are arrayed against us. The devil honors us by howling in pain."

As Dowie saw it, the thousands of sinners who appeared before him repented "the filthy pleasure of sin," so that the devil missed them.

"And so do the saloon-keepers, the drunkard manufacturers. So do also the 'stink-pot makers,' the tobacco vendors. So also do the theaters, the dance rooms, the secret society haunts, and the gambling hells and places of shame. The card table knows them no more in the drawing rooms, and they have no time, taste, nor money for operas, concerts, and lustful music. Hymnbooks have taken the place of the dance and sentimental music. Homes are happy, children are loved, and neglected wives grow young and beautiful again to eyes once bleared with drink and smoke. Howl on ye fiends in every form—your anguish is our joy, and your despair our hope for the captives yet in your dungeons of death."

The battling preacher had no sympathy for those complaining about him, such as medical doctors, surgeons, and druggists selling medications.

"They have rushed to their comrades of the state board of health, they have summoned their henchmen—mercenary lawyers, the policemen, the press, and the pulpit to save them from the wrath to come of a disillusioned people," Dowie wrote in *Leaves of Healing*. "But it is all in vain. The beginning of the end has come."

It was too soon to declare victory. Dowie's opponents thought unfavorable publicity of the kind appearing in the *Chicago Dispatch* would force the minister to shut down his healing houses, but they misjudged. The publicity gave Dowie more prominence.

In January 1895, Dowie and his wife were arrested for practicing medicine without a license, in violation of an Illinois statute, and ordered to appear on January 15 before Judge K. Prineville. The Scotsman hired a lawyer, Anthony Stubblefield, but did most of the talking himself.

You get some idea of Dowie's combative spirit by reading a transcript of the trial, which was printed in the *Leaves of Healing* and described by Dowie's biographer, Gordon Lindsay.

The prosecuting attorney, identified only as Mr. Williams, opens by saying that the minister is charged with treating, operating upon, or prescribing for persons who are under physical disability, without license from the state board of health.

Dowie responds: "We want to know what we are charged with. It is not enough to say that we are guilty of operating or prescribing. We want to know the names of any persons, so that we may be able to deal with specific cases."

When Williams says Dowie has "patients," Dowie interrupts, saying, "We object to the word 'patients.' We call them guests." The guests come to Dowie's home, which he calls Divine Healing Home Number One.

Asked if that's where he treats his guests, Dowie responds: "I don't treat at all." So what does he do?

"I pray for the sick," says Dowie. "I pray to God for the sick in the name of the Lord Jesus Christ. I object to the word 'treatment,'" and later adds, "I pray for recovery." Dowie demonstrates how he lays his hands on his guests.

Williams: "You do it for the purpose of curing the person of that disease?"

Dowie: "No sir. I do not heal anyone. I do it for the purpose of obeying God, who uses me in the healings."

Williams: "You do it for the purpose of effecting a cure?"

Dowie: "Of God effecting a cure. I have never healed anyone, or claimed that I did."

The jousting went back and forth. Dowie testified that divine healing was secondary in his work, that getting people "to give up their sins" came first, and sometimes was necessary before healing took hold. His healing homes, he said, were not hospitals with doctors, nurses, medicine, or treatments.

After the prosecution rested, the preacher commented on the case out of court. He took aim at the prosecutor, saying he was chewing tobacco and spitting on the floor.

"You are sinning by defiling your bodies," Dowie said of the prosecutor and others like him. "I call them 'stinkpots.' I say to them that you may call yourself a Christian, but you do not smell like one. You have no right to ask me to ask God to heal you, whilst you are creating disease by your bad practices."

That message would seem very sensible a generation or so later to a more health-conscious public.

Judge Prineville fined Dowie $100; he refused to pay. The prosecutor filed an order committing Dowie to prison; Dowie appealed. The case against Dowie and his wife was dropped, since a higher court might not see Dowie as much of a nuisance as did Chicago authorities, and there were Constitutional freedoms of religion and speech to consider.

The case before Judge Prineville was the first of many, but it offers insight into Dowie's feisty character and the way he described himself and his ministry. Some said Dowie was arrested 100 times; others said 100 warrants for his arrest were issued. Not exactly the same thing, but they give some idea of the furor public authorities aimed at the Scotsman.

Chicago authorities pressed their attack against Dowie, adopting a Hospital Ordinance aimed directly at him, then charging him with violating it. The ordinance was found unconstitutional.

Meanwhile, testimonials and affidavits of "miraculous healings" poured into Superior Court, praising Dowie, including accounts of people dropping their crutches and walking away restored. One of them involved 10-year-old Willie Esser, a Chicago lad, who had suffered from a withered leg three inches shorter than the other for six and a half years. He went to Dowie, who prayed for him and allegedly pulled the shortened leg down to match the length of the other.

Testimonials from celebrities heightened public interest in Dowie. They included Amanda M. Hicks, a cousin of Abraham Lincoln, and Sadie Cody, niece of wild west figure Buffalo Bill Cody. Remember Sadie? She fell ill while attending her uncle's show during the Chicago World's Fair.

According to a testimonial from Sadie Cody, who lived in Rensselaer, Indiana, for nine months after her visit to Chicago, her condition worsened and she became helpless. Five vertebrae were worse than useless, an abscess as large as her fist grew at the base of her spine, she developed a tumor, and one limb was three inches shorter than the other. Doctors told her they could do nothing except to put her in a plaster body cast.

Upon seeing a copy of *Leaves of Healing*, Ms. Cody asked to be transported to Chicago, where Dowie prayed for her in Healing Home Number Three. All her symptoms disappeared and both limbs were of equal length.

This healing especially delighted Dowie, who considered it revenge for all the suffering Buffalo Bill and his Wild West show caused him during the World's Fair.

"Ah! She is a Cody; a relative of Buffalo Bill Cody, and we have had our revenge on him and the Wild West show," Dowie wrote in *Leaves of Healing*. "He captured Indians and hung their scalps at his belt. We have captured a Cody from the murderous demons of disease, and here she stands as a witness for God, testifying in the very place where Cody's Indians 'massacred Custer' daily."

Dowie thanked his adversaries for causing his believers to flood court records with testimonials. "Not one of these affidavits has been impugned, either by our opponents or by their allies in the Chicago press," he wrote. And they would be preserved in court records. The more Chicago officials fought him, the more famous he became.

Some considered him a fraud, but the numbers of his faithful multiplied. No advertising was needed, since newspapers provided a lot of free publicity.

Chicago authorities were not done with the minister. The Chicago Post Office revoked his second-class mailing privileges, so Dowie traveled to Washington, DC, to talk to the postmaster general, who lifted the revocation. Dowie took the opportunity to complain about "immoral and obscene" publications the Chicago postmaster allows to pass through the mail at second-class rates.

While in Washington, Dowie met briefly with President William McKinley and offered to pray for him, which the president appreciated. Writing about it later, Dowie said he noticed disappointed office-seekers hanging around the White House and thought "President McKinley ought to protect himself a little bit more than he does."

McKinley would be assassinated a few years later in 1901, six months into his second term, while shaking hands with the public, by a disgruntled anarchist. He was the third American president to be assassinated, following Abraham Lincoln in 1865 and James A. Garfield in 1881.

Returning to Chicago, Dowie leased the Chicago Auditorium in the winter of 1895 and filled it with crowds of more than 4,000. In May 1896, he bought the Imperial Hotel on Chicago's Michigan Avenue, close to the heart of Chicago, and turned it into his headquarters. That year, he disbanded the International Divine

Healing Association, formed the "Christian Catholic Church in Zion," and took the title "general overseer."

In February 1897, Dowie moved into an empty church on Michigan Avenue, formerly St. Paul's Episcopal, and called it the Central Tabernacle. Chartered with 500 members, membership soared in central Chicago, where 60,000 people had no church, Protestant or Catholic. Three years later, his church had 6,246 converts; by 1906, there were 23,000.

Dowie preached a social gospel of equality and tolerance, disparaging anti-Semitism and expounding on the plight of African Americans. His Zion Press made a point of publicizing frequent lynchings at the turn of the century. More than 2,500 lynchings, mainly in the South and mostly of African Americans, happened between 1884 and 1900. In 1901 alone, 100 Black people were lynched. The Zion press also noted with approval when Black people were successful in business. The minister called for voting rights for African Americans and was anti-war. In America, he said, "I do detest this hatred of the Jew. It is one of the most shameful and disgraceful things in American life."

In an attack on bigotry, he said: "It is bad to hate the negro because he is black, but if there is a degree of wickedness worse, it is more shameful to hate the Jew because he is rich, when all your salvation comes through the Jew. . . ." (Jesus was a Jew.)

The missionary arm of the church, called the Seventies, organized in 1897. Operating in pairs like Mormons, they spread across Chicago, performing works of kindness in the tenements to scrub floors, cook meals, care for the sick, and distribute food and clothing to the needy.

Dowie criticized Chicago churches for failing to salvage young men before they became drunkards, and the Seventies missionaries visited Chicago saloons and red light districts to speak to those in jeopardy.

An outgrowth of this work was the establishment of the Home of Hope for Erring Women, illustrating Dowie's concern for the practical aspects of the gospel. A building was set aside near the Zion headquarters to rehabilitate homeless girls and women of the street. Although some help for prostitutes had begun nationally, Dowie pioneered in their rehabilitation in an America that was doing little in this line of welfare.

The motto of the Christian Catholic Church in Zion for 1899 was "Go Forward," leading to several important steps. On February 14, Zion College opened its doors. In March, the Zion City Bank and the Zion Land and Investment Association were launched. Concern for young working women gave rise to the "Christian Home for Working Girls," which opened in May 1900. A Bureau of Labor and Relief of the Poor was created to assist employers and employees to care for the needy.

Branch tabernacles were opened in many parts of Chicago, as well as other cities, such as Cincinnati, Cleveland, Philadelphia, and New York and other parts of the world, including Europe, Australia, South Africa, England, and Scotland.

These were golden years for the church.

Michigan Avenue in the vicinity of 12th Street, near downtown Chicago, was becoming Zion territory. It signaled Dowie's interest in laying the groundwork for what he planned to do next, his biggest miracle.

But that did not mean Dowie was finished with fighting the City of Chicago. He declared a three-month "Holy War."

Chapter 10: Holy War

CHICAGO WAS A KIND OF HELL FOR JOHN ALEXANDER DOWIE, THOUGH IT COULD be argued that the torment brought out the best in him.

Now it was payback time. Few clerics of his time strongly challenged the status quo and found it safer to avoid challenging conventional wisdom or the authorities. That was not Dowie's style.

In the fall of 1899, Dowie launched a bitter attack against his "enemies." In a full-page announcement in *Leaves of Healing* on September 30, 1899, he declared his "Three Months Holy War Against the Hosts of Hell in Chicago." Few segments of society were spared his fury. The preacher reminded his followers that the church was supposed to protest wherever sin was found, including "apostate churches," which was sure to incite clerical anger.

In an October 8 pronouncement, Dowie said it was "time that the Baptist Church was utterly smashed." As for Congregationalists, "The lord have mercy on you ... you are living on the Pilgrim Fathers' dust; the brains of a dead theology." For the Presbyterians, "if there is a miserable people on God Almighty's earth, it is you."

Like a bulldog, Dowie attacked and condemned major segments of society like the press, politicians, doctors, freemasonry, and major denominations, especially churches for allowing the use of alcohol, tobacco, and drugs while saying little about social reform.

The Holy War stung these groups and led to violence.

On the evening of October 18, Dowie gave a lecture on "Doctors, Drugs, and Devils," meaning druggists, in a tabernacle on the corner of Madison and Paulina Streets. Advance notices of the lecture were posted by students of Rush Medical

College, suggesting, "We want to give him a hot reception."

More than 2,000 medical students, plus several thousand sympathizers, cursed Dowie and threw bottles of foul-smelling liquids as he stepped from his carriage. An estimated 100 Chicago police officers charged the demonstrators with their batons and arrested many of them, hauling them away in patrol wagons. Some called it a riot.

Officer John D. Shea of the Chicago Police Department assured Dowie that regardless of his views, he had the right of free speech and that he would receive police protection. With all charges against him dropped and the city's Hospital Ordinance declared unconstitutional, essentially all major city opposition against Dowie ended. It was a turning point. Popular sentiment began favoring the battling preacher, and police became friendly. He had battled his opponents to a standstill.

Except for those he newly antagonized by his Holy War.

"Truly, we have drawn upon us the fire of the enemy," Dowie wrote in an October 21, 1899, editorial. "For the doctors of this city, who have subsidized the press, took counsel together and determined that they would stop the rising tide by the most disgraceful and riotous proceedings within their power—hoping, doubtless, that they could in the confusion and darkness seriously injure, or perhaps, destroy our life."

But he also was delighted at getting such a strong response. "Never in all our years of ministry have we felt so supremely joyful and happy, even though sad and sorrowful for those who were doing the devil's work, because we felt beyond all question the lecture had been magnificently illustrated by the facts which all could see, hear, and smell."

Dowie went on to speak in Hammond, Indiana, and Oak Park, Illinois, and faced similar disturbances. A report in the November 4, 1899, *Leaves of Healing* said a "bloodthirsty mob" shouting "kill him" and "do the old fakir up" in Hammond intended to murder Dowie, but he escaped. Accompanied by his wife and son, Dowie spoke at a tabernacle in Oak Park, where about 2,000 to 6,000 high-school-aged youths broke the windows and pelted Dowie guardians with eggs, stale bread, and vegetables.

"In spite of all the pandemonium which was raised by the horns and the crashing of windows," reported *Leaves of Healing*, "the meeting proceeded with scarce an interruption, every word of the general overseer being easily heard by the five hundred present, and listened to with the closest attention."

Much of the crowd dispersed by midnight, but a group of about 100 men was seen lurking nearby, raising fears of an attack. Dowie decided to conduct an all-night vigil of prayer and praise, including testimonials. This continued until about 3 a.m., when the gongs of a police wagon coming up the street were heard.

Unexpectedly, Chicago police sergeant Muldoon appeared at the door, offering the services of a squad of Chicago policemen to escort Dowie home. Followed closely by a police patrol wagon, the Dowie family returned home safely at 5 a.m.

In December, the third and final month of Dowie's insult and hostility onslaught, he aimed his war at the Chicago press, saying its policymakers deserved to be in a penitentiary. The press's politics, religion, and social ethics, he said, were governed by money. In this era of sensational "Yellow Journalism," reformers might have agreed with him. "I do not believe there ever will be a truly honest and God-fearing newspaper in Chicago until Zion prints it," he said, referring to his own church.

Dowie's crusade against Chicago's evils riveted the attention of the public and the press. Newspapers in other cities carried accounts of what the preacher was doing in Chicago.

You could say it was a battlefield smoke screen.

"What have I been doing?" asked Dowie. "I had a holy war for four months at the close of 1899, which the vile press of Chicago will remember. I was so very much engaged in it, day and night, that they never imagined that I could be buying 6,500 acres of land in Lake County as a site for Zion City."

In this telling, the holy war grew from three months to four.

The minister's interest in a Christian utopia was not new. As early as February 8, 1895, he had expressed in *Leaves of Healing* the need for a better location for his work. With a city of his own making, Dowie believed he could finally be able to appeal to the godless masses of large cities, individuals who lacked direction for their lives, especially to the "lost souls" in Chicago.

"My heart was filled with a holy passion for the misguided, ignorant, uncared for and perishing thousands who are in the bondage of Satan in our cities," said the preacher.

With the help of confidantes in 1899, the crafty Scotsman started scouting for a site near Chicago for his dream city. Allegedly dressed as a tramp so he would not be recognized, Dowie looked at farmland north of Chicago with a remarkable cultural and geographic history.

Northern Illinois was home for centuries to many American Indian tribes, the Algonquians, Pottawatomis, Illinois, Iroquois, Sauk, Miamis, and Fox. Under the Treaty of Chicago in 1833, most of them were forcibly removed. At the time of the signing, Potawatomi chief Metea lamented that his people had lost land under a series of treaties.

"You think, perhaps, that I speak in passion; but my heart is good towards you. I speak like one of your own children. I am an Indian, a red-skin and live by hunting and fishing, but my country is already too small; and I do not know how to bring up my children, if I give it all away.... Our land has been wasting away ever since the white people became our neighbors, and we have now hardly enough left to cover the bones of our tribe."

In the wake of that treaty, the Federal Land Grant Law of 1851 granted 2.5 million acres of Illinois public land to the new Illinois Central Railroad for a rail line that would span the entire length of the state. The railroad could sell land to finance the project for $8 to $12 an acre. It was advertised as "the finest farming lands equal to any in the world!!!" and called Illinois "the garden state of America."

The black soil of Northern Illinois is some of the most fertile land in the world. Dark and unusually rich, it's hundreds of feet deep in some places. Water underlies all of Illinois in natural underground reservoirs, a ready source of well water.

Soon German immigrants and Yankee settlers from New England moved in. The German immigrants farmed the land, putting up log houses and split rail fences around crop fields with rooting pigs and chickens in the farmyard. The Yankees were real estate operators, buying and selling land. Farmers often found arrowheads in their fields, mute testimony of earlier residents.

Into this landscape, Dowie and his operatives sought to create a holy city. After rejecting other possible sites, including one in Indiana, Dowie focused on a tract of land about 40 miles north of Chicago. He wanted about 10 square miles.

The minister and his cronies came to believe the site was chosen by God.

In early December 1898, Deacons H. Worthington Judd and Daniel Sloan hired a horse and buggy in Waukegan and surveyed the area about six miles north of the city on an overcast day. When they reached a summit about two and a half miles west of Lake Michigan, the clouds parted and sunshine beamed down on the site for 10 minutes.

It was an omen. "The thought came to us at once, God approves of the selection of this land and how wonderfully plain he has shown it to us," said Judd. The same "omen" happened again during two later visits to the site. The men took 30 snap-shots of the site and told Dowie about what they had seen. He believed the Lord himself had reserved this spot for Zion.

A real estate salesman, E.D. Wheelock, had the job of getting options on 115 parcels of farmland. Dowie and his associates tried to keep their interest as quiet as possible to keep prices for the land from rising.

Many of the farmers were Methodists and balked at selling their land to an outsider. But one by one, they sold. One of the staunchest holdouts was the pastor of the East Benton Methodist Church. Dowie himself approached the minister and offered to buy the church and its land.

The minister refused, until he had only five families left in his flock. Many of them had moved to Utah. The minister reconsidered and sold.

Almost immediately, one of Dowie's associates, Burton J. Ashley, a civil engineer, wrote to more than 50 community leaders, asking for details about successful urban planning that would shape Zion into a model city: drainage, sewage and refuse disposal, landscaping, ideal alignment of streets and boulevards, sidewalks, utilities, alleys, and parks for outdoor recreation and relaxation.

The City of Zion, like Washington, DC, is one of the few cities in the United States that was planned in detail before development began.

Always a showman, Dowie called his followers to a late-night service on the eve of the New Year, 1900, in a Chicago tabernacle. Upon entering the building, they saw a canvas, 25 feet high and wide, veiled by a curtain. As the religious service concluded and the hour drew close to midnight, Dowie grabbed a cord and waited until the clock struck 12 as the city outside celebrated the incoming New Year with horns and whistles.

Dowie jerked the cord, and the curtain fell from the canvas, revealing a map of the proposed City of Zion, where drugs, tobacco, liquor, theaters, brothels, dance halls, and other sinful activities would be banned. It would be heaven on earth.

The audience gasped. But Dowie was not finished. He pulled another cord, and unveiled a huge, detailed painting of what the future City of Zion would look like. Judd, Sloan, and Ashley described the features of the location. The crowd talked until dawn about how this vision would become real. It would cost millions.

On Saturday, January 6, Dowie and 90 of his followers boarded a special train to the vicinity, then explored the property by carriage and wagons.

"Inside of five years," said Dowie, "you will see a city of 25,000 inhabitants here, and in 20 years there will be 200,000 people."

Giving three cheers for the holy city, the group joined Dowie in singing, "Go Forward, O Zion." With a parting blessing to the farmers watching nearby, the minister led a caravan back to the train to return to Chicago.

Chapter 11: A Holy City

Now Dowie's stage for spectacles was as big as a city, Zion, named after Mount Zion in Israel.

The crowning event for the summer of 1900 was the consecration of the 16,000-seat Zion Temple site, which attracted a crowd of about 10,000 church members and friends transported by trains from Chicago.

They marched three times around the temple site, accompanied by a choir wearing white robes. Dowie turned the first spade of sod on July 14, 1900, amid elaborate pageantry, speeches, and prayers.

A bugle called the group to attention, and Dowie made clear who or what would rule Zion. "The rule of the people, by God and for God, is the right rule," he said.

Until winter set in, surveyors and road crews did preliminary work, which resumed in the spring with grading streets, working on the drainage system, and planting trees, adding to the natural groves of maple, oak, and hickory on the west. Teamsters arrived to pull stumps out of the ground and level the surface.

To the east, Lake Michigan's sandy beaches promised future summertime recreation, with boardwalks leading from the city to the lake. Zion workers planned a harbor for boating. By 1900, the business district included a harness shop, a cobbler's shop, coal and lime yards, a barber shop, and a bank.

Via covered wagon, bicycle, horseback, and rail, worshippers came to Zion to be among its first occupants. By July 13, about 2,500 people were camped nearby. On Sunday, July 14, preliminary worship services were held, along with the First Feast of Tabernacles. Some 8,000 people assembled, many brought by Chicago excursion trains.

Fifteen hundred visitors attended a conference presided by Dowie, who explained land sales and how the land should be used. The Land and Investment Association prepared for lot sales. Land was allotted according to the number on stocks. Those who bought shares early would have first choice. Dowie told the listeners to be patient, there would be no favoritism.

In Zion, said Dowie, "we must pool our interests, every man, every dollar, must stand together."

On July 15, 1901, several subdivisions opened and 6,000 lots were offered, triggering an expected stampede at the Zion Land Office. The city consisted of 800 blocks.

An unusual feature aided the land rush: two 75-foot-tall wooden land survey towers topped by platforms, from which prospective landowners could see for miles. They climbed the towers and looked for inviting home sites. Then they waited nearby under the shade of an oak tree. At the sound of a gunshot, they raced to their chosen spot to claim squatter's rights. These first residents claimed hundreds of lots, and investors poured tens of thousands of dollars into the Zion Bank to get the promised high rate of interest.

Building started immediately. Early life in Zion was rugged. Men pitched tents or built temporary shacks for housing. Some moved into a boarding house, or bought homes from nearby farmers who sold their land and moved away. A few workers brought their families with them. The first girl and the first boy born in Zion arrived in October 1900.

The boom was on. A chorus of hammers and saws sounded all hours during the summer months. Officially, Zion's first house was completed in August 1901. Others, some beautiful and sturdy, replaced the temporary shacks.

Everything about the new city was unique. Land could not be sold. It was leased for 1,100 years. Dowie reasoned that Christ would return in 100 years, followed by a 1,000-year reign of Christ after his second coming.

People had a feeling that they were part of a great crusade that was starting a new era on earth, and they were willing to work hard and sweat for that. And they must agree to restrictions imposed by Dowie.

The "Zion City Lease" forbade gambling, theaters, and circuses, including the manufacture and sale of alcohol and tobacco. The lease also banned pork, dancing, swearing, spitting, politicians, doctors, oysters, and tan-colored shoes. Whistling on Sunday was punishable by jail time. Violators of these terms would lose their leases.

Many of the original settlers in Zion City, primarily of Dutch, German, and Irish origin, were attracted to the community because of Dowie's reputation as a faith-healer and to promises that Zion would be corruption-free and an ideal place to raise a family, away from the crime and corruption of major cities.

The second winter forced a lull, but building resumed in the spring of 1902, the year Zion was incorporated. At the first city council meeting on May 6, 1902, Dowie presented a proposed Zion corporate seal, which was adopted. It was a shield with a dove, sword, and crown, all religious symbols, at the top and the word "Zion" on the bottom of the shield. A banner containing the words "God reigns" appeared above the shield.

In his presentation, Dowie said he wanted it to be clear that "God shall rule in every department of family, industry, commercial, educational, ecclesiastical, and political life."

The seal appeared on the city flag, city letterhead, the city council chambers, city vehicle stickers, the shoulder patches of city police officers and fire fighters, the city's water tower, all city street signs, and all city-owned vehicles.

Under Dowie's vision, Zion would not be some kind of bedroom suburb, for residents only. He wanted a self-supporting city with commerce and industry, including a 180-acre farm to feed the growing the community.

That vision quickly took shape.

Zion's first industry was a lace factory, brought to Zion in 1901. English lace manufacturer Samuel Stevenson of Beeston, Nottingham, England, became interested in Dowie after reading a copy of *Leaves of Healing*. He came to the United States, was interested in the idea of a Christian city, and decided to bring his business to America. Stevenson returned to England, bought new lace-making machinery that was sent to Zion, and convinced about 100 of his skilled lace-making workers and their families to come to America with him. Stevenson became a Zion deacon in the spring of 1900.

Administration buildings were among the first to spring up, followed by the Elijah Hospice, a massive four-story frame structure with 350 rooms to house men pouring into Zion to build their own homes. Built in 1902, the structure, later called the Zion Hotel, was 340 feet long and 130 feet wide, making it one of the largest of its kind in the nation. The hotel was demolished in 1979, and only a bandstand with a domed roof remains on the site.

Next in 1902 came the Shiloh Tabernacle, heart of the Christian Catholic

Apostolic Church, Dowie's church, the only church in town and geographical centerpiece of the city. Like many things in Zion, the tabernacle was outsized. It had almost 8,000 seats and an electric pipe organ with 5,000 pipes. The music must have been thunderous, and the singing loud and joyous.

Displayed on the walls of the tabernacle were crutches, canes, braces, flasks, pistols, tobacco pouches, medicine bottles, and brass knuckles surrendered by the healed faithful, souvenirs of winning battles with Satan. Above the souvenirs, in giant stylized letters, are the words "Christ is all and in all," and "I am the lord that healeth thee." An arsonist burned the structure down in 1937.

Also among the early structures in 1902 was Shiloh House, Dowie's $90,000 personal residence, an enormous amount for the time. A three-story red brick and Indiana limestone structure of Swiss chalet design, it is a 25-room mansion now used by the Zion Historical Society as a museum featuring Dowie and the city's early history.

The colorful preacher left his mark on Zion in many lasting ways, including personally naming the streets and boulevards in Zion. Most are named for biblical figures or places. Exceptions are a street named Caledonia, the Roman word for Scotland, Dowie's native country, and Edina Boulevard, an old Roman abbreviation for the City of Edinburgh, Dowie's birthplace.

Construction finished with amazing speed. The Elijah Hospice, for example, containing three million feet of lumber, was done in four months.

The Zion Printing and Publishing House, started in Chicago, moved to Zion. Zion Bank and the Zion Land and Investment Association followed.

Zion Bakery was known nationally for its Zion Fig Bars, still produced today by another bakery. Zion had a candy factory. The massive Zion City General Store offered the latest in clothing, fine pottery and china, and groceries.

Up went the Zion City Power, Plumbing, Lighting, and Heating Association building. The Zion Lumber Yard, the Zion Brick Yard, a railroad depot, and the Zion City Fresh Food Supply appeared. Raised board sidewalks kept residents from walking in the dust or mud. A Zion Radio Station appeared years later to broadcast sermons and religious messages, taking advantage of new technology to broadcast Zion's ideology.

The Zion Building Industry Office kept construction activity organized. All Zion institutions and industries were owned and operated by the Christian Catholic Church, in theory.

Zion residents paid a 10th of their incomes to the church. At the end of each fiscal year, according to some accounts, profits from the flourishing Zion industries and stores were divided among them. Others said the profits actually supported the Christian Catholic Church and its activities, which were controlled by Dowie, along with tithes and offerings.

Dowie's critics now looked differently at the odd little man in flowing robes. It looked like the little miracle worker pulled off an urban development miracle. Meanwhile, Dowie had visions of Zion reaching a population of 200,000 and spreading his ideology worldwide.

Zion was a company town, and the company was the church, and Dowie.

Just as it looked as though things could not get any better, on May 14, 1902, a personal tragedy struck Dowie. His only daughter, Esther, at the age of 21, was curling her hair using a curling iron heated with an alcohol lamp in her bedroom in the Dowie Chicago residence. Dowie had forbidden any form of alcohol in the house.

Esther locked the bedroom door because she did not want her father to know she was using an alcohol-fueled lamp. She accidentally knocked the lamp over, spilling alcohol on herself and turning her into a human torch. She screamed, but the locked door delayed help from reaching her until she could unlock the door. She suffered severe burns over three-quarters of her body and was lucid, talking to her father, asking his forgiveness for disobeying him, until she died later that day.

A crowd of about 7,000 attended Esther's funeral, where people were reminded that it took only a single act of disobedience and "the devil struck her."

Dowie's luck was changing.

Chapter 12: Dowie's Reign and Fall

Zion City brought new meaning to the words "God's country."

Holy living meant all activity stopped twice a day for two minutes of prayer when a whistle blew. People stopped walking in the middle of the street. Teams of horses stood while drivers bowed their heads. Activity resumed when the whistle sounded again. The whistles blew at 9 a.m. and at 9 p.m.

In their homes, families prayed at mealtime and before retiring at night. Each employee in the city, whether in an office, a factory, or in the fields, spent 15 minutes in prayer and worship services before beginning the daily routine. A church official conducted the service.

Residents greeted each other with "Peace to thee," and responded with "Peace to thee be multiplied."

Zion had its own police force, known as the Zion Guard. Dressed in black uniforms with visored caps, they carried Bibles in their holsters. It was probably the only city in the country where law officers began their day on bended knees.

Zion adopted "blue laws" in 1902 that outlawed drunkenness, alcohol, profanity, spitting, gambling, and smoking. The guards mostly nabbed smokers, arresting them and taking them to court to be fined. Jail was mostly for outsiders who might "attempt to disturb the peace." When an old church deacon absent-mindedly spit on the street, police and the court took his age into account and sentenced him to 10 days at home. Guards also made sure everyone went to church on Sunday and served as church ushers.

Rigid adherence to rules and overzealous authoritarianism had their downsides. Families spied on each other and would rise at prayer meetings to denounce and shame a neighbor seen going to a movie in nearby Waukegan, or engaging in other forbidden activities. Dowie encouraged them to report "hidden iniquities." Usually, repentance brought forgiveness.

Overall, though, the joys of feeling they were living in heaven on earth brought a sense of solidarity and unity, enforced by religious processions and celebrations that tied them all together. Thousands of residents wearing Zion's colors—white, blue, and gold—paraded together in the streets, and filled the massive Shiloh Tabernacle from wall to wall to hear Dowie's exhortations. These were spectacles on a grand scale.

Odd thing, though. By 1900, the number of visible miraculous healings began declining. This might have been because Dowie was busy promoting other interests, such as building a holy city. Some of his followers wondered if the minister was straying from his true calling, the healing ministry.

But who could quarrel with the fact that a new city was taking shape about three miles south of the Wisconsin state line? By the end of September 1901, 200 new buildings dotted the landscape near the lakeshore, a city of thousands protected by Bible-packing guards.

Then Dowie did something startling in June 1901. "Zion is to be a theocracy, not a democracy," he declared, thereby becoming the absolute ruler of the City of Zion. In this theocracy, the Bible was the supreme law.

The general overseer began attaching new titles to himself. He decided that he was the "messenger of the Covenant," and the successor to Elijah, a miracle worker and perhaps the most beloved prophet in the Bible, ranking himself among the celestial favorites. Eventually, he called himself John Alexander, first apostle of the Lord Jesus, the Christ, in the Christian Catholic Apostolic Church in Zion, who is Elijah, the prophet of the restoration of all things.

Dowie changed the name of his church in 1903 from the Christian Catholic Church in Zion to the Christian Catholic Apostolic Church in Zion. That might seem to be a small change, but Dowie was exercising power by renaming the church to his personal satisfaction as part of his new, grandiose image of himself.

His church generally is described as Evangelical Protestant. Some said he had visions of becoming something like the pope.

The minister's flowing robes became more flamboyant too, more ornate, multi-layered, and colorful in gold, scarlet, and purple with a turban-like hat. They were similar to the robes of many colors worn by high priests in ancient theocracies.

World religion leaders denounced him, calling him an imposter and a mountebank. Hundreds of ministers lambasted him from their pulpits. Church periodicals carried attacks against him and his ministry of prayer for the sick.

Always sensitive to criticism, Dowie furiously denounced them for denying God's power of healing. He became a master of invective, railing against those who opposed him.

"Thus the broad scope of his ministry which had attracted worldwide attention narrowed down to a strongly sectarian character, practically restricting God's program to his own projects," wrote his biographer. "Subsequently, his preaching gradually deteriorated into a denunciation of his enemies, lectures on political views, exhortations to invest more liberally in Zion's business projects, etc."

In the spring of 1902, Zion headquarters were officially moved from Chicago to the City of Zion, where Dowie's new home was his 25-room mansion, Shiloh House. With their prophet living among them, residents felt a new source of inspiration.

Zion was growing into a boomtown, adding industries and homes at a fast clip, convincing Dowie that he indeed was Elijah the Restorer, and everything must fall in line with his programs or be destroyed.

Feeling invincible, Dowie accepted an invitation to be interviewed by a New York editor, assuming he would be friendly. The story appearing in the *Century Magazine*, written by that editor, described Dowie as a "flamboyant mixture of flesh and spirit" who believed he was the Restorer. If Dowie believes that, wrote the editor, "he is in the moonlit borderland of insanity." Believe it or not, "he is but another imposter."

Stung by what he considered a betrayal, Dowie decided to rent New York's Madison Square Garden and demonstrate to the world the soundness of his mind and mental capacities. No one could attack the Restorer with impunity and get away with it, including that New York editor. He vowed to "restore New York" and bring his message of healing and holy living to the Gothamites.

Hearing that Dowie was coming to New York, the Pittsburgh *Leader* said it was like "the Sodom of Chicago" coming to the "Gomorrah of New York."

On Saturday, October 17, 1903, eight train cars carrying 3,000 members of the Dowie party arrived in New York.

In a story headlined "The sketchy faith healer who tried to save New York from vice," a journalist wrote:

"Dowie's evangelists were instantly recognizable by the black leather satchels they carried and the greeting they habitually uttered: 'Peace to thee.'

"They canvassed tens of thousands of homes in the city, as well as countless saloons, gambling halls, and brothels. They preached on street corners and handed out pamphlets promoting rallies and healing services at Madison Square Garden, which Dowie had rented for two weeks.

"But dissolute New Yorkers were in no mood to be lectured by a horde of bible-thumping hicks from the Middle West. Almost everywhere they went, the Dowieites were greeted with jeers."

The next day, an estimated 14,000 people filled Madison Square Garden for the opening service, the first of several rallies. The processional of robed officers and 500 adult and junior choir members sang "Open Now Thy Gates of Beauty." Then the general overseer began speaking with 1,000 of his congregation sitting behind him as part of the colorful pageant.

Gothamites did not give Dowie the kind of reception he got in the City of Zion.

"The rallies at Madison Square Garden were a catastrophe," the article continued. "Dowie's longwinded and incoherent sermons were frequently interrupted by catcalls and hisses. One especially unruly meeting was cut short by police who feared a riot would erupt. Amid cries of 'Blasphemer' and 'Imposter!' Dowie was forced to flee the arena for his own safety."

The New York *Examiner* dismissed Dowie as "a coarse-grained, low-minded, shame-bereft, money-greedy adventurer," and his followers as "weak-framed, dull-witted creatures who crave a master as a dog does."

Lashing back, Dowie denounced his critics in the press, mainstream ministry, and the audience as dogs, flies, rats, maggots, lice, and pigs.

The New York crusade cost somewhere between $250,000 to $350,000 and was a disaster. Something had gone terribly wrong, and it was not getting better.

Upon his return from New York, Dowie got 14 summonses from Zion's creditors trying to collect on overdue bills, which was reported by the Waukegan *Daily Sun* on November 13, 1903. Dowie immediately assured his flock that the city was financially sound. But that was not true, and the demand for payment revealed the first clue of a fatal flaw in the religious utopia.

The tragic error committed by Dowie at the very beginning of the Zion City venture was that he controlled everything but had no experience in municipal administration or money management and took advice from no one. His followers, known as Dowieites, tended to believe he was overzealous or maybe naïve, but they accepted what he did without question.

Others, like University of South Australia researcher Barry Morton, see something far more sinister in Dowie's insistence on keeping everything under his personal control, away from prying eyes.

"A vital element is missing from our understanding of Zionism," writes Morton. "Put quite simply, the entire Zionist enterprise was founded by a professional con man—John Alexander Dowie—a man who perfected the dark art of 'faith healing' to attract followers, after which he fleeced them for as much money as he could." The City of Zion itself, maintains Morton, "was a carefully devised large-scale platform for securities fraud requiring significant organizational, legal and propagandistic preparation to carry out."

Dowie sold worthless stock in Zion industries, described as legally incorporated when they were not. All members of Dowie's church were told to deposit their funds in the Zion City Bank. Everything in Zion, including the bank, was owned and controlled by the prophet.

Money coming in from Zion industries was deposited in the bank's general fund and spent, instead of being turned over to the industries to pay their employees and buy raw materials and equipment needed to keep those businesses operating. They were going broke. Trying to make up for the deficits, Dowie issued coupons to workers instead of money. These quickly dropped in value as crooks counterfeited the coupons.

Zion as early as 1903 began to get into serious financial trouble. It went into receivership, but since Dowie did nothing to correct the problems, they went from bad to worse.

Major reasons for this financial failure were twofold: the Dowies lived lavishly and the minister was using the Zion City Bank general fund as his personal bank account, spending the money deposited there by Zion residents and by Zion industries. Some inclined to be charitable might call that foolish or unwise, while others might call it criminal.

Dowie and his wife went on world tours, took the most expensive suites in the finest hotels, entertained and bought costly clothes and other merchandise. The minister spent $2,000 a month on personal expenses, which would be $70,550 in purchasing power in 2022. The Dowies also owned a summer home in Montague, Michigan, and entertained hundreds at a time.

This did not go unnoticed.

"The one incomprehensible element in the man's gigantic success is the personal luxury in which he lived, and his superb refusal at the same time to account for any of the sums of money entrusted to him," caustically wrote T.P. O'Connor, an Irish member of the Australian parliament, in Australia's *The Bulletin.*

"His horses are worth a fortune in themselves; his carriages are emblazoned with armorial bearings; his wife is said to dress with the gorgeous extravagance of an empress. When he travels, hemmed round with a little army of servants, the prophet of humility and self-denial has a special train chartered, and whenever the spiritual burdens become too great a tax there is a delightful country residence belonging to him in which to retreat from the clamour and importunate appeals of the faithful."

Always considered a model of rectitude and probity, Jeanie Dowie also was taking some lumps. It was noticed that she had developed a taste for finery and fashion and was going to Paris to shop for her dresses. Some measure of decline in her spiritual life was noticed too, along with influencing her husband to spend $50,000 to furnish Shiloh House, including a replica of a table owned by King Edward VII costing $1,400. Light fixtures and bathroom fixtures designed as silver swans were brought from Europe.

By some accounts, Dowie had amassed a fortune in excess of $10 million, and was getting an annual income of $250,000 from tithes alone. Morton writes that he no longer had to rely on petty swindling.

Dowie made no apologies for his wealth. One of the reasons people were attracted to his ministry was Dowie's gospel of wealth. The minister preached that "poverty is a curse" and that Jesus came to make people rich, not just in the hereafter but on earth too. That was in harmony with American ideals expressed in Horatio Alger books about people of humble beginnings rising to the top. "I am simply a businessman in the ministry," Dowie often said.

In 1904, Dowie took a six-month world tour and returned from that trip with a young deaconess, Ruth Hofer, of Zurich, Switzerland, whose mother supported the Zion church and wanted her daughter educated in the Zion utopia. Dowie built an apartment for Ms. Hofer next to his residence, Shiloh House, and she was seen riding with the First Apostle in his carriage. His attentions toward the young lady were noticed.

This was happening as Dowie appeared to be changing his mind about polygamy. At first opposed to it, the minister reasoned that in a thousand years, morals and manners likely would change. Several of his officers said Dowie in private conversations spoke in favor of polygamy.

Upon his return to Zion, the Scotsman became deeply involved in planning his next major project, a Zion Plantation Paradise in the state of Tamaulipas, Mexico, requiring millions of acres. With a large party, Dowie traveled to Mexico and was received by President Porfirio Diaz, who was interested in development. Dowie saw the project as a chance to enhance his reputation by converting millions in that country.

Also on Dowie's agenda were more Zion cities around the world—four, five, six, seven, or more. And to buy Jerusalem from "the infidel."

Gossip about Dowie's changing attitude toward polygamy led to rumors that he was planning a polygamous colony in Mexico, and that he had even started to collect a harem. A Chicago newspaper reported it with the headline "Harem in Mexico."

Jeanie Dowie was jealous, saying she found the general overseer and the young lady together on several occasions under questionable circumstances. The prophet's wife at first said her husband did advocate polygamy, but later changed her mind and said he didn't. Son Gladestone said his father was a rascal. Ms. Hofer returned to Europe in 1905 to do missionary work. By that time, the Dowies seemed like characters in a French bedroom farce. The damage to Dowie's reputation was irreparable.

In September 1905, the First Apostle was preaching his farewell message in the Shiloh Tabernacle before heading back to Mexico when, according to an account, "suddenly their leader shook his right hand as if some foul thing clung to it. He beat it upon the arm of his chair. Those near him saw him sway." The sudden numbness Dowie tried to beat off signaled that he was suffering a stroke, which would paralyze half his body. He went to Jamaica to recover.

In the aftermath, Dowie denied the allegations regarding Ms. Hofer. His attentions to her were paternal, and his wife and son were disgruntled that he cut their shares in his will. He admitted that he should have taken advice, and had wrongly interpreted the Bible. Finances were mishandled, including Dowie's overdrawn bank account, to the tune of around $475,000. His followers saw poor judgment but no intent to defraud.

Zion officials barely got through the winter without financial failure. Creditors were hounding them for payment of bills long overdue. They pleaded with Dowie to recognize the gravity of the situation, but recognized he was seriously incapacitated.

Back in Mexico now, Dowie, fearing what might happen to Zion if he died, sent a telegram to a man he trusted, Wilbur Glenn Voliva, who was working as the overseer of the church's Australian branch. Dowie told Voliva to move immediately with his family to Zion and take over its administration. He traveled 22 days by water and six days by land to reach Zion, arriving on February 12, 1906.

Beaten by hard times, Zion already was a changed town. Residents were turning away from habits that identified them as Dowie followers. Instead of saying "Peace to Thee," they were greeting friends with "Howdy do."

Voliva, at the age of 36, was heavy-browed with slicked-back dark hair and resembled Napoleon Bonaparte. He had a reputation for being a man of promptness and decision, with management skills. It didn't take long for him to realize that money that should have been invested in Zion industries went to Dowie's personal use. Voliva and investigators contended that $2.5 million to $3.4 million were unaccounted for.

Voliva immediately began reorganizing the administration hierarchy.

Offered Shiloh House as his residence, Voliva preferred the Zion Hospice, saying he wanted to be near the people. Dowie was allowed to live in the landmark mansion.

Meeting with city fathers, Voliva called a public meeting on April 1, 1906, in the Shiloh Tabernacle, which was attended by 3,500 people. Voliva explained why drastic change was needed. Investors were furious at losing their money, and people could not access their savings accounts. Workers were paid with worthless coupons and faced losing their homes.

Revolting against Dowie, some of the congregation accused him of corruption and polygamy. By standing vote, 95 percent of those present backed their new leader, Voliva, who sent Dowie a telegram on April 2. It charged the prophet with "extravagance, hypocrisy, misrepresentations, exaggerations, misuse of investments, tyranny, and injustice." It further read:

"You are hereby suspended from office and membership for polygamous teaching and other grave charges. . . . You must answer these satisfactorily to officers and people. Quietly retire. Further interference will precipitate complete exposure, rebellion, legal proceedings. Your statement of stupendously magnificent financial outlook is extremely foolish in view of the thousands suffering through your shameful mismanagement. Zion creditors will be protected at all costs."

They kicked Dowie out of the organization he created.

Stunned at first, Dowie wired back, suspending all the rebellious overseers involved and canceling authority he'd given to Voliva. He left Mexico and arrived in Chicago on April 10.

Weeks later, Dowie faced Voliva in the Chicago courtroom of Judge Kenesaw Landis, of the U.S. federal court for the Northern District of Illinois. Two weeks of hearings ended with a receiver owning all of Zion's money and property, including its industries, stores, and schools. The church owned nothing. Some property was sold to pay debts. Judge Landis asked why no charges were filed against Dowie, a question that apparently did not occur to Voliva and other Zion leaders.

The court ordered an election by secret ballot to decide who should be Zion's general overseer. Dowie refused to run, realizing sentiment was against him. Voliva was the only one considered shrewd enough and with enough business savvy to save Zion. He won the election.

Dowie felt betrayed, writing at the time that he was aghast "at the strange conduct of my faithless officers who have betrayed my confidence and sought my ruin at a time when I needed their loyal support. In return for my kindness to them, calling them from obscurity to positions of trust, they have led my people in revolt and imperiled the very foundations of Zion, while I was absent and weak in body through excessive toil."

Writing in August 1906, Dowie admitted he was aware that his days as the leader of Zion possibly were drawing to a close. Like Moses, said Dowie, his people turned against him.

A Chicago newspaper at the time wrote about Dowie and the city he had created:

"It is a city hopelessly bankrupt, facing an indebtedness of six million dollars—a city built upon sand. The dream of Zion as conceived by Dowie is gone forever. The vision has faded. A receiver from the courts held the keys to the administration building, the factories, the hospice, the bank. The venture into commercial enterprises, a field in which Dowie was an amateur, overthrew the church. Dowie, sick, suffering from hallucinations, still sat in the Shiloh House. The fire of ambition still burned in his eyes, and a note of defiance still sounded in his voice. But his eyes were deeply sunken, and his voice quavered in disappointment."

In his final days, Dowie was in questionable mental health. He was seen lying on his bed near the second-floor window in Shiloh House, watching crowds pass, saying, "Oh, my people, I love you, though sometimes you are naughty children."

In his final hours, lying on his death bed, Dowie reportedly became delirious around 1 a.m. and began preaching, as he did before throngs of followers in his prime, denouncing sinners with vigor and ordering guards to throw out disturbers. He gradually became weaker. A friend watching his last minutes said Dowie's last words were "The millennium has come: I will be back for a thousand years."

Dowie died March 9, 1907, at the age of 59. He looked much older.

Mrs. Dowie was living in White Lake, Michigan, when informed of her husband's death. She rushed back to Zion to find the estate bankrupt; her husband had left her destitute. Friends bought furniture she did not want, and Mrs. Dowie loaded a railroad car with belongings she wanted taken to Texas, but found she did not have money to pay the freight. She sold more furniture to raise the amount needed.

The widow had hoped she would become Zion's next general overseer, but the job already belonged to Wilbur Glenn Voliva.

Chapter 13: Voliva Turmoil

FOLLOWING IN JOHN ALEXANDER DOWIE'S FOOTSTEPS, WILBUR GLENN VOLIVA declared himself a theocrat and took control of the city as its population dwindled and split into Theocrat and Independent factions, which fought ferociously.

Five competing Christian Catholic Apostolic churches in Zion claimed to be the true church.

In 1911, Voliva got support from his followers to float a $950,000 bond issue to buy back some of the property lost through sales. Once he became sole owner of church property, Voliva launched a colorful campaign of using billboards to proclaim Zion a city strongly opposed to any form of sin. Voliva said it was a battle between the forces of light and darkness.

The new theocrat said there was no room for people with certain undesirable habits, and his sermons, which some said could "take the hide off" a foe, and his billboards let everyone know in plain English what that meant. He called opponents "dirty dogs" and "swine," but ended his sermons with love, joy, and peace.

One of the billboards, for example, said, "No one except a low-down scoundrel, a person lower than the dirtiest dog, yes, lower down than a skunk, would chew or smoke tobacco in Zion City." The language was vividly against sin and vice, stating the founding principles of the city. Zion, said a sign at the city border, was "the only place where it is easy to do right and difficult to do wrong."

The signs also made it clear the Zion was for Zion families only: "A clean city for a clean people." Another mentioned restrictions against "tobacco, whisky, beer, theaters, doctors, drugs, pork, oysters and all other evils." Another said, "dancing, gambling, unclean foods and all other vices strictly prohibited." Some of them bore Voliva's name.

Billboards favored by Voliva's theocrats were placed throughout Zion. To show their resentment, the opposing independents burned them or tore them down, leaving them in scattered pieces. Those two factions brawled over the billboards, their skirmishes becoming known as the "Sign Board Wars."

Most of the bans were dropped in 1921 after a war of words, fists, and billboards.

A row of stores on 27th Street, operated by independents, became known locally as "Rat Row," another sign of intense hostility between neighbors. The intersection of 27th Street and Sheridan Road became a thriving shopping district, but the business community was sharply divided by church affiliations and economic rivalry. Both parties claimed to be Dowie's heirs. The independents wanted more democratic procedure in city politics.

Instead of a city where neighbors lived peacefully, as Dowie envisioned, Zion was a tumultuous battleground of quarrels, violence, disagreement, anger, bitterness, and chaos. It was not heaven on earth.

But now it was Voliva's turn to rule Zion. One of the major differences was that Voliva took advice from a cabinet on commercial affairs and never called himself a prophet. He shunned robes, calling them women's clothing. Born on a farm in Indiana, Voliva was taller than Dowie, clean-shaven and brawny. He wore suits and sported 10-gallon hats.

Voliva's main claim to national fame was his vigorous advocacy of the flat earth doctrine. The earth, said Voliva, was round and flat as a plate, with the North Pole in the center and the earth stationary with other planets moving around it. He tried to prove that with maps, charts, and photographs. Zion promoted the flat earth theory and church schools taught it until the 1940s.

"The idea of a sun millions of miles in diameter and 91 million miles away is silly," he insisted. "The sun is only 32 miles across and not more than 3,000 miles from the earth. It stands to reason it must be so. God made the sun to light the earth, and therefore must have placed it close to the task it was designed to do. What would you think of a man who built a house in Zion and put the lamp to light it in Kenosha, Wisconsin?"

If the earth was round, he said, the ocean's water would flow off into space. Voliva offered $5,000 to anyone who could prove him wrong. The Bible, he argued, was all he needed to know about the earth. The new overseer focused on destroying the "trinity of evils": modern astronomy, evolution, and critical examination of the

Bible and scripture, called "higher criticism."

Voliva was anti-science. He predicted the world would end in 1923, 1927, 1930, 1934, and 1935. He lived on a diet of Brazil nuts and buttermilk and predicted he would live to the age of 120. In 1923, he became the first evangelical preacher in the world to own his own radio station, WCBD, which could be heard as far away as Australia.

Like his predecessor, Voliva developed a lavish lifestyle, amassing a $5 million personal fortune by 1927, which began to alienate his followers.

Once Voliva became sole owner of church property in 1911, he was more determined than ever to return the city to the old, founding ideals. By 1922, Zion was an active community again and industries under Voliva's management were thriving, including the bakery and candy factories. Voliva estimated the value of church and industrial property at $11 million.

Then the Great Depression of 1929 struck and took its toll. Zion industries, owned by Voliva, declared bankruptcy and were put in receivership again. Industries that were not profitable were sold and others were closed. He failed to return the church and the city to their former glory. By 1937, Voliva was bankrupt. He was general overseer of Zion from 1906 until his death in 1942 in a Chicago hospital of a heart ailment at the age of 72.

By then, another generation of Zion dwellers was influencing the course of the city's future. The staunch theocracy ended long before Voliva's death. New overseers still clung to Dowie's ideals, but without the unique theocratic and prophetic views. They were more conciliatory and worked to unite residents with a community-minded philosophy. They ended the wars between battling neighbors and civic strife.

Dowie and his "saints" did end up creating a model community with high family ideals and a spirit of brotherhood and sharing.

Make what you will of Dowie's ministry of healing. Author Philip L. Cook might have said it best: "Finally, Zion is a prime example of that American characteristic we call freedom of religion. The city was free to develop in a country that prides itself on its tolerance and understanding of diversity of thought." It is interesting, says Cook, that American religious historians have given relatively little attention to the healing ministries, those on the fringes of "respectable" denominations.

One of Dowie's legacies might be a closer look at what is considered respectable or traditional in religion. One scholar accuses Western culture of theological imperialism—views of religion not supported by scripture or earlier Christian tradition that modern Western biblical scholars and theologians impose on others.

As theological reflection becomes more global, it's also becoming more supernaturalistic. The reason for changing attitudes in Western societies is the growing influence of "majority world" countries, once known as the "third world." The new description highlights that the developing countries of Asia, Africa, and Latin America have a majority of the world's population, some with local cultures linked to the supernatural. The knee-jerk rejection of miracles by many Western scholars strikes them as a mistake.

Skepticism remains, but there has been a widespread trend since the 1960s toward a greater acceptance of "miracles" and the scientifically unexplainable.

"Christianity is actually moving toward supernaturalism... a vision of Jesus as the embodiment of power, who overcomes the evil forces that inflict calamity and sickness upon the human race," argues Philip Jenkins, professor of history and religious studies at Penn State University. He calls the astounding growth of Christianity in the Southern Hemisphere over the last century "The Coming Global Christianity."

It would be interesting to see what kind of a supernatural utopia a John Alexander Dowie of the future would make of that.

Dowie was an innovator, described as a man who thought out of the box and created many ministries. By the 1950s, the City of Zion was beyond its brother-against-brother struggles, beyond its lunatic fringe image for thinking the world is flat, and beyond multiple bankruptcies.

A man like Dowie would be looking for the next big thing that promised a rebirth in a time of emerging big technology. He would grab it with both hands. But what could that be?

It would be nuclear power.

Part II: The Zion Nuclear Power Plant

Chapter 14: Nuclear Power Comes to Town

THE ONLY THING THAT SEPARATES TODAY'S RESIDENTS OF ZION, ILLINOIS, FROM John Alexander Dowie is time.

Time moves on and Dowie is still in Zion, buried in Lake Mount Cemetery along with the rest of his family. What he created still exists. The City of Zion exists, as do all the streets he named personally. His home, Shiloh House, still stands. Evidence of his time and thoughts remain. In a sense, we shake hands with the past every day.

We also add to the landscape Dowie created; that's called progress.

It's fair to say the biggest thing to happen in Zion since Dowie's days, both physically and ideologically, was a giant, double-domed, twin nuclear reactor plant big enough to produce electric power for about 1.5 million homes.

It was a monument to modern science in a town where its leaders once scoffed at science.

Zion City was the perfect place for that power station, according to the engineering calculus of the time when it was dreamed up in the 1960s. It had available land near Lake Michigan, an ideal water supply, and, as they say in the real estate business, location, location, location.

In this case, that meant a new power plant in Northeastern Illinois would be close to the Chicago metropolitan area, which consumed 80 to 85 percent of the electricity produced by Commonwealth Edison Co., the Chicago-based utility bent on producing electricity by splitting the atom, a promising new technology.

City officials were eager partners, as most would be. The idea of attracting new industry that would produce more tax revenue and create jobs was a winning

combination as far as they were concerned—without considering future consequences like what happens when the plant shuts down, as all eventually do.

"In 1968, nuclear power was a new technology that was to provide low-cost electric power," recalls Al Hill, Zion's mayor from 2015 to 2019. "This was good for Zion,
good for Lake County, good for Illinois and good for the entire country. The City
of Zion cooperated with Commonwealth Edison on this exciting new adventure."

Notice how Hill describes nuclear power as "new," and it was. Westinghouse
Electric Corporation, manufacturer of the two pressurized water reactors built at
the Zion station, boasted of "over 40 years of actual reactor operating experience."
Forty years. The commercial nuclear power industry was still in its infancy.

The City of Zion and Commonwealth Edison struck a bargain. Zion surrendered 331 acres of sand dunes and 2,000 feet of Lake Michigan shoreline
in the far northeast corner of Illinois, about three miles south of the Illinois-
Wisconsin border. The site is about 40 miles north of Chicago and about 42
miles south of Milwaukee.

And it's located in the middle of Illinois Beach State Park, a 4,160-acre lakeshore wilderness of windswept beaches and one of the most popular state parks in
Illinois. Located on the Lake Michigan shoreline, the park is divided into north
and south units, with the Zion station site between them. It is part of a state
coastal management area extending about three miles north of the power station
site and three miles south of it. The northern section of the park was expanded
after the power station already was in place.

Prized as a natural resource, the state park attracts close to 3 million visitors
each year from the Chicago and Milwaukee metropolitan areas to the six-mile
shoreline for boating, hiking, camping, fishing, and swimming.

The Midwest has a talent for putting industry in the middle of natural resources.
About 80 miles east of Zion, the Indiana Dunes National Lakeshore has two steel
mills and the Port of Indiana-Burns Harbor in the middle of it. Political figures who created this marriage of heavy industry and delicate natural resources
argued it would be a showcase demonstrating how the two could exist peacefully
together—setting off decades of environmental battles.

That's another story, but it's a familiar story. Heavy industry and nature lovers
want the same things: open space and lots of clean water. They often end up fighting each other over it.

Commonwealth Edison began eyeing and studying the Zion area in the early

1960s, said Irene Johnson, a Commonwealth Edison spokeswoman in the 1980s.

Land, water, and location were key, she said, and "it's ideal to locate the point of generation close to the point of utilization." To meet the power demands of the Chicago metropolitan area and northern Illinois, Commonwealth Edison had to import power from other generating stations, one 120 miles away, because Edison considered northern Illinois deficient from a transmission standpoint.

"You are talking of bringing in power long distances," explained Johnson, who holds an environmental engineering degree from the Illinois Institute of Technology. Long distance power lines fall victim to storms and accidents, reducing their reliability.

"We only had to build a few miles of transmission systems to link with southern Wisconsin," she said, and part of the national power grid for sharing electricity. "That gave us better reliability."

In the 1960s, most power generating stations burned coal. State and federal environment and safety regulations had not yet caught up with this new thing called nuclear energy. Essentially, the Zion generating station was built to meet standards for coal-burning stations. Public safety often is an afterthought as new technology hits the ground running.

"These were criteria in the 1960s for power plants, not a nuclear plant," Johnson pointed out. "Today, Zion could not meet new criteria. In the 1960s, this was not a major consideration."

Edison announced its intention to construct the first Zion nuclear reactor on February 10, 1967, and announced its intention to build the second reactor on July 11, 1967. Reactor one began operating on December 31, 1973, and reactor two on September 17, 1974.

Built at a cost of $583 million, the Zion station was one of the first of its size in the world, and the first twin-reactor nuclear station in the Edison network featuring two identical pressurized water power reactors. It took 360,000 tons of concrete, 18,000 tons of reinforcing steel rods, and 10,000 tons of structural steel.

The plant was designed to produce enough electricity for a city with a population of about 1.5 million.

Because of changes in governmental regulations and guidelines, the Zion plant would not be built in its chosen location today because it's too close to two large cities, Chicago and Milwaukee. Nuclear disaster drills typically call for public notification and evacuating nearby residents along designated routes, depending on the severity of the accident, or asking people to remain sheltered in their homes.

This would be a challenge for large cities.

From its beginning, commercial nuclear power came under harsh criticism for lax safety standards from environmentalists like Barry Commoner, who pointed to such shortcomings in his 1971 book, *The Closing Circle: Nature, Man & Technology*, only 10 years after the first commercial nuclear power station went into operation.

At the outset, the U.S. Atomic Energy Commission promoted the domestic use of nuclear energy, while also setting safety standards for the construction and operation of nuclear plants. To overcome this governmental conflict of interests, the U.S. Nuclear Regulatory Commission was created in 1975 to protect public health and safety from nuclear hazards.

As a result, governmental regulators began considering what might happen if a nuclear plant accident released radioactive gases and particles, which could be carried long distances by the wind. In this way, such an accident potentially could expose millions of people to dangerously radioactive debris. About 225,000 people live within 10 miles of the Zion station.

"Zion is allowed to operate," Johnson emphasized in the early '80s, "because of the increased safety factor put into the design. But it would not meet today's siting requirements."

As regulators caught up with the potential hazards of nuclear power, they recognized the need to protect large bodies of water that both nature lovers and industry covet.

"Now, no power plant can be built on Lake Michigan, fossil fuel or nuclear," explained Johnson. Before the new safety rules went into effect, six nuclear plants housing nine nuclear reactors were built on the Lake Michigan shoreline, including Zion.

In addition to the recreational charms of Lake Michigan, it is a source of drinking water for millions of people in Illinois, Indiana, Wisconsin, and Michigan. Twelve million people live along Lake Michigan's shores, mainly in the Chicago and Milwaukee metropolitan areas. The U.S. Environmental Protection Agency reports that 9.5 million of those residents get their drinking water from Lake Michigan. The breakdown is 6.6 million in Illinois, 1.5 million in Wisconsin, 888,760 in Michigan, and 485,520 in Indiana. It's a waterway for commercial shipping too. These activities caused governmental regulators to think harder about safety.

Technologically speaking, the lakefront nuclear power plants became lame ducks no longer suited to their locations. Safety restrictions that came after they were built are reminders that enthusiasm for new technology can race ahead of the foresight needed to consider their consequences. It depends on whether society *can* build something or whether it *should*.

Of course it should, said Howard Everline, Zion's feisty mayor from 1983 to 1987.

"The only problem we have are the calamity-howlers from the outside," he said, offering a city official's viewpoint in 1983. "I live very close to that plant. I've lived here 22 years, and I don't get nervous about it at all."

Some local residents accuse Edison of grabbing lakefront property, denying them a chance of having sandy beaches for recreation. That was just a rumor, said the mayor, and he denied it.

"We have no city beach at all, nor did we ever," said Everline. "The City of Zion did not extend all the way to the beach." There was lake access at the end of 25th Street, but he described it as "sort of a trash hole" that was not easily accessible or desirable. Gone, however, was a small lakefront subsection known as Hosea Beach. "When I came here, we had a road back there and homes all through there, and it was called Hosea Beach. There were houses on the shoreline."

In 1968, the City of Zion made way for the power plant by vacating the streets in Hosea Beach, meaning the city no longer owned the streets or maintained them. They were torn up to facilitate power plant construction roughly a mile and a half from the center of Zion.

"It was privately owned land there," said the mayor. "The state, along with Commonwealth Edison, came in and bought up all those houses. Then Commonwealth Edison sold them back to their owners for a dollar to make it legal, on condition they would be moved immediately."

The state of Illinois took away the beaches, insisted the mayor. "The state has taken all the land that ComEd is not using. It has become state park, except for one small area that was owned by the City of Zion." A previous administration gave that land to the Zion Park District.

Everline praised the utility for bringing jobs to Zion and a Westinghouse Power Corporation training center that gave Zion world recognition, with "people coming from all over the world to learn about nuclear power." And he was confident about nuclear safety.

"There is no fear," said the mayor. "We take precautions. My civil defense can evacuate this city like you would not believe. The state gave us a plus rating on their mock drill. They are going to use Zion as an example."

Although emergency drills require public awareness of what to do in an emergency, Everline said, "I think you would find that 95 percent of our people are oblivious to that plant being there. And the other 5 percent are fishermen who like to fish in the warm waters [flushed from the power plant] in the winter, and they don't complain because they are catching fish." The warm water present near the plant in all seasons attracts fish.

One of the perks of having Commonwealth Edison in town, according to the mayor, was that the utility provided electricity to municipal buildings free of charge.

All in all, Everline considered the Commonwealth Edison generating station a success story. It was a step into a new age, the atomic age. Everline believed Zion showed how it can be done.

Nuclear power also brought a new kind of working world to Zion, though that was not foreseen. It brought a cultural shift. The Zion station staff worked around-the-clock shifts in a bluenose community, where they could not find an alcoholic drink in the city until 1999. The Zion Park District golf course got Zion's first liquor license.

Gradually, those constraints loosened, but not much. Even today, Zion does not have a bar or tavern. The city allows the local Piggly Wiggly and Jewel Foods stores to sell liquor. Restaurants may sell alcoholic drinks when served with food. Otherwise, you'd have to go outside the city limits to find a liquor store.

Which is to say, the City of Zion still is in the grip of its temperance-minded beginnings.

Chapter 15: Nuclear Flagship Rises, Stumbles

A<small>NY MAJOR CONSTRUCTION PROJECT IS LIKELY TO HAVE FATALITIES, AND THE</small> massive Zion Nuclear Power Station was no exception.

This was in the late 1960s and early 1970s. Six or seven construction workers met their deaths while building the lakefront plant, according to a Zion supervisor. Commonwealth Edison could not confirm that, saying its records do not go back that far. The Zion station was constructed between 1968 and 1973.

Building the Zion station employed about 1,000 workers at the peak of construction. The Zion station had two nuclear reactors, which Edison described as "nuclear boilers." Each was housed in a concrete and steel structure resembling a silo 150 feet in diameter and 215 feet high. They're called containment buildings, designed to prevent radioactive contamination from escaping in case of an accident.

Some of those workers walking high up near the top of those containment buildings reportedly fell to their deaths through open grating. Occupational safety standards have improved since then, requiring better safeguards when working at great heights.

The work went on, until one of the reactors began commercial operations in 1973 and the other in 1974. The station was Edison's biggest and best—the flagship of Edison's growing nuclear fleet.

Almost immediately, the Zion nuclear station began creating an unusual history. After operating only 18 months, it began getting into serious trouble with federal nuclear regulators.

Then it got worse, causing problems the stunned regulators had never seen at any nuclear power generating station in the United States. The U.S. Nuclear

Regulatory Commission (NRC) fined Commonwealth Edison repeatedly for safety violations caused by human error, but it did no good.

NRC is the federal agency that enforces safe operations at U.S. nuclear power plants and is tasked with protecting public health and safety related to nuclear energy. It began operating on January 19, 1975, as one of two successor agencies to the United States Atomic Energy Commission.

The events that drew the NRC's attention and ire often were peculiar, even bizarre. The first happened in June 1975. An open valve at the Zion station allowed 15,000 gallons of mildly radioactive water to leak into the unoccupied Unit One reactor containment building, forcing an emergency shutdown of the reactor. NRC cited Edison for two other radioactive water spills at its Dresden nuclear station earlier that year and in 1974. In all cases, the spills were traced to operator errors.

NRC officials decided they needed to get Edison's attention by taking strong action. The federal agency called Edison's top executives, Thomas G. Ayers, chairman and president, and Wallace Behnke, executive vice president, to a meeting with James Keppler, the NRC's Midwest regional director, to review Edison's record of "abnormal occurrences" at its nuclear plants.

Keppler warned Ayers and Behnke that continued accidents caused by operator errors would lead to fines or even revocation of permits to operate their nuclear generating stations.

"The tone of the meeting was to inform Edison that we have seen a number of problems happening at the plants and we want them to stop," explained Jan Strasma, an NRC spokesman. "I would not call it a routine meeting. We were carrying the message to the top of the company." Abnormal occurrences were seen as "on the high side," compared with other utilities.

In response, Edison vice president Byron Lee Jr. said, "On the face of it, the operations this year have been doing very well." Abnormal events leveled off, he said, and problems associated with starting up new reactors and bringing them into service were expected.

The next year, NRC fined Edison $13,000 for the radiation overexposure of a Zion supervisor who ignored warnings against entering a highly radioactive area in the Zion station. The unidentified supervisor entered an area under the Unit One reactor while searching for a water leak when the reactor was not operating.

"The man ignored the advice of the radiation protection staff who advised him not to go into the area of the reactor," said NRC spokesman Strasma. In four minutes, the supervisor suffered a radiation dose that was two and a half times higher than the allowable limit for a three-month period. The exposure—8 REMs—was recorded by a radiation detection film badge clipped to his clothing.

The federal agency limits radiation exposures to 3 REMs over a three-month period. In the arcane language of nuclear energy, REM stands for Roentgen Equivalent Man. It is the most common unit used to measure health effects of radiation.

A radiation survey is required before anyone enters a high-radiation area. No such survey was conducted before the supervisor entered the restricted area, another violation of safety procedures in that incident.

Suffering no apparent health effects, the supervisor returned to work the next day. Edison noted the supervisor "took individual action in a violation of established procedures in a well-intentioned attempt to locate a minor . . . leak in a known high radiation area."

Stop and think for a moment. Why would anyone ignore advice against going into a highly radioactive area? Nuclear workers can become indifferent to the hazards of radioactivity, as smokers become indifferent to the hazards of tobacco. Health consequences are not immediate.

It's a mindset worth noticing if you want to understand the Zion station's history and work culture. And the worst was yet to come. It was a blunder that set the NRC's nerves on edge.

In July 1977, operator errors in Zion's reactor Unit Two caused a sudden water pressure surge known as a "water hammer," a loud thud and quake in the plumbing system of the kind that happens in homes when flowing water turns off suddenly. It shook the reactor, causing minor damage to pipe supports and the reactor to trip off.

Trips are part of the automatic safety system that shuts a reactor off when it senses something unusual electronically. Edison does not like to see part of its power generating network go out of service unexpectedly.

While the reactor was shut down as a result of the water hammer incident, station personnel started to perform a routine surveillance test on the reactor safety circuits. They inserted false signals into the system to conduct the test, but neglected to tell reactor operators in the control room.

This resulted in the true condition of the reactor cooling system to be masked by the test signals. Control room operators did not know a test was being conducted and that the conditions they were seeing reported on instrument panels in the control room were not real.

This fakery went on for 40 minutes. It was a classic case of failure to communicate.

The false signals indicated that the water level in the pressurizer, a large tank that maintains pressure in the reactor cooling system, was slightly above normal. Automatic systems began draining water from the pressurizer. Since the false signals indicated the tank water level remained high, automatic draining continued until the tank emptied.

This is another stop-and-think moment. The nuclear industry has long boasted of redundant, backup safety systems in reactor power plants. They say if one system fails, another will jump in to take its place and keep the plant operating safely. It's true that nuclear power plants are amazingly complicated and designed to perform safely, as long as humans do not interfere.

"Had this continued, it could have gotten to the point where there could have been equipment damage to the pumps, but it would not have jeopardized the safety of the reactor," said an NRC spokesperson. The reactor cooling system kept the hot fuel core covered with water during the entire incident, which eventually was detected and corrected.

NRC counted three safety violations in this event, the most serious being failure to communicate to control room operators that false signs were fed into the reactor circuitry as a test. NRC fined Edison $21,000 for that.

The dummy signal event shook NRC's regional officials in Lisle, Illinois, which oversaw 20 nuclear plants at the time in Illinois, Iowa, Michigan, Minnesota, Ohio, and Wisconsin.

"That event was very serious," said regional director Keppler, the worst in his region. "That was the straw that broke the camel's back."

It was shocking because it defied logic, not to mention all the training and meticulous procedures that power plant operators are expected to understand and follow.

Keppler admitted that Edison was making NRC nervous, maybe even desperate.

"Not one of these events jeopardized public health and safety," Keppler emphasized, "but we were nervous because the company was not taking corrective action and something serious could happen if management didn't do something quick."

NRC officials ordered Edison executives to attend a top-level meeting in Washington, DC, in November 1977.

"That was a very candid meeting," recalled Keppler, "in which the company was told that the regulatory performance had to improve or we would take stronger action. I was there, and that was a tough meeting, I'm telling you."

Stronger action could mean tougher penalties or ordering Edison to cut its reactor power levels in half, like reducing the speed of a car to make them safer to operate. NRC officials also threatened to temporarily suspend Edison's licenses to operate reactors.

No U.S. utility had ever faced such punitive action by the NRC, said Keppler. They were actions that could jeopardize Edison's future as the country's leading nuclear utility.

NRC spelled out its concerns: repetitive personnel errors. Deficient procedures for addressing them. Inadequate communication. Failure to take action on problems in a timely manner.

Paradoxically, Edison had more experience operating nuclear plants than any power utility in the world. It pioneered commercial nuclear energy beginning with the Dresden nuclear station in 1960. Yet Edison was compiling one of the worst records for compliance with NRC safety and performance standards.

By this time, Edison had seven reactors operating at three power stations, and plans on the drawing boards for six more reactors at a cost of $3.1 billion, expecting to operate all those new reactors by 1982. NRC fined Edison $105,500 from 1974 to 1978 for safety violations, more than any utility in the country. The fines did not improve Edison's performance.

NRC's Keppler explained what was at stake for Edison. "If the company can't improve the performance of the seven reactors it has now, why should we give them licenses to operate six more? Commonwealth Edison's performance must be improved before they get [operating] licenses for those plants."

This threat to derail $3.1 billion in Edison investments prompted the utility to hire a management consulting firm, Booz Allen Hamilton Inc., of Chicago, at a cost of $350,000 for a six-month program.

Keppler said his agency expected Edison to focus on management of the corporate office in downtown Chicago and on the Zion nuclear station.

"Why limit to Zion?" asked Keppler. "It was felt to be more helpful to be specific at Zion and determine the applicability of their findings at other nuclear plants." He wanted a close look at Edison's organizational structure, its procedures, its methods of handling problems, and its support of nuclear plants.

Booz Allen management experts began interviewing Edison personnel at all levels, from the CEO down to shop foremen and reactor operators. They were exploring NRC's suspicion that organizational problems were causing confusion inside the generating stations, between stations and the downtown general offices in supporting the company's departments.

Byron Lee took part in some of the high-level negotiations with the NRC. He was Edison's vice president for construction, operation and licensing of all generating stations, nuclear and fossil-fueled.

"They're trying to understand how our system works, how we think it's supposed to work, and determine if it is working that way," Lee said. "If they come across any significant deficiencies, they'd let us know about those immediately. We haven't heard of any yet."

What did Lee think about the beating Edison was taking from NRC?

"We don't think things are as bad as NRC thinks," he said. "They are the regulatory agency. We've got to comply with what they perceive. They are the agency and they perceive a problem, so we'd better do something to correct the problem. This consultant is trying to help us and give us some direction."

These problems seem to imply, I said to Lee, that the utility with the most experience with nuclear power didn't know how to operate nuclear power plants.

"No, I don't think that's the case," he responded. "Maybe our size is more the problem. It's more complex to run three stations and three under construction. More is going on. We're subject to more problems. The possibility for error is great in our situation.

"Nobody comes close to the number of units we have. The more you have, the more chance of somebody making an error. But nevertheless, they are concerned. So we're working hard to try to understand where we can make improvements. I think our performance in the last year and a half—operation output, equipment problems—our performance has been very, very good."

I asked Lee if he was saying what some critics believe, that nuclear power is too complicated for humans to operate trouble-free, that there's always a problem waiting to happen. Especially as utilities construct power stations with two or three reactors onsite.

"No, not at all," said Lee. "We are catching 99 percent of the things that occur. It's not that we're not catching things. Our system is set up to catch them. The problem is people are making mistakes, in spite of our training. And people aren't adhering absolutely to procedures. Thousands and thousands of operations occur every day at these things. We're talking about a few instances a week or a month."

As an example, Lee said, one problem was the lack of absolute adherence to procedure, like filling out an "out of service" ticket. It might not be filled out completely, or somebody would forget to sign it. "Those are the type of things people are not adhering to."

Then was Lee saying federal regulations were too complicated and demanding?

"Federal regulations are very demanding," he answered. "And the NRC feels we're not meeting all those requirements 100 percent. That's the basis of our problem."

Lee said he had no doubts the problems would be solved, but thought it unlikely Edison would never be cited for violations.

"There is nobody in the world who can operate without violations," he insisted. "But they want us to cut down on the number of them. Our efforts at Zion in the last months have been excellent."

While Edison bore down on its problems with the help of Booz Allen, NRC bore down on Edison.

NRC stepped up monitoring at all three Edison nuclear stations. Typically, NRC safety inspectors visit nuclear stations once every three to four weeks.

"We are inspecting weekly at all three [Edison] stations, two or three days a week, and many times with more than one [inspector]," explained NRC director Keppler. The beefed-up inspections began shortly after the November meeting in Washington with Edison's top brass.

Edison was making a sincere effort to improve, acknowledged the director, and "we are seeing a change in attitude and performance." But he was not fully satisfied, and thought the company could do better.

"Otherwise, I wouldn't be after them the way I am," he said. "Other utilities demonstrated they can do better. If you take a look at utilities in general, many of the problems that occur at Commonwealth Edison have not occurred at other stations. We're talking about a situation that can and should be improved.

"That doesn't mean everyone can operate with zero mistakes. These plants are complex, and there are going to be mistakes. The key element in my view is management establish controls that assure that they learn from their mistakes, and they take broad corrective action to minimize recurrence."

Then something strange happened in the Zion plant in February 1978. NRC wanted Edison to stop repeating mistakes. But Zion had a history of making mistakes of a kind NRC had never seen before.

It was 2:30 a.m. when a horn alarm blared in the control room while reactor Unit One was at full power. The primary cooling water system flow suddenly increased, and control room instruments showed that flow control valve number 121 was causing the problem.

Three reactor operators ran to the valve, located in the auxiliary building next to the reactor building. Finding the valve completely shut, the operators cranked it open and the reactor cooling water flow returned to normal.

Who closed the valve, and why?

Two men were in the immediate area when the valve closed in that early morning hour. One was a station man, an entry-level job, changing floor pads. The other was a 21-year-old security guard on the job just two months.

"His function was to inform people they were entering a high-radiation area and should exercise the necessary precautions," explained Jack Leider, the Zion plant's assistant superintendent. Not all highly radioactive areas inside a nuclear plant can be locked off, so guards are posted there to warn of the danger.

The guard repeatedly denied closing the valve. "Then he admitted after further questioning that he in fact was fooling around with it," said Leider. "For that reason, he was fired."

The NRC wrote an extensive report on that incident because it was unusual and warranted some special emphasis. The report said the guard denied at least three times that he manipulated the valve, until he was told it could not close "accidentally," but only by deliberately turning the valve at least six hard turns.

There was no apparent conscious attempt to create an incident maliciously, the NRC concluded, "but rather his actions were a result of boredom and lack of realization of what the consequences of turning the valve would be."

The guard was walking around to relieve his boredom. He was curious about some gauges behind his chair and leaned over to watch what they were doing. He claimed that he leaned his right arm on a small valve and, on turning around, must have brushed his arm across the valve and accidently closed it. The guard insisted it was "sheer accident" and that he did not know the purpose of the valve or that anything was wrong "until three operators rushed in the area to check the position of the valve."

This is yet another stop-and-think moment. People know that nuclear power

plants are highly technical and complicated. What they don't know is that boredom and curiosity are hazardous if a person is inclined to tinker to pass the time.
Once again, it was human interference with the exquisite mechanical and electrical
balances that make a power plant work.

The guard, said Leider, was "passing time in the wrong way. We change them frequently because it can be a boring job. For that reason they're changed every couple
of hours. They can walk around, but not operate equipment or get close to it."

After the incident, a memorandum to all guards said, "at no time will a security
officer operate, adjust or tamper with any plant equipment, including valves and
electrical switches. The only exemption is security equipment."

The valve involved in the incident was small, about two inches by two inches. It
controlled a quarter-inch air line involved in taking off small amounts of radioactive cooling water to purify it and return it to the reactor cooling system.

Leider said the small valve could have remained closed for about another hour
before reactor safety systems automatically shut off the reactor.

"The ultimate end would have been a unit trip and loss of power until we identified the cause and corrected the trip," said Leider. "There really weren't any safety
implications, but operation implications." The reactor had plenty of primary cooling water, about 100,000 gallons. If the reactor had tripped off, it would have taken
about four hours to get it back into service, which would have cost Edison about
$50,000 in lost power production.

That was a near miss. A year later, in May 1979, a reactor operator accidentally
tripped a reactor, causing a 24-hour power outage that cost Edison about $300,000
for power replacement. While testing the electrical safety circuits, a reactor operator inserted signals into the system to see if it was working.

"The operator skipped a page of the test procedure and thereby inserted signals
into the safety circuits that simulated the conditions that would exist if a steam
pipe broke," explained NRC spokesman Jan Strasma.

The emergency core cooling system went into operation and the reactor automatically shut off. A review of the shutdown sequence revealed that three pieces of
equipment failed to operate. Two valves failed to close and a pump failed to start,
although backup systems were available.

It was the second accident at Zion in a month. Earlier, 700 gallons of mildly
radioactive water had spilled onto the floor of the auxiliary building between the
two reactor containment buildings, releasing radioactive gases.

Another pause for reflection. The public normally hears about big nuclear accidents. The smaller kinds, like spills and releases of radioactive gases, tend to go unreported or unnoticed. The public seldom realizes that nuclear power plants routinely release radioactive gases and water into the environment.

An exception was an accidental radioactive gas leak from the Zion station in March 1980, causing residents of Waukegan, only five miles from the station, to question the safety of living so close to the plant.

Waukegan comes from a Potawatomi word meaning "fort" or "fortress," and has the distinction of being the only place where Abraham Lincoln failed to finish a speech while campaigning in the town in 1860. A fire alarm rang, and the man who would soon become president was interrupted. Today, Waukegan is the county seat of Lake County, Illinois, and an industrial suburb of Chicago. The predominantly working-class community has a distinct industrial character, with three federal Superfund sites of hazardous substances such as toxic PCB chemicals and asbestos.

Radioactive gases billowed from the station for 20 minutes when a Zion worker turned the wrong valve. NRC officials were concerned because human error was at fault and Edison failed to report the accident to the agency within one hour, as required. Edison notified NRC 18 hours later.

Waukegan's mayor, Bill Morris, was incensed, demanding that the Zion station stop operating pending an investigation. He asked NRC to assign additional inspectors at the plant for 24-hour surveillance. He also sent letters to NRC officials and Edison's president, calling for various actions. He wanted the Lake County state's attorney to consider filing charges against Edison and the Zion station superintendent.

"This accident has called to our attention that Edison occasionally releases contaminated gases into the atmosphere without any public warning under the sanctions of the NRC," Morris said at a press conference in his office. "We demand that all future releases be done only after proper public notice so that citizens have the opportunity to take precautionary action."

An Edison spokesperson accused the mayor of "gross overreaction." The amount of gas released was not considered a public health hazard.

Roger Harrison, Waukegan's director of environmental control, said city officials learned of the accidental gas release from news reports, and many of the city's 65,000 residents were shocked to learn that the Zion plant routinely emitted radioactive gas and water under NRC guidelines.

"I'm getting calls from people asking if it's dangerous and should they move out of town," said Harrison. Local officials were concerned about the Zion plant for some time, but the accidental leak "broke the camel's back."

"I wasn't aware there were scheduled releases of radioactive gas," he said. "I'd like to know who schedules them. Many citizens in Waukegan want to know more about it so they can keep their children indoors when it's happening. Commonwealth Edison owes us that.

"We in Waukegan don't intend to sit quietly by and wait for a Three Mile Island accident before we demand accountability, honesty, and competence from Commonwealth Edison and NRC," Harrison said. "We have a right to be informed and to make our own decisions about what is safe or unsafe."

Now it was the public's turn to apply pressure, on Edison and the NRC.

Concern over the airborne gas marked an early example of local public resistance to the Zion plant, and the beginning of local public awareness of what it means to live near such a plant. Like many residents, Mayor Morris believed that radioactive water and gases were bottled up inside the power plant.

Several days after protests to the accident, NRC officials tried to address public concerns and explain the nuclear facts of life. Don Miller, an NRC radiation specialist responsible for inspecting the Zion station, said all nuclear power plants routinely give off radioactive gas and water while operating.

The gases are byproducts of the nuclear chain reaction inside the highly radioactive fuel core, and any water that comes in contact with the atom-splitting process becomes radioactive. Water cools the intensely hot fuel core. Reactors also give off so-called "noble gases" that are radioactive but don't react with other matter.

"The major source of [radioactive gas emissions] into the air is leakage in the system," Miller explains, estimating that "roughly 90 percent" of all radioactive gases coming from the Zion plant simply leak out. The rest is held in a tank until the radioactivity decays and loses strength over 45 days before it is released slowly into the atmosphere.

In 1979, said Miller, the plant released 34,083 curies of radioactivity into the atmosphere, which was 11 percent below the annual allowable limit. A curie is a basic unit of measurement of the intensity of radiation.

And, said Miller, about 35,000 gallons of radioactive water were flushed into Lake Michigan from the plant about once a week, or roughly 3 million gallons in

1979. Its radioactive content, he said, was 1.44 curies, which he called "insignificant." About a month later, NRC corrected Miller, saying he had greatly underestimated the amount of liquid radioactivity discharged in 1979 by the Zion plant.

The water flushed into the lake that year contained 601 curies of tritium, a form of radioactive water. Miller had failed to take tritium into account in his assessment.

Critics were aghast, saying tritium is by far the largest source of radioactive liquid discharged by nuclear plants. How could it be overlooked? Still, NRC said it was far below allowable limits and not a hazard.

Miller saw no public health dangers from radioactivity released from the Zion station.

Miller's mistake in overlooking or ignoring the tritium content of water discharged from the Zion plant points to the difficulty in communicating technical information about nuclear energy. Even the experts stumble on the jargon and the details, and what is considered important or insignificant. Miller was a 20-year veteran in nuclear safety. Upon being corrected, he said he didn't know why he didn't mention tritium in his calculations.

Miller's oversight "points up the vulnerability of the public when it must depend on vested interests such as the utilities and the NRC to give them facts necessary to protect its health and that of future generations," asserted Catherine Quigg, an antinuclear activist from Barrington, Illinois.

Quigg called for the creation of an independent radiation monitoring and reporting network at the Zion plant, saying the NRC could not be trusted.

Miller's oversight also points to a blasé attitude toward radioactivity by some nuclear power workers, one of whom caused Edison to be fined a whopping $100,000 in 1982.

While Zion reactor Unit One was shut down for refueling, a shift foreman climbed down a ladder into a steel-lined concrete compartment under the reactor, known as the reactor cavity, to search for water leaks. The cavity is highly radioactive. The foreman reported that his personal radiation monitoring equipment went "off scale" because radiation levels were so intense as he climbed down the ladder and moved closer to the bottom of the reactor.

The next day, a shift engineer did the same thing, "fully aware that exposure rates would increase significantly as he approached the reactor vessel," said the NRC in a report.

"During the entry, which only took about 70 seconds, the shift engineer received a whole-body radiation dose of approximately five REMs," said the NRC. That is high for a single exposure.

Once again, federal regulators were dismayed, saying: "Of all the personnel directly involved in the two cavity entries, these two managers were the most knowledgeable of the specific cavity radiological hazards." In other words, they should have known better.

Their actions violated a slew of safety rules. Nuclear safety is based on preplanning and radiation surveys before work is carried out, so hazards are understood in advance to keep radiation exposures as low as reasonably achievable. Nothing should come as a surprise in a nuclear plant.

In this case, NRC pointed out, no radiological work permit was prepared in advance, which would have defined the actions intended by the two managers, the allowed "stay time" in a highly radioactive area, and what safety precautions should be taken. NRC noted that two radiation safety specialists on duty at the time each assumed the other had discussed precautions with the shift engineer before he descended into the radioactive cavity.

Astonishingly, those two examples of managers flouting radiation safety rules duplicate the 1976 incident for which Edison was fined $13,000, when a supervisor went searching for a water leak under a nuclear reactor without taking precautions.

They are examples of what NRC had been complaining about for years: Edison and the Zion staff failed to learn from their mistakes and kept making them, despite threats and fines.

Radiation protection personnel were present at those two recent incidents of rule-breaking, but they did not do what their training required them to do. They "had a general lack of understanding of the reactor cavity's specific radiation hazards," said the NRC, and "were not familiar with the nature and strength of the radiation sources present."

An NRC investigation showed that those radiation protection specialists wrongly thought that radiation strength was uniformly distributed along the length of the reactor cavity and "did not warn the engineer to stop advancing into higher radiation fields."

The bottom of the reactor cavity registered 50 REMs an hour, which is an intense and dangerous radiation field. NRC was worried about a growing trend toward nuclear power plant workers entering highly radioactive cavity areas. Radiation levels of thousands of REMs are possible in reactor cavities. "Entry into radiation fields of this magnitude seriously jeopardizes the health and safety of personnel," the NRC warned.

So what happened after all that? Zion station was troublesome to the end of its days. But this litany of Zion station blunders says something about NRC scrutiny of power plants, and the obligation of utilities to report infractions to the NRC when they happen. We know about mistakes in nuclear plants because utilities must report them. They form a record of the kind of mistakes people make in nuclear plants, and help us to understand how to confront them, to learn from them. In some ways, the Zion plant's performance history was unique. Repeated federal fines and threats seemed to have questionable impact.

It also says something about what is known as the human/machine interface—known as the man/machine interface in earlier politically incorrect times. It's about how humans interact with complicated technology. It helps to answer whether nuclear technology is beyond the abilities of power plant workers to comprehend and control.

One of the disturbing, perplexing aspects of the human interface with nuclear technology is the tendency seen repeatedly to ignore radiation hazards. We might ask what kind of lunacy allows someone to knowingly enter a highly radioactive area without taking precautions? Those two Zion station managers should have known better than to enter the highly radioactive reactor cavity so carelessly. They were veterans of nuclear power who over time became insensitive to the dangers, even contemptuous of fears most people would feel. Some might call that bravado, or plain stupidity, but it says something about the nuclear mindset.

Does it lead to ignoring the rules? Or was it something about the Zion station staff in particular? Because, after all those gaffes, blunders, mistakes, violations, fines, and reprimands, the Zion station did something truly shocking and outrageous—off the charts and over the top for any nuclear plant. Nothing like it happened before, or since.

Chapter 16: Sex, Drugs, and Alcohol

IT STARTED LIKE A SCENARIO FROM A SCRIPT FOR A TELEVISION COP DRAMA.

Agents worked undercover for a year following a tip that more than 20 employees at the Zion Nuclear Power plant were using drugs, some of them during working hours.

Their investigation into drug trafficking in the lakeside community led the agents to an isolated road surrounded by cattail marshes and tall trees, where two men were waiting in the shadows near the power plant to sell them cocaine. They quickly found themselves in handcuffs.

Instead of a sale, it was a drug bust. Two power plant employees, Jeffrey J. Kostroski, 23, of Waukegan and Scott A. Klepzig, 26, of Kenosha, Wisconsin, were arrested on drug charges. Both had worked as equipment attendants for the past three years. Undercover agents had bought cocaine from Kostroski twice before making the arrest during a third attempted sale.

It looked like a fairly standard outcome of good police work on February 24, 1981, until further investigation revealed that Kostroski was undergoing training as a nuclear reactor operator at the Zion station.

It was the first case of its kind in the nation, with serious implications for nuclear safety. The very idea that a drug-addled employee might get his hands on the controls of a nuclear reactor rattled U.S. Nuclear Regulatory Commission officials. Not only that, but as equipment attendants, Kostroski and Klepzig routinely handled vital safety-related machinery inside the station.

"It's very unsettling to us, and we're very concerned about it," said Russ Marabito, an NRC spokesperson. "We can't recall any incident of this kind."

The federal agency immediately assigned a third full-time inspector to the Zion plant to see if it was operating according to NRC requirements. Marabito pointed out that NRC inspectors in the past "observed no abnormalities on the site relating to drugs."

After one month of heightened inspections and spot checks, NRC officials decided the station was operating safely and according to requirements, and discontinued increased surveillance, thinking they had put this behind them.

Shortly after that, the City of Zion's mayor, city attorney, and other municipal officials notified the federal agency that informants working inside the Zion station said between one-fourth to one-third of plant employees were using or selling drugs. On June 1, 1981, Zion city officials met with NRC officials, requesting the federal agency to place an undercover agent inside the plant to check out the allegations of drug use there. NRC said that was beyond its civil jurisdiction and could not mount such an undercover investigation.

But never underestimate the power of television. Security guards working at Edison's Dresden Nuclear Station near Morris, Illinois, approached Chicago television station WMAQ-TV asking if they would be interested in a story about labor and morale problems at the Dresden and Zion stations with the guard forces there. On December 16–18, 1981, the television station broadcast feature stories about allegations of drug use and other irregularities at both the Zion and Dresden stations. Guards were shown silhouetted with their voices distorted to disguise their identities.

This time, the entire Chicago region got wind of concerns that two nearby nuclear plants might be harboring drug-using workers. It was an apparent threat that government officials and Commonwealth Edison could not ignore.

The result was two separate investigations. The biggest and most intensive was by the NRC. The other, with some comic overtones, was an ill-fated undercover investigation by Commonwealth Edison, which was having trouble enough operating its power plants to the federal government's satisfaction. Playing detective was outside its skill set. More about them later.

Everything in government begins with a meeting. In this case, it was a big one gathering every level of law enforcement—federal, state, regional, and local. The meeting was held January 6, 1982, at the request of NRC's Midwest regional office. To name a few, it included the first assistant U.S. attorney for northern Illinois, Lake County's state's attorney, the Zion police chief, the North Shore Metropolitan

Enforcement Group, and the U.S. Drug Enforcement Administration.

All parties concluded that the issue of primary concern was the possible impact of drug use on the safe operation of the Zion plant. Dresden plant issues centered mainly on security guard staff dissatisfactions. All agreed that NRC should take the lead in the investigation.

Taking allegations raised by those who appeared on the television broadcast as well as others, the NRC compiled a list that fairly well spelled out the agenda for its investigation at the Zion plant:

· Widespread use of drugs such as marijuana, cocaine, amphetamines, nitrous oxide, and LSD on station premises.

· Widespread use, sale, or possession of various drugs outside the workplace.

· Commonwealth Edison management's cover-up of known drug problems.

· Security guards not properly trained or capable of properly protecting the facility.

· Errors in operations caused by drug use.

· Control room operators asleep while on duty.

· Employees engaged in sexual activities inside the station that were detrimental to their duty performance, including sexual intercourse and oral sex.

· Alcohol consumed on plant premises and employees reporting to work drunk.

No commercial nuclear power generating station in the United States had ever been subjected to such an investigation. It was another first added to the Zion station's remarkable history.

Almost immediately, the federal investigation was bogged down in legal technicalities because of possible criminal charges for drug possession. Eleven lawyers from seven law firms took part in the investigation, employed by Edison, Zion employees, and their union.

Some workers refused to cooperate because of possible self-incrimination. Others believed their drug use during off-duty hours was not relevant to the investigation or was an invasion of their privacy. Some questioned the reliability of eyewitness accounts and whether the NRC had any the authority to ask these questions.

As a result, agreements had to be reached to get beyond this deadlock, including to forego criminal prosecution for certain drug use or revoking a reactor operator's license for past drug use outside the workplace. Identities of employees questioned would be kept confidential on a "best efforts" basis.

NRC also consulted medical authorities on the lingering effects of using certain drugs, like marijuana, amphetamines, and cocaine, for fitness for duty. Under established guidelines, they should not be taken within eight hours of reporting for work. The guidelines also define occasional drug use versus chronic use, and effects of using combinations of drugs. The federal agency was expected to answer questions about the effects of drug use on duty performance.

These clarifications about legal liability or acceptable drug use broke the impasse. Three teams of NRC investigators went to work, interviewing 215 individuals, some of them more than once.

Commonwealth Edison's investigation began when it decided to place its own private undercover investigator in the Zion station after several law enforcement agencies said they did not have the manpower or the money to conduct such an operation to check into allegations of widespread drug use there.

An Edison lawyer contacted a private detective firm, W. E. Davern and Associates, which had previously worked for Edison. On April 9, 1981, the lawyer met with the owner of the detective agency and with the designated undercover agent, Janice Michal. She had three years of experience as a detective, mostly on divorce cases.

At that meeting, Michal was told to coordinate her activities with the Metropolitan Enforcement Group, known as MEG, a regional drug enforcement agency. MEG was supposed to be her primary law enforcement contact for the undercover operation. She got MEG contact information and 24-hour phone numbers to call before making drug purchases or if her safety was jeopardized.

On May 18, 1981, Michal began working at the Zion station as a stationman, an entry-level job title later changed to "station laborer" to ensure gender neutrality. She submitted weekly activity reports to Davern, who edited them and turned them over to Edison's division vice president for nuclear stations.

In these reports, Michal told what she observed outside the Zion station, conversations about drug use, and eyewitness accounts of Zion employees using marijuana at locations outside the station. She also mentioned conversations about drug use she overhead.

The Edison division vice president decided that Michal's reports "contained mostly gossip and innuendo about the personal lives of plant personnel and were not oriented to the undercover operation." A second undercover operation by the Burns International Security Services, Inc., reported no observations of drug use on the Zion premises or outside, and did not get as much attention as private-eye Michal.

Michal reported purchasing marijuana outside the Zion plant in amounts varying from less than half an ounce to up to half a pound. She reported the purchases to Davern, who told her to tell the Zion station personnel supervisor so he could pick it up and give it to the Zion police chief.

Completely forgotten was the original plan to coordinate all these activities with MEG. According to an NRC report on Michal's activities, the Zion station personnel supervisor told the Zion police chief that the marijuana turned up by Michal "was found in a corner of the plant and was not connected with any individual and that the division vice president, nuclear services, would be in later to see him on the matter." The Zion police chief said he never got that visit and was surprised to learn from newscasts that the marijuana in question was gathered in the undercover operation.

The Zion police chief also told the NRC that he received about 23 grams of marijuana from Edison during the undercover operation, so what happened to the half-pound of marijuana Michal obtained? That question was never answered.

Here is the NRC's assessment of Edison's undercover operation: "Apparently, confusion and mismanagement by CECO concerning the undercover operation led to a deviation from the original plan [to work with MEG]. This resulted in no law enforcement agency being advised of alleged drug purchases made by the undercover operative until approximately five months after the purchases were completed."

In short, Edison mismanaged its undercover operation as badly as it was managing the Zion plant to the NRC's satisfaction.

Michal's career as an undercover agent lasted three months, ending almost in a comical way. In mid-August, she went to a Zion dress shop and applied for credit to buy a wedding dress. During the fitting, she told the store associate she was working at the Zion station as a stationman, but that she actually was there as an undercover agent looking for drug violations. Michal never explained why she unmasked herself to the shop saleswoman. Maybe it was a moment of hubris or a bid for a better credit rating by mentioning her James Bond gig.

What happened next showed how tightly knit the Zion community was with the nuclear station. The shop sales associate's husband was working at the station. She told him about Michal's confession, and he told the assistant plant superintendent, who told the personnel manager who was working with Michal on the undercover operation.

When the attorney monitoring the Zion undercover operation heard that Michal revealed her clandestine role in the dress shop, he contacted the detective agency owner. They agreed to remove Janice Michal from the operation for her own safety.

But that is not the last we hear from the Inspector Clouseau-like agent. She appears again when Chicago television station WMAQ-TV broadcast a half-hour prime time special on March 27, 1982, on the ongoing drug investigation at the Zion power plant. She accuses Edison of covering up her identification of criminal activities at the Zion station and misleading police about the sources of marijuana she had purchased.

Her allegations of an Edison cover-up were added to the many issues being investigated by the NRC, which were becoming a heavy load. NRC's final report on the Zion station investigation came out on November 2, 1982. Right at the top, it said:

"We did not substantiate the allegations of widespread use of drugs either onsite or offsite by station and or guard personnel. However, we did conclude that marijuana was smoked onsite. We also have concerns about certain station personnel who admitted using drugs offsite within the time frames and frequency determined as likely to have a potentially detrimental effect on jobs performance, and certain guards who admitted to offsite use of drugs."

The federal investigation began with the understanding that safe operation of the Zion facility was the primary concern. Investigators focused heavily on reactor operators in the control room, the nerve center where all reactor control and safety systems were operated.

"The conclusion was there was no evidence of onsite drug use by licensed personnel," said NRC spokesman Jan Strasma, meaning the licensed reactor operators, "and there was some evidence of offsite drug use."

"Our concern was drug usage by guys with hands on controls that run the plant," he explained.

As for the allegations that reactor operators were sleeping on duty, the investigation found no reason to believe they were asleep, but there had been occasions when they "nodded off" or placed their heads on the desk for five to 10 minutes. Their attentiveness "is an area of concern developed during the investigation," said the federal report. Constant attentiveness is expected from control room operators, and NRC told Edison managers to address this matter.

When a power plant is operating at full power and normally, there is not a lot going on in the control room. Control room operators monitor gauges and dials.

They are there to deal with routine changes, like startup and shutdown, or dealing with a problem, a malfunction of equipment, or some error made in the operation of the plant that requires a response. It can be dull and boring waiting until something happens.

The NRC report shows reactor operators speaking candidly about their struggles to be vigilant during long and irregular hours, a very human dilemma. The control room essentially is a windowless dungeon. Operators must be vigilant in the face of stifling boredom.

Reactor operators cannot occupy themselves with card games, novels, or watching television. Those are forbidden. The only reading material allowed is professional, job-related literature provided by the station superintendent. Newspapers, non-professional magazines, and other non-job-related material are not allowed. No commercial radio or television broadcasts are permitted in the plant, except those authorized by the station superintendent. Violations of these rules could be met with severe disciplinary action, including discharge.

"The confidence of the general public in our ability to operate our nuclear stations can be severely damaged by any failure to adhere to a professional code of conduct," warned Edison in a 1980 directive on conduct of operations.

Other key findings:

- Sex: Federal investigators found no one with knowledge of sexual activities, such as intercourse and oral sex described by one accuser who appeared in a television exposé. Looking at it somewhat bureaucratically, the report concluded that the investigation "did not obtain any information to conclude that any onsite sexual activities resulted in unsafe operation of the plant."

- Alcohol: The presence of alcohol containers, such as empty beer cans and whiskey bottles, indicated that alcohol had been consumed onsite. The exact time frame and identity of individuals involved was not determined. Five individuals were found to be under the influence of alcohol when reporting for work and were prevented from working until they sobered up. Alcoholic abuse did not affect safe operation of the plant.

- Marijuana and other drugs: 30 station employees were accused of using marijuana and/or other drugs on station premises. Widespread onsite drug abuse was not substantiated. Based on the number of persons who smelled the odor of marijuana and the termination of one contract employee for smoking marijuana onsite, it was concluded that marijuana was smoked onsite. No information was developed to identify those people, except for the terminated employee. Forty-three station employees were accused of using marijuana and/or other drugs while away from work. Some of them would be checked for fitness for duty.

· Edison cover-up: Utility management was accused of failing to act when informed of widespread use of drugs. NRC concluded that no action was taken against individuals mentioned in the undercover agent's reports because there was no hard evidence of onsite drug use. The investigation did not conclude that Edison misled law enforcement officials about the undercover agent's drug purchases, but decided a miscommunication occurred.

· Guards: Security guards were adequately trained and properly armed to repel a terrorist assault on the station, and there was no reason to believe the guards would run away if attacked. However, the frequency of drills was an issue of concern. Typically, about two dozen guards are on duty in a shift and they are expected to defend the plant against five armed invaders.

Two guards were fired for refusing to cooperate with the NRC investigation. The guard staff figured in one violation of NRC requirements. Edison was cited for failure to follow proper search procedures for some hand-carried items. The investigation revealed that security guards were not carefully checking thermos bottles, foil-wrapped food, or the contents in Tupperware-type containers for alcohol, drugs, or explosives as employees entered the station through the guardhouse entry.

Guards thought it was sufficient to pass such items through X-ray machines, although this resulted in poor images of what was inside a thermos. Owners of such items sometimes objected to their food being X-rayed, and guards complied or gave quick visual inspections. NRC ordered corrective action, saying guards should tell the owners of hand-carried items to open them for close inspection, including drinks in paper cups.

Responding to the federal investigation, Commonwealth Edison Co. adopted a new "Company Policy Regarding Drug Abuse" and ordered all 17,000 company employees to attend meetings to review the policy. Under a new drug awareness surveillance initiative, supervisors made personal contact with all plant reactor operators, radiation control technicians, stationmen, and instrument technicians in their first hour of coming on duty to assess their fitness.

This should be the end of the story, but not yet. We have not heard the last of Janice Michal, the failed undercover agent. She was in the news again in January

1982, by this time married and known as Janice Michal King.

She told a *Chicago Tribune* reporter that she was assaulted in her apartment building by a man wearing a ski mask and a Zion plant company jacket. She was treated at a Waukegan hospital for bruised ribs, face cuts, whiplash, and a possible concussion.

"I'm petrified," she told the *Tribune.* "I'm considering leaving the state. Our life has been in chaos since this whole thing began. It just has been hell on wheels," she said of her accusations against Edison on WMAQ-TV.

The attack on her happened several hours after someone apparently tampered with an electrical circuit breaker in her building, twice cutting off electricity to the apartment she shared with her husband, Dennis, an employee of the Zion station. The first time the electricity went out, her husband claimed a notice left in the mailbox said, "Ha, Ha. See what happens when you screw with CECO."

According to the *Tribune* account, Mrs. King was in her building's laundry room when she apparently surprised a man wearing a ski mask and a Zion station jacket.

"He asked me if I was Janice King and then he called me every foul name in the book," she said. "He told me to leave ComEd and the people at Zion alone. He beat me on the head and face with his fists. I ran toward the door, and he hit me again. My head flew back and he hit me five or six more times. Then I went down. I just couldn't handle it anymore."

The couple said they had been harassed since Mrs. King appeared on television making accusations against Edison. After the televised interview, many of her husband's fellow employees gave him the "cold shoulder," he said. Others tacked notes on the company bulletin boards referring to him as "spy King" or "the spy's husband."

"Things had settled down over the last couple weeks," he said. "And then this happens," referring to the attack on his wife.

John Alexander Dowie's religious utopia ended with neighbors fighting neighbors. A new generation who arrived with a modern superpower plant had their own issues to fight about.

This is not the way Dowie would have liked it, and neither did the Zion station superintendent who thought the power plant's reputation was being tarnished unfairly.

Chapter 17: The Invitation

NUCLEAR PEOPLE ARE PARANOID EVEN IN THE BEST OF TIMES, BELIEVING THAT the media are out to get them and the public doesn't like them. They're partly right.

They also are proud of their work and defensive toward criticism. They might even feel some elation when astonished strangers say: "You work in a nuclear power plant?" It's not the usual line of work like selling shoes. It's an odd mixture of pride and punishment.

So when news reports headlined the unusual sex and drugs federal investigation at the Zion station, it didn't look good, even though it largely absolved Commonwealth Edison. The *Chicago Tribune*'s headline on the story I wrote in November 1982 read: "U.S. study backs safety at Zion."

Still, the nuclear industry doesn't like to get that kind of attention. Which probably explains why I got a call months later from Ken Graesser, the Zion plant's superintendent, asking me to meet him in his office at the generating station. I don't recall ever meeting the man before that and was glad for a chance to hear what he had to say.

We set a date and I drove the 38 miles from my home in Evanston to Zion, got permission to enter the plant at the guard house, and walked escorted to the service building where Graesser's office is located, then up three flights of concrete steps. My first trip up those stairs.

The first thing I noticed while climbing the stairs was that the air smelled fresh, which is not surprising since the plant site is washed by Lake Michigan breezes. Graesser's office was in the southeast corner of the building overlooking the lake. It had a magnificent view of lake and sky. The lake was peaceful that day, less agitated than the plant superintendent.

After a cordial greeting, Graesser settled behind his desk and I took a seat in front of his desk. The plant superintendent was a bear of a man, not tall but husky with broad shoulders. I would learn later that he grew up on a dairy farm in central Wisconsin, milking cows until the age of 18. He had 15 years of nuclear power experience in a career spanning 24 years in the electricity business.

Graesser was smiling and polite, but there was a clear undercurrent of agitation. He had something he needed to get off his chest. He didn't like all that unfavorable publicity about Zion. He believed it gave a false impression of his workforce as unruly and unprofessional. He believed the Zion station had been maligned, giving the public wrong ideas about what went on in the facility.

Graesser wanted to set the record straight. He complained that the media gave a warped view of the nuclear industry because reporters focused only on accidents or other unusual events that would be ignored in other industries, even at other generating stations that burned coal or natural gas. They all got a pass, except for nuclear energy.

And another thing, he said: reporters don't have the slightest idea of how a nuclear plant works, not even the most basic rules and regulations that govern nuclear power.

Jabbing a finger in my direction, his voice rising, Graesser said forcefully: "I wish you could work here and see what it's REALLY like." To which I replied, "Okay, I accept. When can I start?"

Judging from his startled expression, Graesser was not expecting that. After recovering from the shock of my acceptance of an invitation more wishful than intended, Superintendent Graesser proved to be a man of his word.

I wish I could have heard the reaction from Edison's top management when he told them of his invitation to me, a reporter. Consternation? Disbelief? Brilliant but definitely unconventional? I don't know. I had interviewed many Edison bosses during the course of my career as a *Tribune* environment and energy writer, so they knew me and my work. I don't know if that had anything to do with it, or if they believed it necessary to support Graesser.

Maybe they simply were willing to give me a chance and hoped that a fair report by the *Chicago Tribune* might help to restore some public confidence in the Zion station.

It set in motion a unique chain of events in journalism/nuclear power relationships: an up-close and in-depth view of what goes on in a nuclear generating station, and getting to know the people who work there.

The next time Graesser and I talked, it was clear that he and Edison management had thought it over and had come to some ground rules. The first was that it made no sense to hire me for some kind of entry-level job in the plant. That would confine me to certain areas of the plant. And there could be union problems. Of the 496 workers in the Zion station, 284 were members of the International Brotherhood of Electrical Workers.

Graesser didn't want me pushing a broom. He wanted me to get a big picture of the dynamic human and mechanical forces that harness atomic power to produce electricity safely. And the timing was right.

In late 1983, Edison scheduled a $7.2 million overhaul of Zion's Unit One reactor after its first 10 years of operation. That side of the plant would be shut down and every major system would be inspected, dismantled or tested, then restored to like-new condition, and upgraded to comply with new federal safety regulations.

I would be escorted and given free run of the station to see the operations firsthand. It would be a rare, inside view of an atomic power plant overhaul lasting five months, and a close look at the people who run and maintain nuclear machines.

My escort was David Smith, a Zion staff maintenance manager. He was my guide and mentor almost every time I set foot in the station. His job was to explain what I was seeing, and to keep me out of trouble. We would be going into radioactive areas, where mistakes can be harmful.

This was an experiment in trust and journalism ethics. Workers in nuclear power generating facilities generally despise and don't trust reporters. Edison's willingness to allow me to witness everything happening at the plant, from the inside and for an unlimited time, was fairly unique.

From a journalist's perspective, no attempt was made to go undercover or try to get information by indirect routes. There would be no attempt to sensationalize. This was a story that had to be experienced and told clearly because of the highly technical nature of nuclear power. No restrictions were placed on me, except those dictated by safety in hazardous areas.

The assignment involved risks for the reporter and for the utility. I believe Edison did something that was characteristic of its pioneering spirit and tradition.

Commonwealth Edison has sole authority to decide who enters its power plants, but everyone in the plant must follow rules and regulations imposed and enforced by the U.S. Nuclear Regulatory Commission. Safety is a big deal in a nuclear power plant. Even though I had Commonwealth Edison's permission to be inside the station, I had to take radiation safety training first, just like anyone else who enters the plant.

Take the training and pass the test, I'll get in. Fail the test, I won't.

Chapter 18: Radiation Training

PAST THE LOW-SLUNG, HIGH-TENSION POWER LINES, THE ZION STATION'S ONE-story training center sits just inside the main gate but outside the double chain-link fence topped with barbed wire that surrounds the main power plant buildings.

They are separated for a reason. You have to qualify to get into a nuclear power plant.

The training building's remote location says something about why I'm here. I need a special kind of training to get past that barbed wire and that double line of high chain-link fences.

It's been years since I attended classes at Northwestern University and I'm a little nervous. A lot hangs on what happens in the next few hours. It's called Nuclear General Employee Training—known as N-GET in nuclear parlance—required at all commercial nuclear power plants in the United States. Employees must take the training course once a year and pass a test to keep working there. Me too.

It's 7:15 a.m. on September 26, 1983, and I enter the training center in the shadow of the towering domed reactor containment buildings. I'm walking in a corridor toward my classroom when I hear Peter LeBlond, a training supervisor, tell someone in one of the other classrooms that a *Chicago Tribune* reporter will be on the premises and the reaction is "mixed." The person he's talking to says he will be "brushing up on song and dance." Overhearing that conversation tells me I'm not entirely welcome.

The class of 11 students, including me, starts at 7:30 a.m. They're all men, mainly pipefitters, boilermakers, and laborers in denim jeans and flannel shirts.

The classroom walls are decorated with various kinds of warning signs colored in yellow and magenta, many of them with the three-bladed symbol for radiation. "Danger, high radiation area. Authorized entry only," reads one. Another says, "Caution, airborne radioactivity area." White canvas coveralls hang from coat hooks on the wall.

School-style chairs with attached desks occupy the front of the classroom. A yellow booklet on radiation protection guidelines and helpful hints rests on each desktop.

Our trainer, Clifton Nehmer, is slim and decisive. He gets down to business immediately without any breezy chatter. We start by filling out forms, then Nehmer gives an overview of what he will be talking about, including radiation protection and the biological effects of radiation and "what we can do to be a safe worker."

The class will end with a test of 25 questions, with a passing score of 70 percent. "It's a very difficult test," Nehmer warns. "Pay attention and listen up." Those who fail might not get a second chance.

A slide presentation on the Zion station layout shows buildings inside "the protected area" surrounded by the double security fences. In an emergency, we should go to the machine shop in the service building. We get the phone number for the security office.

Hard hats are required in the plant, along with "sensible shoes" and proper clothing. Eye and hearing protection are used where needed.

Get familiar with the color combinations for warnings: yellow and magenta for radiation protection, black and yellow for general safety.

Then Nehmer gets to the main reason we are all sitting there: radiation protection. He'll talk about the types of radiation "and we'll talk a little about biological effects."

Radiation, explains Nehmer, "is nothing more than energy from an unstable atom," an atom being the smallest unit of matter and consisting of protons, neutrons, and electrons. "It needs to get rid of excess energy."

Atoms of uranium-235, the fuel of a nuclear power reactor, are stable in their natural form. But when a neutron crashes into a uranium atom, the atom splits apart, releasing energy in the form of intense heat, two or three neutrons, and about 200 types of radioactive "fission fragments" that account for most of the radioactivity in a reactor.

Splitting atoms is called fission. Like crashing balls on a billiard table, neutrons released in fission crash into other atoms, breaking them apart and releasing more heat and more neutrons and more radioactive debris. It's a chain reaction used in nuclear power plants to turn water into steam.

Nuclear plants cannot explode like a bomb. Uranium fuel is enriched enough to sustain a nuclear chain reaction, but not enough to pack the explosive force of a bomb.

A chain reaction produces high-energy waves traveling at the speed of light and high-speed subatomic particles. They're called ionizing radiation because they can break molecular bonds and change the living cells of humans, animals, and plants. That makes them dangerous.

It's Nehmer's job to explain those dangers, name them, and offer ways to avoid them. He starts by getting into the language and terminology of radioactivity and the four types of radioactive hazards typically found in a nuclear generating station that are products of the nuclear chain reactor inside a reactor.

Alpha particles are positively charged and react with material. Normally alpha particles are not found in a nuclear plant except in uranium fuel-handling areas. Alpha particles are a weak kind of radioactivity that can be stopped by a thin sheet of paper.

But if swallowed, warns Nehmer, it "could destroy the lining of the intestinal tract and lead to an infection of the blood."

Beta particles are stronger, negatively charged subatomic particles that could penetrate a quarter of an inch into a person's body but could be blocked by a sheet of aluminum or plastic. They can cause radiation burns on a person's skin and eye cataracts.

Gamma radiation is extremely dangerous, a high-energy electromagnetic radiation traveling at the speed of light emitted by the nuclei of radioactive atoms. It can penetrate steel. But shields of two inches of lead, 10 inches of concrete, or 24 inches of water can reduce its penetrating force to a 10th of its full power.

"Gamma ray is a major concern to us," says Nehmer. "All vital organs of the body can be exposed. We're very concerned about gamma radiation."

None of those three types of radioactivity can make a person radioactive. The danger is from inhaling, swallowing, or smearing material contaminated with those radioactive particles on human bodies, or being exposed to gamma rays.

A fourth kind of radiation *can* make a person radioactive, and anything else that comes in contact with it. It's neutron radiation, a subatomic particle with no electric charge that shoots out from the nuclei of atoms in fission inside the reactor while it's operating. Neutrons make cooling water passing through the reactor fuel core radioactive, which is why the power plant staff must guard against cooling water leaks. Workers walking through radioactive water puddles could spread contamination around the plant.

After describing those radioactive hazards in a nuclear plant, Nehmer turns to how nuclear workers defend themselves against radiation, including ways to describe it and measure it, and allowable exposure limits in the workplace.

Among the first things the class hears from Nehmer and that little yellow booklet lying on their desktops is this message: "The ultimate responsibility for your radiation protection, safety, and security is in your hands."

Which means workers in radioactive areas must learn to protect themselves. This includes learning the language of measuring radioactivity, using terms that most people never hear. They include REM, RAD, and millirem. REM stands for Roentgen Equivalent Man, a unit of radiation measurement describing the energy absorbed by the human body. A millirem is one-thousandth of a REM, also a unit of radiation dosage that a wise nuclear power plant worker keeps in mind to be aware of his or her exposures.

REM and millirem are the most common measuring terms used in United States nuclear plants. There are others, including RAD, standing for radiation absorbed dose, a unit of measurement used to describe the amount of energy absorbed in any material from any ionizing radiation that can impact human health.

Such terms become more understandable and usable when workers learn that Commonwealth Edison's whole-body radiation exposure limits are 1,250 millirems for a calendar quarter, 300 millirems in a week, and 50 millirems in a day. Federal limits are higher. Limits for hands can be higher because they are not vital organs, and Edison's radiation safety specialists may approve higher weekly or daily exposure limits.

Nuclear plant workers encounter various degrees of radioactivity, ranging from low to very high. To work in those areas safely, radioactive work permits (RWP) specify time and radiation exposure limits for work assignments in radioactive areas.

Nehmer explains the importance of knowing the meaning of those radiation measurement terms, and what they mean to human health. He pulls no punches.

With a 100 REM radiation exposure, which is extremely high, "You'll be vomiting, nausea, diarrhea, and your hair will be falling out and your skin will be turning red. That's radiation sickness."

I'm taking notes, but look up to see how the class is taking this. They're still and Nehmer has their complete attention. They are riveted.

Dose is the amount of radiation received, the trainer continues. Dose rate is the amount of radiation received over time. They need to know how radiation affects their bodies. Radiation bombarding the human body smashes into living cells, killing or damaging them to some extent. Explanations of what that means to health may differ, depending on the source of information.

"Radiation hits you, and then it's over," says Nehmer. "Then your body repairs the damage. It is nothing you give off. It is like getting an X-ray." Radiation doses at the Zion station usually are small, "and usually effects are small. Cells are damaged, but you need massive amounts of cell damage before an organ will start to malfunction. Usually, when cells are damaged, they repair themselves. Like a cut, cells repair themselves."

Cell damage can appear early or late in life, including genetic changes and "the chance of something happening to my children and grandchildren. Both are valid concerns," the trainer says.

"Overexposures do occur, and we've had some fatalities from it," says Nehmer, but not in nuclear power plants. The first signs of a radiation overdose in the 25 to 50 REM range would be changes in blood cell counts. "You would not have observable effects under 100,000 millirems," he says. "Over 100,000 millirems, and things start to happen," including vomiting, diarrhea, and nausea. "Lethal doses take weeks or months to kill a person."

But not everyone is affected the same way, like a cold. "Some people are susceptible and die quickly. Some people are resistant to it."

Without medical treatment, 100 percent of people would die from a 600,000 millirem or 600 REM exposure. Treatment would include blood transfusions and bone marrow transplants.

Radiation is not stored in the body, he says. "Radiation hits you and then it's over. Then your body repairs the damage." (An exception would be inhaled or swallowed radioactive material, although the human body has ways to flush it out or the material loses its radioactivity over time like a fire that burns itself out.)

A 500,000 millirem exposure in 24 hours would be dangerous, but it might be harmless if spread over a lifetime, said the trainer. This brought the conversation to chronic, low-level exposures of 40 or 50 millirems over 50 years.

"The main concern is cancer," says Nehmer. This is a potential risk of working in a nuclear power plant. If 10,000 people were exposed to 1,000 millirems, he explains, one might die of cancer. But if those 10,000 people were exposed to all types of cancer-causing agents, 1,640 of them would be expected to die of cancer. It would be impossible to identify the radiation death from the others. To complicate matters, radiation from our natural surroundings exposes most people in the United States to 100 to 250 millirems every year, more at higher elevations like Denver, Colorado.

This is the first time Nehmer mentions cancer. It is a word the nuclear industry uses gingerly, and defensively. I'm curious how much he'll say about it.

Despite what the public might think, Nehmer says, "we know more about the effects of radiation than anybody does… We know what's going on."

To prove his point, the trainer puts radiation hazards in some context. People increase their risk of death by one in a million in lots of ways: smoking 1.4 cigarettes, drinking a half liter of wine, living with a smoker for two months, a bicycle ride of 10 miles, an auto ride of 300 miles, and a 10 millirem whole-body exposure to radiation.

Nuclear people often compare radiation exposures in power plants to medical chest X-rays, but it's a poor comparison. Medical X-rays usually are focused on specific parts of the body. The entire body is exposed to radiation in a nuclear power plant.

"Even the lowest radiation level causes some danger," says Nehmer. "There's always something done when you are exposed to radiation."

Nehmer compares radiation damage to sunburn, caused by radiation from the sun, a giant ball of nuclear activity. "Even that causes biological danger," he says, but adds, "the cells repair themselves and there is no lasting damage."

Allow me to deviate from Nehmer's lessons. I've added some description of a nuclear chain reaction, but purposely described Nehmer's class in detail to show what you would have seen and heard if you were sitting in that very important class with me. Nehmer largely gives the nuclear power industry viewpoint, but there are some in the industry who differ.

In my time covering Commonwealth Edison, I encountered James Toscas, an Edison spokesperson who also taught radiation safety classes. He believed there was no safe level of radiation exposure, a view shared by antinuclear activists and some medical authorities. But that did not mean Toscas was opposed to commercial nuclear power, only that he had a more nuanced point of view. He was more blunt and outspoken about cancer.

"If a radiation-damaged cell does not die," Toscas explained, "it may become a mutant cell or cancer. Even 1 millirem has a theoretical chance of causing cancer, but compared with workers in other industrial plants, [nuclear workers] have fewer accidents, so they seem to be better off." He saw the power of trade-offs.

Toscas also pointed to the overall safety record of the commercial nuclear power industry, compared with others. The U.S. nuclear power industry in 1984, for example, reported no on-the-job fatalities. But accidents claimed 124 lives in the nation's coal mines that year, when the reporting for this part of the story was done.

By 2021, the hazards of coal became much clearer. It is the world's most dangerous conventional energy source, measured by deaths from accidents and air pollution, according to Our World in Data, a scientific research organization in England. Coal kills roughly 25 people per terawatt-hour of electricity, largely because of particulate air pollution. The death toll for nuclear electricity was 0.03 per terawatt-hour, including deaths from the Chernobyl and Fukushima disasters. The World Health Organization says air pollution from burning coal causes one million deaths a year, including 30,000 deaths a year in the United States. Brown coal, also known as lignite, causes about 33 deaths per terawatt-hour of energy production.

Back to Nehmer and the radiation safety class.

Radiation is energy that travels through the air like light, heat, and radio waves. It is invisible, silent, tasteless and odorless. But it can be measured by instruments like the ones used and worn by nuclear workers to detect radiation and contamination.

Those tools include a Geiger Mueller detector, the RAD Owl, friskers, and portal monitors, the last being a metal housing the size of a doorway.

The Geiger Mueller detector comes with a probe attached by a cable to a housing with a dial. It makes a metallic clicking sound when the probe encounters radioactivity, ranging from a few lazy clicks to a storm of clicks that sounds like an angry buzz if the radiation is intense. When leaving the plant, a worker sweeps the probe over his shoes, clothes, or hair to check for contamination.

A RAD Owl, resembling a camera counted on a handle, serves the same purpose. A device known as a "frisker" looks like a big metal lollypop, with a round and flattened head on a handle. It is attached by a wire to a box that chatters like radio static when excited by radioactivity, and a dial shows how intense the radioactivity is.

If it is silent when you pass the device over your feet, hands, and body—you're clean. If it chatters, you're contaminated. You might get a click or two from background radiation; nothing to worry about.

Portal monitors resemble airport metal detectors. When starting the workday and ending it, nuclear workers walk through the portal monitor which scans their entire bodies. It's a "whole-body counter" that detects both interior and exterior body contamination. Step in, wait for the green light, and step out. If the light is red, you're contaminated. Portal monitors also have slots to check your hands for contamination. Everyone leaving a power plant walks through a portal monitor.

"If tripped," advises Nehmer, "get a paper suit and a plastic bag. Then walk each article of your clothing through the walk-through monitor. You're going to have to go home in a paper suit if your clothing is contaminated," or if the source of the contamination can't be found. It happens once in a while, says the trainer, but "don't get the idea you go home in a paper suit every day."

Nehmer explains allowable exposure limits set by Commonwealth Edison and by the Nuclear Regulatory Commission, for the whole body and for extremities. Extreme caution is taken with pregnant women because of hazards to an embryo or an unborn child.

"We take action to be sure she doesn't stay on a radiation job," explains Nehmer. "Once we find she is pregnant, she is out of a radiation exposure job. She doesn't lose tenure. After birth, she can return to the job."

Everyone has a right to report safety violations to the NRC. Forms are in the radiation safety office.

Work in radioactive areas is guided by a policy known by its acronym ALARA— as low as reasonably achievable. But the trainer reminds the class, "it falls on each individual to reduce his exposure as much as he can." Work that involves a radiation exposure above 50 millirems requires a radiation work permit (RWP) that specifies a time limit for the job.

Typically, there are three ways to reduce exposures. The first is the amount of **time** exposed to a radioactive area or material. The second is **distance**, staying as far away as possible from a radioactive area or substance. The third is **shielding,** which can include thick layers of lead, concrete or water.

Contamination, says Nehmer, is "radioactive material that is someplace we don't want it." Over time even the dust and grime around a nuclear reactor becomes radioactive, as well as the water that courses through a reactor to keep it cool.

Contamination can settle on a worker's body, or he can inhale or swallow it. "Think of contamination as invisible dirt," the trainer advises. When in a radioactive area, "assume all surfaces and equipment are contaminated. Assume your hands are contaminated." Don't scratch your nose or face, or smoke, drink, eat, or chew tobacco. "Working at a plant like this," he says, "there is always a chance of airborne radio-activity, always a chance to ingest something," which can lead to contaminating a body's interior, including glands, bones, muscles, and intestinal tract.

Holding up a length of magenta-and-yellow-colored rope, Nehmer warns: "You never cross this rope without checking with rad protection," referring to radiation safety specialists on duty. Colored ropes are safety boundaries.

Then Nehmer introduces the class to two personal radiation detection devices that will be their best friends in a nuclear plant, the dosimeter and the film badge, which are clipped to shirt pockets while on the job anywhere inside the power plant.

A dosimeter is a metal or plastic tube about 4.5 inches long and half an inch in diameter. Aim it up to a bright light, look into it with one eye, and you see a scale from zero to 200 millirems. A line on the scale moves to the right, showing increasing radiation exposures a worker gets throughout the day. A dosimeter can be reset to zero at the end of the day so the worker can keep track of daily expo-sures. It is sensitive to gamma rays.

The other personal monitoring device is a film badge, a plastic housing measur-ing 1.75 inches wide, 3.25 inches long, and a quarter of an inch thick. It has a slot in the middle to insert a strip of black and white photographic film that gradually fogs from radiation exposure, indicating the extent of on-the-job exposure. The film is developed every two weeks and replaced. It is sensitive to gamma rays and beta particles.

Commonwealth Edison sends workers quarterly reports on their radiation expo-sures, which are required by the NRC. Nehmer urges the workers to keep track of these records, especially if their work takes them from one power plant to another.

Workers also are fitted with respirators on their faces if they must work in areas with airborne radioactivity.

Protective clothing is another layer of defense: those baggy, white canvas coveralls we saw hanging on the walls when we came into the classroom. They "keep contam-ination off the body, not to stop radiation," explains Nehmer. "It might stop all alpha and some beta, but gamma will go through you and out the other side of your body."

Nehmer introduces us to the gear that essentially will be our uniforms while inside radioactive areas. Along with the coveralls, which have chest-high pockets for film badges and dosimeters, are yellow hoods that cover our heads down to our shoulders, cotton glove liners, black rubber gloves, black rubber overshoes and yellow plastic boot covers.

The trainer drills the class in donning the protective gear in exact order: cotton glove liners, yellow plastic booties, canvas jumpsuits with hood, rubber overshoes, rubber gloves and masking tape to fasten the top of the gloves to the jumpsuit arms, closing the openings to make them leak proof. The dosimeter and film badge go into a pocket.

Typically, inside the power plant, before donning this protective gear, a man strips down to his shoes, socks, and underwear. A woman strips to a one-piece bathing suit or body sock.

Now that we're all suited up in loose jumpsuits, we look like dumpling men. The next, and final, stage of our training, is learning how to remove that clothing in exact order as though they might be contaminated.

Clothing removal follows a three-step process with the help of "step-off pads," three plastic floor mats with instructions written on them. Coming from working inside the power plant, a worker encounters the first step-off pad saying: "Remove protective footwear before stepping here." Don't step on it yet. Remove the masking tape at the top of the gloves. Then remove the rubber overshoes. It's harder than it sounds, tugging at the overshoes while standing upright. Drop the overshoes in a barrel and move forward.

The next two step-off pads offer similar advice as a worker removes his hood, overalls, and rubber gloves, dropping them in barrels and standing there in his shorts.

While advancing on the three steps, "never step back," warns Nehmer. Assume that the step-off pad behind you is contaminated, and that you are advancing onto clean step-off pads as you remove your outer garments and leave them in barrels. Try not to touch outerwear.

"It's important to turn things inside out and contain contamination" that might be on the garment, says the trainer. Don't just reach up and pull the hood off your head. "You chance dripping contamination on your hair," he says. Instead, slide your fingers inside the hood and lift it off and away, and into the disposal barrel.

While your hands are still covered with cotton glove liners, use your fingertips to remove the film badge and dosimeter from the jumpsuit and place them on the pad in front on you. Unzip and roll the baggy jumpsuit off and down so it's inside out and step out of it, then drop it into a barrel. Dispose of the cotton glove liners.

Retrieve the dosimeter and the film badge.

It's a good idea to run the process through your mind and memorize it: the precise order of donning the clothing, and the precise order of taking them off in a way to avoid smearing any radioactive contamination on the clothing on yourself.

We're not done yet. The final step is scanning ourselves and the two radiation detection devices before leaving the step-off pad area. We pass a frisker over our feet, hands, body, and the devices. Take a full two minutes to scan yourself, warns Nehmer.

If you're clean, he says, "pick up your stuff and go."

When entering the power plant, a worker's belongings are X-rayed, including their lunch containers. Workers worry that the X-ray will cause their food to become radioactive. Nehmer says that can't happen.

The test of 25 questions takes 45 minutes. It includes responsibility for safety, radiation, where to go in case of an emergency, radiation limits for pregnant women, protective clothing, radiation exposure limits and guidelines, reporting safety violations, radiation work permits, and types of radiation warnings found in the workplace.

Later company guidelines for plant access would become more stringent, calling for two tests covering security and safety and fitness for duty, and a separate test for radiation worker training.

I get one question wrong and a score of 96. Passing is 70. I stumbled on question seven, which asked what a yellow and black rope means. The correct answer is "a hazardous area." I said a contaminated area.

I get a card saying that I am N-GET qualified. I ask training supervisor LeBlond what that entitles me to do, in real terms.

"That means you've successfully completed Commonwealth Edison's general employee nuclear training program," he answers. "You're now qualified to be issued a film badge and work inside the station and to work in a radiation area.

"If we said later, go and take a look at the containment building [where the reactor is located] you could put on the clothes and walk the step-off pads and do what you did for real."

LeBlond tells me I will meet David Smith at 8:15 a.m. tomorrow at the station gatehouse. "He's sort of like your social director," LeBlond quips.

I leave the training building at 3:05 p.m., almost eight hours after arriving there, and write in my notebook: "Today, I am qualified to be a nuclear worker."

Chapter 19: Meeting My Nuclear Family

FOG SWIRLS AROUND THE ZION POWER STATION AS I REPORT FOR MY FIRST DAY of an assignment that lasts 28 months, fully qualified and trained to go anywhere inside that radioactive facility.

It's early and the sun is a bright red ball rising over the rushing Lake Michigan waters as I pull into the employee parking lot with a feeling of excitement and nervousness like the first day of school. Two reactor containment buildings standing 200 feet high are dark silhouettes in the gray mist. A high-pitched whine sings from the nearby turbine building.

The 38-mile drive from my home in Evanston, Illinois, soon becomes routine as I repeatedly visit this mammoth power plant, sitting there in a web of power lines. Passing motorists on Sheridan Road, in the City of Zion only two miles away, might wonder what's happening there. And who are those people who work in radioactivity? I wonder too, and I'm there to find out.

Although I don't realize it at the time, I will form an attachment to this nuclear generating station. It will become my nuclear home, my nuclear family of the fission kind. Five hundred people work there. They form attachments like any family, they hug and bicker. I want to get to know them. Like all family relationships, they will teach me the facts of life—nuclear life.

It starts when I join hundreds of workers on that misty morning to walk through the front gate and past two chain-link fences running parallel to each other, topped with barbed wire. We run the guardhouse gauntlet at the main entrance to the power plant. Although 500 workers normally staff the plant, that number doubles as contract workers come in to help during this outage.

One at a time, they walk through bomb and metal detectors, putting their lunch-boxes and toolboxes on conveyor belts passing under X-ray units. Three armed female guards in dark uniforms stand beside two banks of scanners, watching the workers pass or searching their belongings, alert for weapons, drugs, or alcohol.

I go to a window marked for visitors and get a visitor's pass, then sign a waiver acknowledging that I'm aware of hazards in the facility. I empty my pockets and get pat-searched. Eventually, I get a photo identity card allowing me to enter that plant like one of the workers.

Dave Smith and I immediately identify each other. Among the passing tide of hurrying workers, we're the only two looking around for somebody we've never seen before. Smith, jokingly called my "social director," is a maintenance staff manager with a lot of power plant savvy. Young and trim, deliberate with a talent for details.

He's my escort and guide. Under agreement with Commonwealth Edison, I may go anywhere in the plant as long as I have an escort, and it's usually Smith. It turns out to be a good arrangement.

Smith has a firm grasp of nuclear engineering and what is going on inside the Zion station. My periodic visits to the plant come at an opportune time, when reactor Unit One is shut down for a system-wide overhaul after operating 10 years. Some of those maintenance operations take days and weeks. To save time, Smith will call me to alert me when a major operation is about to start, or reaches an important point, so I can go to the plant and watch.

First, my guide leads me to a trailer parked behind the turbine building, near the lakefront. It is our temporary office, where I get a desk and a chair. In the coming weeks and months, Smith and I will start our day here, where he'll brief me on what I am going to see that day, and we will often return to the office at the end of the day to hash out what we saw and heard together. Smith explains technical points and I ask questions for clarification, including the meaning of technical terms or how to describe equipment being dismantled. Sometimes the exchange gets personal, Smith speaking as a member of the nuclear community and I as a journalist.

I will not report to the plant every day, but my plan is to see some part of every major step in the outage.

The agenda today, explains Smith, is to meet some of the senior managers who are supervising the outage. They probably want to look me over, and to explain some ground rules before I actually make my way into the center of activity.

On our way to that meeting, Smith explains something he thinks I should know: "Some people don't understand why we change the plant." Some changes in plant design are mandated by the Nuclear Regulatory Commission based on what it learned from accidents or operating experience at other plants, like the Three Mile Island accident. That includes fire safety upgrades.

Because Commonwealth Edison has been cited often for safety violations, people wonder how well the utility relates with NRC "on a day-to-day basis." Smith says I might be able to sit in on a meeting between Edison and the NRC to see how they get along. "We'd like to have you attend those meetings," he says.

Smith leads me to the central office building for the orientation session. I'm introduced to the top managers for operations, maintenance, safety, radiation safety, and training. They are a mix of young and middle-aged men, some wearing suits in a power plant that prides itself on flannel shirt informality.

Ed Fuerst, the station's assistant superintendent for operations, speaks first. He's a broad-shouldered man with a thick, black mustache and hair, wearing a suit and tie. A veteran of the nuclear navy, like many nuclear power people, Fuerst is gruff, with the military bearing of an impatient drill instructor. But his greeting is cordial, and direct.

"You're not imposing," says Fuerst. It's important for the public to see the side of nuclear energy that I'm going to see. "It's vital the public realize we're not a bunch of necessary bastards. My time is yours, when you want it."

It's a generous offer from one of the busiest men at the Zion station. There's something else on his mind, and he expresses it bluntly: "We always have suspicions about reporters. We have had bad experiences with them."

It's a fair warning, and a fair description of how people in the nuclear industry, and in the Zion power station in particular, regard journalists. It won't be the last time somebody in the station expresses doubts about having a reporter in their midst.

Though outwardly stern, Fuerst has a sense of humor. "It's kind of ridiculous," he says during one of our encounters. "You come to an electric generating plant and the first thing they do is give you a hard hat and a flashlight." On another occasion, I will catch a glimpse of the passion, almost poetic, which Fuerst feels for his work as we enter the turbine building together and hear the piercing, high-pitched whine that's almost deafening as the turbine blades spin 1,800 revolutions a minute. It's one of those areas where workers should wear hearing protection.

"I open a door and hear that sound," says Fuerst. "It's the sound of a thousand megawatts being generated. I get a chill even now."

Back at the meeting, Fuerst says the 10-year anniversary outage is a little bigger, involving a little more work, than other periodic maintenance outages. "Anything at a power plant can be broken down to something that is simple and recognizable," he asserts. Peter LeBlond, training supervisor, agrees, saying, "It's like working in a factory."

George Pliml, assistant superintendent of administration and support services, adds: "Most people here are proud to work at Zion." It's a pride that comes from working with nuclear energy. It's far from being an ordinary factory.

The conversation swings to my presence in the station.

"Don't be bashful," advises Kurt Kofron, assistant superintendent for maintenance, advice a reporter usually does not need, but is appreciated. "We are concerned about your radiation safety," he adds.

Remembering basic rules of safety learned in my N–GET class is "a good start," says Pliml. I'm reminded of the color-coded safety rope barriers.

Outwardly jovial and plump, Dick Principe is the plant safety coordinator whose demeanor turns deadly serious when ticking off some rules of safety and things I should not do. "Don't touch my cards!" he says, his voice rising practically to a shout, emphasizing the importance of what he is saying. Principe is referring to cardboard out-of-service tags measuring roughly two and a half inches wide by five inches long that are attached to equipment and devices throughout the plant during an outage, indicating they are not operating.

"It's an honor system, and a sacred system," Principe says of the cards. "They protect people in the plant." It's forbidden to meddle with them. "You will see valves and switches around the plant," he says. "Hands off as far as switches and valves."

He tells me the evacuation alarm is a continuous, steady blast of a horn. "Believe me, you will hear it," Principe says. I'm told to go to the assembly area in the machine shop and am given a card printed with a map of the Zion station, showing the location of the machine shop.

I'll be in areas where an overhead bridge crane is operating, hauling loads of equipment or material.

"Don't stand under loads and look," warns Principe. Mindless gawking can be dangerous. Things fall. "There are hazards in the plant. Be aware of what is going on around you. If you have a problem, come to me."

I'm told to wear my dosimeter and film badge, and to keep records of my radiation exposures.

"We are going to take all of the fuel out of [reactor] Unit One, and all support mechanisms" inside the reactor vessel which holds 140 tons of uranium fuel, explains LeBlond. "We'll do a very detailed inspection of the condition of the reactor." Removing the 193 fuel assemblies containing the uranium will take three to four days.

Terry Rieck, a radiation safety supervisor, explains that four or five of his safety specialists will be watching these operations in person or via video monitors. His crew is "ready to step in if there is a problem," says Rieck.

Typically, only a third of the highly radioactive uranium fuel is removed and exchanged periodically as it "burns out" in the nuclear fire of an atomic chain reaction. But for this outage, all of the fuel comes out of the reactor core so the steel reactor vessel in which it sits can be inspected for defects.

"We're looking for any changes in the vessel from one that is brand new," says Smith. "Hairline cracks that could cause trouble later." The reactor vessel would be called a "boiler" in non-nuclear generating stations.

LeBlond and Rieck watch me as I take notes and ask questions. This is technical information. They want to see if I'm keeping up. At this point, some of this information is theoretical and hypothetical. It will become very real and personal in the next few months.

As the orientation meeting winds down, I tell the managers that I promised Graesser to report what is usual and ordinary about the power station, and that I will be guided by their enthusiasm, which I've already noticed.

After the meeting, I'm handed a yellow plastic hardhat and a yellow flashlight. The hardhat has my name "Bukro" spelled out in bold red letters. But it disappears a few days later and is replaced with a plain yellow helmet like the ones most plant workers wear. Maybe someone thought it was not a good idea to make me so conspicuous, like a target. Maybe someone is allowing me to do what I'm trying to do, which is to blend in and do my work like everyone else.

As Smith walks me to our next stop, the auxiliary building, he says, "think of us as being manufacturers of electricity." The auxiliary building is a warehouse of tools and equipment, and headquarters for the radiation safety department.

I get a dosimeter and a film badge, and paperwork to keep records of my daily radiation doses. I sign a form detailing the requirements to enter radioactive areas. It comes with a note on the latest survey of radiation readings in areas where people are working, including the auxiliary building and the containment building.

I meet Bob Cascarano, the outage coordinator, youngish and wearing a shirt and tie, his dosimeter and film badge clipped to the shirt pocket, and carrying a clip-board. His shirt sleeves are rolled up. There are 493 "critical path jobs" during the outage, explains Cascarano, and he must keep track of all of it, as his job title implies.

A five-page printout shows every step scheduled, beginning on September 3, 1983, when reactor Unit One stopped producing power, to December 10, 1983, when the reactor is scheduled to be operating again.

On the day of shutdown, the reactor temperature was 547 degrees Fahrenheit. That drops to 140 degrees in a day or so, cool enough to perform some tests, then to 70 degrees soon after.

About 60 plant modifications are scheduled. Thirty major systems will be discon-nected, dismantled, repaired, and evaluated. This includes a close look at the hun-dreds of miles of pipes that make up the vital primary, secondary, and emergency core cooling systems needed to keep the superhot nuclear reactor from overheating.

Dating to 1973, the reactor Unit One complex was built with asbestos, which now must be removed under new safety regulations. During the shutdown, the Zion staff will update and modify the plant to meet new NRC safety standards for fire protection and for "environmental qualification," an example of arcane nuclear vocabulary requiring some explanation.

Carl Schultz, a staff technician, explains the meaning of "environmental qualifi-cation." If the main steam line breaks inside the containment building, the interior environment of that building will be very hot and wet. "We have to show activities will function in that environment," explains Schultz.

Diesel generators must be tested for reliability. "We have to show a diesel will start at a certain time and reach load in a given time," he says. This includes a 24-hour diesel endurance run, which stresses the diesels. Some people balk at stressing diesels like that, but NRC requires it.

Hard wiring will replace temporary wiring.

A major item during the outage, says Schultz, is pressurizing all the pipes on 26 water supply systems, including reactor coolant. Water pipes and tubes of all sizes run through the power plant. They will be pressurized to 102 percent of normal operation to see how well they withstand the higher pressure and don't burst.

The goal is to test piping integrity. "You are looking for weld problems and leaks from bad welds," explains Schultz. Water lines are flushed with hydrogen oxide to wash away mineral deposits coating the interior pipe walls, producing what work-ers call a "crud burst."

Keeping the reactor vessel from overheating is critical in a nuclear generating station. The four reactor coolant pumps will get special attention during the outage. Each of them is 26 feet long and six feet, four inches in diameter. Each pump can circulate 88,000 gallons of water a minute through the reactor and to the four steam generators at a temperature of 590 degrees Fahrenheit and pressures of 2,235 pounds per square inch.

Every system in the plant has operated or been tested over the course of the last 10 years, except for one. That, says Schultz, is the overhead spray system at the top of the reactor containment building. It consists of five concentric rings of pipe.

The system is designed to douse high heat and pressure in a major reactor accident. "We've never had to spray the containment with sodium hydroxide," says Schultz. "It's there for the very worst accidents. That's probably the only system that hasn't come into play." Sodium hydroxide scrubs radioactive iodine out of the boiling air in such an accident and flushes it onto the floor of the building, where it could be washed away.

I'm told that Edison once tried to hire a contractor to test the overhead sprinkler system. An advertisement of a suburban company said, "No job is too big." When a representative of the company looked at the sprinkler system, he said, "This job is too big."

In another key test, technicians will determine how much radioactive air or gas might escape from the containment building in a major accident.

"As the technology changes, we find ourselves back-fitting the plant to upgrade to the present state of the art," says John Johnson, an outage coordinator. This includes installing new radiation monitors because previous monitors in the fuel-handling area were not reliable. Another modification is relocating valves to areas where they are easier to reach for maintenance.

It becomes clear that human beings are not the only "actors" in a nuclear power plant. The machinery deserves recognition, too. The size of the Zion plant and its components are truly remarkable. A small car could drive through some of the massive water pipes that guzzle cooling water from Lake Michigan and into the plant.

The size and complexity of its parts are impressive, beginning with the reactor vessel containing the fuel core. It's 43 feet, 10 inches tall and made of carbon steel, bigger than a school bus. It's connected to four reactor cooling pumps, each 26 feet long. Those shoot water to four 67-foot-tall steam generators, where water is converted to steam that goes to the turbine to produce the electricity that goes out to homes and industry via transmission lines. The turbine housing is long and streamlined, about the size and shape of a modern railroad locomotive. A condenser turns steam back into water, which is cooled and sent to Lake Michigan.

One more vital piece of giant equipment needed in this high-temperature balancing act is the pressurizer, 52 feet tall, nine inches in diameter.

Zion's Unit One reactor is a pressurized water reactor, meaning the cooling water coming from the reactor at 590-degree-Fahrenheit temperature must be kept under high pressure to prevent the water from boiling at 212 degrees Fahrenheit. The pressurizer does that by exerting 2,500 pounds per square inch of pressure on the cooling water system. When full of water, it weighs 1.1 million pounds.

At the end of my first day inside the plant, Smith wants me to see one more thing.

The Zion generating station is a water processing plant, Smith tells me, and invites me to see the lakefront water-intake structures. Cooling water for the Zion station comes from the lake at a point about 2,600 feet from shore and 22 feet below the surface.

Two discharge pipes extending 760 feet from shore and 154 feet north and south of the intake pipe discharge water that is 20 degrees warmer than lake water. These pipes are 16 feet in diameter. Two pumps circulating 250,000 gallons a minute each are enough to supply all the water the plant needs.

"It uses millions and millions of gallons an hour," says Smith. Both pumps working together can propel 60 million gallons an hour if needed, he says. "I can't imagine that much water."

It's the first time all day that Smith is stumped. Otherwise, he's had an answer for every question.

Before leaving the plant, I check my dosimeter for the first of many times. It says I had a 3 millirem radiation exposure that day.

Chapter 20: The Real Thing

THE DAY STARTS WITH DAVE SMITH INVITING ME TO ATTEND A "GATHERING OF the eagles."

The "eagles" are a select group of young supervisors in charge of various parts of the outage. They are troubleshooters and decision-makers, expected to answer problems springing up. They meet every morning for problem-solving, information gathering, and updating the outage schedule. "Surprise prevention" is a major goal, says Smith, who also describes the young men as "Johnnies-on-the-spot."

They meet in a corner room overlooking Lake Michigan and Illinois Beach State Park. Twenty-seven men clad in plaid and short-sleeved shirts sit there, their dosimeters and film badges clipped to their shirt pockets. They are a macho-looking group. Open-necked. No ties or suits. They wear boots, cowboy and engineer's.

Outage coordinator Bob Cascarano starts the meeting, saying the top of reactor Unit One might be lifted off that night. It's called a head lift, a major event in the outage. Other men give their reports. One of them says 3,100 tubes in a steam generator were inspected; three were found to be defective and plugged. The mood is relaxed and informal.

"Okay, thanks," says Cascarano, ending the meeting after 26 minutes.

"Outages are real interesting because you see things opened up you've never seen before," explains Smith, while leading me to our next stop, the auxiliary building. On the way, a mouse scampers across the floor. Smith calls rodent control, worried that it might be contaminated and get out of the plant. It did not occur to me that wildlife was in the plant. Further on, we see oil burning in a pit and employees engaged in firefighting training.

The auxiliary building tucked between the two reactor containment buildings is a radioactive crossroads of many systems connected to the reactors. They involve safety, radioactive wastes, chemicals, cooling water, and controls for those systems. Expect 1 to 2 millirems of exposure from our tour of the building, Smith tells me. It's my second visit, but the first was short and on the edges. This time, it will be longer.

Inside the building, I get a radiation work permit (RWP) allowing me to enter a radioactive area for a limited radiation exposure. A technician at the check-in window wants to see my N-GET card to be sure I'm trained in radiation safety, then says "no problem." I sign a card with my time of entry, my name, and my badge number. The card goes into a rack until I sign out. The card allows me to be in the auxiliary and the Unit One reactor buildings.

"We are on our own to make sure we don't get into trouble," says Smith as we walk further into the auxiliary building. "We'll check [our dosimeters] on the way out to see if we get anything."

When do we get into the radioactive area? I ask Smith. "We're in it now," he answers. We pass laboratories, barrels of radioactive waste, a decontamination room with a shower, some medical supplies, heavy-duty cleaners, and a room equipped with a stretcher and a gurney, which do "not get a lot of business," Smith says. Stored here are radiological emergency kits in a red metal box.

Contamination is the major concern, says my guide, including kinds that can be swallowed or inhaled. "We are not equipped for radiological emergencies," he says. They can do first aid. "People decontaminate by taking a shower—the normal washing process. Occasionally people get their clothes contaminated, and they lose the clothes."

"We're in a low-dose area," Smith says. The auxiliary building floor is painted white with a special paint that is easily decontaminated.

Some workers in the auxiliary building are wearing canvas anti-contamination clothing, some are not. A green box called the "hot box" is marked "contaminated tools." Tools are reused and decontaminated as much as possible. Plastic bags full of radioactive trash—cloth, paper, rags, wooden planks, glove liners, and gloves— await disposal. They are compacted into drums and hauled away to disposal sites.

The auxiliary building contains reactor control panels, used in case the control room is not usable.

A frisker sits near a doorway for people who want to check themselves for contamination.

Pumps jut from the floor in one part of the building. One has radiation warning rope strung around it, and a sign reading: "Caution—contaminated area." Clusters of warning signs and ropes hang nearby.

"We'd go crazy trying to stop all the leaks," says Smith, "so we contain them and try to prevent problems."

Then we see a problem.

A man wearing street clothes hands a bundle of bolts to a worker suited up in anti-contamination togs and standing behind a radiation warning rope. Reaching to give the bolts, the man in street clothes leans over the yellow and magenta warning rope and into the potentially contaminated area where the other man is working.

The worker in anti-contamination gear also is trying to measure something. He gives the end of a measuring tape to the man outside the warning tape, who then leans over the rope and into the contaminated area again while the man in garb measures. They violate all the rules of contamination prevention.

"What he did was a no-no," says Smith. We approach the man in street clothes, identified as George by his badge. I won't mention his last name. Smith stops him as he is walking away and asks: "Are you going to get that tested?" referring to the measuring tape the man is holding. Smith explains that the man in the contaminated area might have contaminated his gloves, which could have contaminated the measuring tape, "and now it's on you."

George looks surprised, saying that he believed what he did was okay, "as long as I stay in the building." I don't know what George did after that. The incident reminds me that although nuclear power demands precision and careful planning, humans are unpredictable.

Smith and I leave the auxiliary building. My dosimeter shows that I had a 1 millirem radiation exposure during my tour.

At midday, I get a surprise. We encounter Terry Rieck, Zion station's radiation-chemistry supervisor. That's a technical way of saying he's in charge of radiation safety. He invites me to join him on a tour of the reactor containment building, where much of the key outage work is planned. It's the nuclear reactor's lair.

For the second time that day, I fill out cards and get a work permit spelling out the allowable radiation exposure limits for what I am about to do, and for where I am going.

Rieck becomes my escort. He leads the way to the auxiliary building where workers suit up in anti-contamination gear. Canvas coveralls of different sizes are neatly stacked on shelves, along with hoods, gloves, glove liners, and yellow plastic booties. Rubber overshoes of various sizes fill barrels. This will be the first time I suit up in protective gear for real.

A moment of panic strikes as I wonder if I'll remember those precise steps for donning that gear. I settle down and concentrate, and Cliff Nehmer's drill from two days earlier comes to me. Rieck reminds me what to take. Gathering our gear, we go to a room for Edison employees and strip down to socks, shoes, and underwear. Then we suit up in our anti-contamination clothing.

"The real thing," I scribble in my notebook quickly, remembering to take notes. I'm about to take a plunge into a highly radioactive world. It seemed like going from gentle basic training to the rigors of combat.

Rieck and I walk to a uniformed security guard who takes our identification tags and puts them into a rack until we return. In an emergency, I'm told, unclaimed name tags would identify missing persons.

A giant doorway leads into the Unit One reactor containment building. It's built to withstand tremendous forces. The reinforced steel door, now open, is 20 feet in diameter. It looks like a door to a bank vault, but much bigger. The concrete wall at the base of the containment building is three and a half feet thick. It's massive, intended to keep the mighty force of the nuclear genie cooped up inside, where it belongs.

Rieck and I walk through that doorway, into a world that is both alien and familiar, the inner sanctum of a nuclear power plant. Some of the machinery is common to all power generating stations. But nuclear energy is a different breed, requiring special precautions and constant warnings of radioactivity. It takes vigilance and a level of awareness not found in most industrial activities.

Once inside the reactor building, my face seems to tingle. I know the sensation is only my imagination. I'm thinking about being bombarded with radioactivity, but I know you can't feel this radiation. Still, it's a bit unsettling.

The air is cool and smells clean and fresh. The containment building is pressurized so that air flows inward, not outward. This is a safety precaution to keep airborne contamination from escaping the building.

The floors are covered with yellow plastic sheets. Rieck explains that it's easier to dispose of the sheets when the work is finished, rather than scrub contaminated floors.

A bell suddenly rings and a siren shrieks a warning. I startle. Rieck, who always wears a leprechaun's impish grin, explains: "The overhead crane is moving."

We walk to what is known as the refueling cavity, where the nuclear reactor lives. The cavity, 45 feet deep, looks like a stainless steel swimming pool surrounded by guardrails. Rieck and I step to the guardrails and look down into the cavity and see a tall bell-shaped dome to our right. It's the top of the 43-foot-high nuclear reactor, known as the reactor head, and it covers the control rods that govern the reactor. The rest of the reactor is below the cavity floor.

This is my first look at the yawning space where much of the outage work will focus in the coming weeks. I'll learn a few things and see some things even veteran nuclear power people never see.

After a few minutes, Rieck says, "Let's not stand here and talk." He's reminding me that we are standing in the reactor's intense radiation glare. A radiation scan over the cavity registers about 25 millirems an hour.

Three men wearing yellow plastic suits over their coveralls are on the cavity floor below, loosening the bolts that fasten the reactor head to the cavity floor. They're wearing breathing apparatus to avoid inhaling airborne radioactive dust or debris. Rieck says they are in a "smearable area" of contamination that demands extra precaution.

The stubby reactor, surrounded by pumps and machinery, squats in the center of the containment building. Equipment rests on a multileveled maze of concrete and steel-grate landings connected by steel ladders—like the interior of a ship.

Suddenly, Nehmer's warnings against scratching my nose while in a radioactive zone hits me full force. My nose and face itch, and I can't do anything about it. The urge to scratch grows from a nuisance to an obsession. I try to ignore it. The warning seemed a bit silly at the time, but think about it. How many times a day do we scratch or touch our faces? We do it without thinking. But in a contaminated area in the nuclear world, you *must* think about it. Later, I will find an answer to this dilemma.

"If you want to climb," says Rieck, "I'll show you the valve that went wrong at TMI [Three Mile Island]." We climb 40 feet up a steel ladder to the top of a pressurizer and look at a saucer-shaped pressure relief valve. A valve like that got stuck in the open position at TMI, causing a loss of cooling water and touching off a series of operator errors that caused a partial fuel breakdown.

"Let's not discuss it up here," cautions Rieck. We're standing in a 50-millirems-an-hour radiation field from the reactor below. We climb down the ladder and discuss TMI. "If they closed that block valve," says Rieck, pointing to a nearby valve, "they could have stopped the leakage" and the accident.

Next, Rieck leads the way toward the basement, past signs warning of a "hot spot" and "160 m/r/hour" on a pipe. We walk between concrete walls four feet to six feet thick, serving as protective missile barriers against flying metal parts in case some of the nearby tanks and equipment under high pressure rupture.

We stop at a basement doorway through one of those concrete walls, where radiation safety technicians set up a checkpoint. The doorway leads into an area of high radiation. It's tightly restricted. Rieck gives his personal approval for me to enter with him. Technicians record my name, badge number, and current dosimeter reading.

Rieck asks the radiation safety foreman about radiation readings where we are going. The foreman says they range from 300 millirems an hour in the "hot spots" to a low of 5 millirems an hour and identifies their locations.

While inside that area, passing warning signs, we encounter a group directing work via a television monitor on a steam generator, a highly radioactive component in a nuclear plant. Workers dart through an open hatch at the bottom of the steam generator, looking for leaking tubes, plugging them, and darting out again to reduce their exposures. A worker hands me a headset and I listen to a supervisor in a trailer parked outside the plant directing the operation.

Rieck and I go to another highly radioactive area, the pump deck. This is where the four giant reactor coolant pumps are located. Cooling water coursing through the reactor becomes radioactive and so do the coolant pumps. We come upon one of those pumps, the size of a car, lying in pieces as though it's being dissected.

"I attack those pumps like a surgeon," says Phillip "Ski" Stachelski, holding up hands encased in thick rubber gloves proudly. A seal specialist, he is the mechanical maintenance foreman in charge of reactor coolant pump maintenance.

Twenty-six workers have been laboring on the four coolant pumps in the last six weeks in rotation, so none of them reach their exposure limits and get "burned out."

Ski apologizes that the pump desk is dirty, saying it's normally spic-and-span. And he cautions me against getting too close to an open pump.

Radiation protection specialists like Rieck wear orange hoods instead of the standard yellow hoods. That makes them easy to identify, especially if someone has questions about radiation. I see two of them in the pump deck area using scanners to check radiation levels.

"They're always watching," Stachelski says. "Their job is, if somebody dumb comes and sits on a [radioactive] pipe, he gets kicked out." They can order workers to leave the area if radiation rises higher than expected and help keep exposure levels as low as possible.

Workers seem antlike in size, next to some of the massive machinery.

When leaving the high-radiation area, we check our dosimeters. Mine reads 12 millirems.

Walking while clad in the canvas coveralls is a new experience. It feels like shuffling along while encased in a baggy sock.

Rieck and I return to the refueling cavity area and meet Larry Thorsen, the fuel-handling foreman.

"If you want to see fuel moved," jokes Thorsen, "it'll cost you. We sell popcorn." Why is it so popular? "The whole idea of a fuel outage is to remove that fuel, so everybody is interested in fuel movement." The sight is fascinating in part because so few people in the nuclear business get a chance to see it. And it's all done underwater.

This tour is ending. Rieck leads the way out of the containment building and into the auxiliary building, to the step-off pads where we must disrobe in the prescribed manner, including the little tricks for avoiding contaminating clean parts. This is easier because the three step-off pads have written instructions. I remember the procedure, starting with the rubber shoe coverings.

While standing on one foot and pulling off a shoe covering, I teeter toward a barrel of discarded boots before regaining my balance. With my glove liners, I lift the hood from my head from the inside, then slip off the coveralls, remembering to turn them inside out to avoid contaminating clean parts.

After removing my coveralls, I absentmindedly tug at my T-shirt while still wearing glove liners. Rieck frowns but does not say anything. I can tell by his expression that he is not pleased. It was a careless breach of training and he reacted to it. I should not touch clean underclothing with clothing that might be contaminated.

The frisker, though, says I'm clean.

One thing has changed since we went into the containment building hours ago. A female security guard has replaced the male security guard at the entrance to the containment building. I walk to her in my shorts to reclaim my identity tag before going to the room with my street clothes. She looks me straight in the eyes.

I follow Rieck to his office for a debriefing and for an expert's view of what we saw and did in the containment building. Why did he take me there?

"I wanted you to see major pieces of equipment the operations would be involved with in the next few days," he says, major equipment like the steam generator. "I especially wanted you to see the care people go to, to protect each other. The radiation monitoring. I wanted you to begin feeling comfortable going in and out of a radiation area. I wanted you to talk to Ski and Thorsen to see the pride they have in their jobs. We saw woman experts in different areas. A lot of people have expertise in different areas. We work together as a team. They individually know their jobs. They're good."

I ask Rieck what he thought about his radiation control specialists as we toured the plant. His answer: "People working for me were concerned about where we were going. When I asked about radiation levels where we were going, they were aware of what they were. I saw two of my people checking things out. They were watching and checking people. I felt good about what I saw. The people who work for me were doing their job."

That was the end of a busy day, unlike any I'd ever experienced. But it was why I was there, to see nuclear power from the inside.

I had spent 88 minutes in the auxiliary building in the first part of the day, and my dosimeter tells me I got a 1 millirem radiation exposure. The tour with Rieck in the reactor containment building lasted two hours and 35 minutes, with a radiation exposure of 17 millirems, for a total of 18 millirems.

As I head for the exit and the portal monitor, I remember something Nehmer said after the training class had ended. "How long will you frisk in the plant?" he asked. I drew a blank. "Two minutes," he said.

It was a gentle but firm reminder I never forgot. As I prepare to exit, I pick up a frisker and pass it over my head and face, my shoulders and arms, my hands, over my chest and back and down my legs and over the soles of my shoes.

Slowly, I take the whole two minutes and get no chatter from the frisker. I'm clean. Considering where I had been that day, I'm glad to know it.

Chapter 21: How to Open a Nuclear Reactor— Very Carefully

JOHN MILLER IS A SLIGHT MAN WITH THE BOOMING BARITONE VOICE OF A country-western singer, and he wears the orange hood of a radiation protection specialist.

"If I jump up and down and scream and holler—run!" he says only half-joking, jabbing a finger at the open doorway of the reactor containment building. These are his final instructions as he and about 20 other hooded figures wait for a major step in reactor Unit One's maintenance outage.

They're going to lift the top off the reactor—called a head lift—all 350,000 pounds of it, exposing the highly radioactive fuel core where the nuclear chain reaction sizzles. When that happens, radiation levels in the reactor building will soar. It's like lifting a lampshade off a bright light, only in this case the radiation is invisible and potentially deadly.

Miller is holding a radiation monitoring device. He knows radiation levels will jump during the head lift. But if they unexpectedly go dangerously high, he will order everyone out of the building.

It proves to be a waiting game. My guide, Dave Smith, and I speak several times on the telephone the night before, trying to figure out when the head lift is likely to occur. It's a major event in the outage, and I want to be there to see it. The first estimate was 1 a.m. the next morning, but then management rejected that, fearing workers might be tired at that time of night. They should be fresh and alert on a job that takes their full concentration and skill.

Smith and I agree to meet at 7 a.m. the next day in the Zion station. Upon arrival, we head to a foreman's office. "If you are there by 8 a.m., you'll have plenty of time," he says. Smith and I go to the auxiliary building to ask another foreman for his permission to watch the head lift. He approves, but cautions us to stay near the containment building doorway when the overhead crane lifts the reactor head.

At 7:40 a.m., Smith and I sign the paperwork allowing us into a radioactive area and enter the reactor containment building.

"We'll have it up in 45 minutes to an hour. Piece of cake," says Dennis Sheehan, general foreman, who's been working in the Zion station for 11 years.

It starts at 8 a.m. A man in protective clothing stands on top of the reactor head, slowly guiding the crane's hoist, inch by inch, until it connects with a triangular metal frame called a "lifting rig" fitted to the top of the reactor head. The man is dwarfed by the machinery around him.

The crane is equipped with an electronic scale that measures weights up to 400,000 pounds. The crew knows the reactor head weighs 350,000 pounds. If the scale registers that weight when the hoist lifts the reactor top, it means the massive part has cleanly disconnected from where it was bolted to the refueling cavity floor. Too much weight means the reactor head is snagged.

But there's a glitch. The electronic scale is not working. Sheehan's predictions of a "piece of cake" proves optimistic. They must replace the scale, which takes an hour and 45 minutes.

Meanwhile, we wait. Workers kill time by exchanging jokes, banter, and good-natured insults. One man asks busy pump specialist Ski Stachelski if he does anything in the plant beside plan the annual Christmas party.

The hiatus proves to be a blessing for me, providing answers to the vexing problem of an itchy face or nose in a contaminated workplace. I see how the seasoned professionals do it while they loll around waiting to go back to work. Some simply rub the inside fabric of their canvas hoods against their faces. Others remove an outer rubber glove and scratch with the cotton glove liner.

As we wait, I hear Dave Smith telling general foreman Dennis Sheehan that he heard earlier today that some parts of the reactor emit radiation levels of 100,000 REMs an hour. "It makes you wonder how they can ever decommission these things," answers the foreman.

Sheehan explains to me that when the head lift operation resumes, he will go down a ladder 25 feet to the bottom of the refueling cavity. When the reactor head rises exactly 133 inches, Sheehan will look to see if the 61 control rods and drive mechanisms inside the head separate and lift freely from the bottom part of the reactor vessel and its fuel core. Control rods start and stop a reactor.

"I'll take a look and get out of there, fast," says Sheehan. Though most of the workers are idle, the radiation protection specialists wearing orange hoods are not. They prowl like mother hens with handheld radiation monitors, scanning the surroundings. The place where we are standing registers 20 millirems an hour.

The mood turns serious again when a rumble followed by a soft whirr signals that the crane is ready to lift the reactor top. It rises slowly, then stops. Three men dart down to check the separation, then dash back up. Miller uses scissors to cut off their outer yellow plastic suits. All are wearing respirators to prevent inhaling radioactive dust.

The plastic suits go into the trash because they are almost always contaminated in a job like this. "A little piece of dust can have enough activity in it to be a problem," explains Smith, who remarked on how fast those three men did their short inspection tour. "They were moving."

Speed is essential, but everything in a nuclear power plant is planned carefully, and sometimes rehearsed. "It takes twice as long to prepare for it than it does to do it," explains Sheehan. It also must be done according to federal regulations and safety guidelines.

Donning plastic coverings, Sheehan and a radiation protection technician go down the ladder to the floor of the cavity. Sheehan trains a spotlight on the interior of the reactor head, known as the upper internals, to double-check separation from the reactor core. While he's doing that, the radiation specialist takes radiation readings.

Satisfied there is separation, the two men scamper back up and are cut out of their plastic suits, which are presumed to be contaminated. Sheehan says his personal radiation monitor shows he got a 23 millirem dose in less than two minutes; the radiation protection specialist got 38 millirems. A scan showed they were in a radiation field of 3 REMs an hour. It was like standing at the edge of a radioactive volcano.

A cluster of control rods like the spines of a porcupine bristle from the bottom of the reactor head as it hangs suspended by the crane.

"That's probably the hottest job we do and it was the first time we did it," remarks James Ramage, the station's ALARA radiation safety coordinator, referring to the head lift and inspecting the interior of the reactor head as it was lifted. Typically, nuclear workers avoid close encounters with intense radiation exposure.

In this case, it's known as a "hot job."

"I didn't waste any time," says Sheehan.

"We have authority to stop any job in the plant where exposure is a problem or if it is not running smoothly," Ramage explains. Exposure readings of 20,000 REMs an hour were measured where Sheehan went for the close-up inspection.

As the reactor head lifted, I and others retreated to the containment building doorway to get away from the invisible glow that reactor workers call "shine." The radiation dose where we were standing before the head lift was 20 millirems an hour. After the head lift, the measurement was 50 millirems at the same spot, indicating that invisible radiation was "glowing" from the reactor core now that the top was off.

With the reactor top off, the reactor core where 140 tons of fuel assemblies are crowded together is visible. From a distance, the fuel core appears black, its edges a dark rusty red. Some parts look scorched, but I'm told they are corroded.

"I'm getting away from the dose," says Miller as he walks away from the edge of the cavity, which is empty now except for the gaping black hole in which sits the reactor fuel compartment. "I don't have to be there right now, and you don't get any more dose than you have to. I want to work the whole year."

If Miller gets "burned out" by surpassing allowable quarterly radiation dose limits, he will be banned from working in radioactive areas for a while. Increasing the distance from a radioactive source or spending less time in radioactive areas cuts the dose.

We see another way to lower the dose. The crane carries the reactor head to the side of the cavity, and gently settles it on a ring stand. Powerfully radioactive, the bottom part of the reactor head is covered with yellow lead blankets to make it less hazardous to nearby workers. A man climbs to the top of the reactor head and disconnects it from the crane hoist.

Unexpectedly, the musical sounds of a waterfall ring through the reactor building. Clear water is gushing up from the reactor fuel core and filling the cavity. One end is 12 feet deeper than the rest of the pit, creating a cascade of falling water until the cavity fills slowly. That's the usual pace of a nuclear power plant, slow and deliberate.

"This is a lot more interesting than running tests," a worker remarks.

Some had expected a sudden rush of water into the refueling cavity. But the quality control department objected to that, although the procedure was used elsewhere. Quality control said a sudden wave of water slamming into the refueling cavity might cause a spray of radioactive mist in the reactor building.

Once the pool fills with 350,000 gallons of water, the surrounding radiation level drops to 5 millirems an hour, proof that water is a good shield against radiation. Now they are ready to begin lifting 140 tons of uranium fuel out of the reactor's hollow interior.

Workers stream out of the reactor building and to the step-off pads to remove their anti-contamination gear. On the way out of the containment building, Dave Smith asks a radiation technician to scan my notebook to see if it is "crapped up," the term nuclear people use for contaminated. It is clean, but that gets me to thinking: what would I do if my notebook were confiscated?

I was in the containment building for three hours and 40 minutes, taking notes all the time. Losing those notes would have been a major loss.

My radiation dose for the day is 10 millirems.

Chapter 22: The Nuclear Reporter

"Back into the shine," I write in my notebook.

Today, I'll be going into the reactor containment building for the third time. In the auxiliary building, I fill out the entry card and get a radiation work permit that sets my exposure limit at 100 millirems for the day.

A supervisor tells me that "the most important thing is you know what your dose level is," meaning keeping track of how much exposure I'm getting and remembering limits set by work permits. The permits make you conscious of the need to watch your exposure levels.

I'm beginning to feel comfortable about radiation. I'm no longer super-cautious or tense about it. I don't feel imaginary tingles in radioactive areas.

A strange thing happens after the first exposures. Once exposed, knowing radiation exposures can cause cell damage—is causing damage—you want to believe it will cause no harm. You *must* believe it will cause no harm. Otherwise, the uncertainty would be unbearable, even though I know the effects might not be known until 20 years later.

It's easy to trust radiation if the damage is not immediately recognizable, and depends greatly on individual susceptibility.

I'm coming to terms with nuclear power in a personal way, in a way that potentially affects me physically, as most nuclear energy workers must. Over the years, when I've asked nuclear people why they're working in a nuclear facility, they've often said they needed a job. It's a choice that includes a reckoning with radioactivity.

Contamination is on my mind as Dave Smith and I make our way to the auxiliary building to start our day. I tell Smith that I'm worried about getting my notebook contaminated with radioactivity. A day's work could be lost if my notebook is confiscated.

A non-journalist might say just recreate events from memory. But memory is fickle, especially recalling subtle details and technical terms over many hours when talking to many people. A notebook goes where I go and sees and hears what I see and hear. It's recoverable memory in writing. A transcript of sorts. This was before the advent of cell phones. Even so, there is nothing more reliable and accurate than a notebook full of notes.

Until now, when taking notes in a radioactive area, I'd remove the outer rubber glove and hold a pen with my right hand covered with the cotton glove liner, the notebook in my gloved left hand. Then I'd replace the rubber glove until I took notes again. All the while, the notebook and pen are in the open and exposed to potential contamination.

Today, I'm going to do something different. We suit up in anti-contamination clothing in the usual way. But this time, I put a notebook and two pens in a large plastic bag, then put my right hand into the bag, without a rubber glove on my right hand, but wearing a cotton glove liner on that hand. The plastic bag is big enough so that I can flip notebook pages and write with the pen.

"The nuclear reporter," Smith dubs me, chuckling, while sealing the top of the plastic bag around my right forearm with tape. "We dress to allow ourselves to do our job. You've taken it a step further here at Zion," he says. He calls my invention "a technological advance in reporting in radioactive areas."

But with my right hand enclosed in the plastic bag, Smith warns me to be careful since I am physically hindered, which could lead to an accident.

Inside the containment building, we are eager to see nuclear energy's greatest show on earth, visible only if you dare to get close to a red-hot nuclear reactor and its bundles of wildly radioactive fuel. Except that it's blue-hot, a phenomenon known as Cherenkov's radiation. In the intense fire of the nuclear chain reaction, these parts become so hotly radioactive, they glow with a cobalt blue light unlike any other in the world. But the price of admission is a dose of radiation.

"If we didn't have the water there, we'd be dead in five minutes," says Ed Fuerst, Zion's assistant superintendent for operations, who we met shortly after entering the containment building. "It's extremely radioactive."

We entered as a glowing 12-foot-long fuel bundle was pulled up underwater from the refueling cavity through a channel to the spent fuel pool. The water in the refueling cavity is 25 feet deep, so even if a 12-foot-long fuel bundle stands upright, there are 12 to 13 feet of water above it.

"The water acts as a shield," says Fuerst. "There would be millions of millirems with no water here, extremely deadly."

Fuerst orders the lights in the building turned off so we can fully appreciate this phenomenon named for the Soviet physicist who discovered it in 1934. Cherenkov's radiation is caused by charged particles from the reactor fuel traveling faster than the speed of light in water. It's like a sonic boom, a shock wave of sound. But in this case, it's a shock wave of light. It is awesome and seldom seen even by nuclear power workers because few get to enter the containment building and see the ultra-radioactive reactor fuel moved from the underwater fuel compartment.

Cherenkov's radiation is a ghostly cobalt glow radiating from the fuel bundles. Metal parts look black, but spaces between them are filled with the blue-purple light. Together, they look like a blue and black X-ray image.

"Wow," says Ronald Chin, a station nuclear engineer standing next to me. "It's neat, the first time you see it. It's a radiated fuel assembly. Most radiation you don't see. This is actually seeing the results of it."

"Sort of like St. Elmo's Fire," says Fuerst in admiration.

The Zion nuclear reactor's belly is packed with 193 of those bundles, standing upright in racks in a checkerboard pattern. A bundle, also called an assembly, is a box measuring 15 inches by 15 inches, and 12 feet long. Each bundle contains 204 pencil-thin rods, and the rods contain uranium dioxide pellets an inch long and less than half an inch in diameter. There are 39,372 rods in the fuel core. Each fuel bundle weighs 1,381 pounds and costs half a million dollars.

As part of the Unit One reactor overhaul, all 193 fuel bundles are being taken out and moved to the spent fuel pool, so the reactor vessel can be drained and inspected.

But at this moment, the refueling cavity is filled with 350,000 gallons of water, with underwater lights illuminating the fuel bundle removal activity. The telescoping quality of water makes it appear the bundles are only a few feet below the surface, but they are much deeper. Bubbles rise lazily to the surface.

"Looks good enough to swim," remarks Fuerst. "Only one problem," responds Smith, pointing out that the pool contains high concentrations of boric acid, which retard nuclear activity in the fuel bundles.

We watch as a grappling tool more than 20 feet long reaches down from the overhead refueling crane, through the water and into the reactor, grabbing a fuel bundle and lifting it out of the core. Moving on tracks on each side of the refueling cavity, the crane transports it to the spent fuel pool. The scene underwater is distorted by movement of the water in the pool as it ripples.

The radiation field at the edge of the refueling cavity is 8 millirems an hour.

For a better look, Fuerst and I climb up to the refueling crane and look down at the yawning dark reactor core, about 14 feet in diameter with edges a dark rusty red.

It reminds me of a black hole in space, the eerie blue glow of Cherenkov's radiation in its depths heightening the illusion of something outer worldly, like the florescence of a distant galaxy. I ask Fuerst if we are looking at the black hole of a nuclear power plant. He is offended. "I wouldn't call it the black hole of a nuclear power plant," he replies. "It's the heart. Radioactive waste is the black hole."

I ask Fuerst, "does anything else glow that way?" He answers: "Nothing you see on an everyday basis."

We climb aboard the refueling crane operated by Ernest Abbott, a fuel handler, and directed by Larry Thorsen, fuel-handling foreman. Abbott has pulled 120 fuel bundles out of the reactor so far with the telescoping mechanical arm, lifting them one by one, and transporting them to the spent fuel pool.

Each bundle is identified by a letter and a number, like E-10. Abbott and Thorsen operate as a team, with Abbott saying what he intends to do, and Thorsen gives his approval. It is a way to prevent errors before they happen.

"Permission to trolley," says Abbott, and Thorsen agrees, before the crane begins moving. From above, the fuel bundles look like square doughnuts, with the blue glow outlining each bundle and illuminating the hole in the middle.

The bridge crane runs on rails on both sides of the refueling cavity, the rail marked with numbers. Letters appear on a dial on a control board. Together, they locate the exact position of individual fuel bundles. Abbott lowers the mechanical arm and watches a metal measuring tape which is marked with depth indicators. The point at which the arm normally connects with the top of the fuel bundle is marked in black.

At some point, a computer takes over the fuel recovery system. The operator watches to be sure the grabbing arm reaches the depth at which the arm connects with the top of the bundle. Then he observes a scale to see if it registers the correct weight of the fuel bundle. If there is no weight reading, the bundle is not being lifted. If too much weight, the bundle is snagged and should be handled carefully.

The smooth operation suddenly encounters a glitch. In the intense heat of the nuclear reactor, one of the fuel bundles has bowed out of alignment and is stuck in the rack. Bundles normally stand perfectly upright leaning against each other.

"That's the first one we had trouble with," says Thorsen. He peers downward, searching for the reason the fuel bundle is stuck, then finds what he's looking for. At the side of one of the blocks of bundles below, he sees a clear line of blue light revealing a separation between two bundles. That should not be there. It's out of alignment by about a quarter of an inch.

It takes 15 minutes to remove the misshapen bundle by moving the grabbing tool a quarter of an inch to the east and an eighth of an inch to the south. Normally it takes about three minutes to extract a bundle, "but I've had them take almost an hour" when stuck, says Thorsen.

He tells me he wrote the plans for this operation. "It's like seeing your work done like an architect. You see the finished product."

Once the reactor core is empty, workers look for items that fell or washed into it. "You always find junk in the core, but we don't know what," says the foreman.

As we're leaving the containment building, some workers laugh when they see my hand covered in a plastic bag clutching my note pad and a pen.

Then a worker points up at a sparrow fluttering from pipe to pipe 200 feet above, an unexpected intruder.

"He'll be six feet tall before he gets out of here," jokes Thorsen, referring to the myth that radioactivity causes things to grow enormously. It is believed to be the first bird to reach the reactor's inner sanctum, and its chances of getting out are slim. A worker cautions against bird droppings while another suggests setting a trap for it.

If it does get out before the containment building is locked up, "it will be contaminated, although not much," says Thorsen. Management will hear about the bird.

The sparrow and I suffer a similar fate. On my way out of the containment building, I put the plastic bag containing my notebook in another bag, then go through the step-off pad procedure, get dressed and collect the bags and take them to the radiation-chemistry office for a scan.

The bag containing my notes is contaminated. "About 30 counts," says the technician. Technically, it could be taken out of the building because it's below the 100-count limit. But he confiscates the plastic bag, then hands me my notebook and pens after finding them clean. My technological advance in reporting in radioactive areas worked. My notes were not contaminated with radioactivity.

I fill out my exit cards and record my exposure for the day. An hour and 45 minutes in containment. Exposure: 12 millirems.

Chapter 23: Underwater Mechanics

It never occurred to me that when all those radioactive fuel bundles are removed from the reactor vessel's belly, somebody would willingly dive into the thing.

But that's what John Peterson is getting ready to do on my next visit to the Zion Unit One reactor containment building. An underwater mechanic, he's donning a diver's suit and helmet weighing 165 pounds—like the ones you see in movies of deep-sea divers.

I never expected to see anyone looking like a deep-sea diver in a nuclear power plant.

"It's a fun job," says Peterson, at 59 a white-haired, hawk-nosed man wearing horn-rimmed glasses and an easy smile. "It's a shame to take the pay," which is $500 a day for him and his tender, John Geschrey, plus insurance.

"I enjoy my work and the conditions in a nuclear power plant are absolutely pleasant," says Peterson, who owns a Morton Grove, Illinois, diving company, with five divers. "Clarity of the water is fantastic, and the jobs are always interesting."

By comparison, water in rivers usually is murky and he does that work largely by touch because visibility is so poor. Peterson has been doing underwater work for Commonwealth Edison for 37 years.

Peterson sits calmly at poolside, discussing his dive with two Edison radiation safety specialists while his tender helps him put on the gear in what appears almost like a ritual. The reddish-brown rubber diving suit goes over a canvas anti-contamination suit underneath. A metal plate with bolts goes over his head and shoulders and is fastened to the suit. The tender fixes weights to Peterson's boots.

A helmet with a black rubber hose attached sits at poolside. All the equipment is conventional deep-sea diver's equipment, except for a special adaptation to the helmet to prevent water from entering it.

Like almost everything that happens in a nuclear power, every move was planned in advance.

"Last night, I spent a couple of hours planning the job," he says, for a task he believes will take an hour. "We feel we have all the bases covered." The preparation is "twice as long to get to the job, putting the wrench to the nut." But he agrees it's worth the time. In a radioactive area, "you don't stand there deciding what to do next."

Like other workers in the plant, Peterson is conscious of his allowable exposure limits. If exceeded, he would be banned from working for a time in radioactive areas.

"The job," in this case, calls for Peterson to place an inflatable rubber plug in a 29-inch diameter opening in the side of the reactor vessel called a nozzle, which is radiating a 250-millirem-an-hour field of radiation. It's 35 feet below the surface of the refueling pool. The reactor vessel has eight nozzles that flush cooling water in and out of the reactor while it is operating.

The reason for plugging that nozzle is to reach a leaking valve about 50 feet inside the cooling water pipe from the nozzle. Once plugged, workers can drain that section of pipe and repair or replace the leaking valve, which is radiating a 300- to 400-millirem-an-hour field. The work will require a rotation of workers to spread out the dosage.

Peterson spent two hours underwater the day before, inspecting the interior of the reactor vessel and taking photos. He got a 225 millirem radiation dose while doing that. Then he drafted a plan describing how he intended to insert the plug.

That seemed like a lot of millirems, but Peterson tells me, "It's low for this type of work we are doing. If it was in the dry, people would get a higher dose faster." I asked about his highest radiation dose, and he says it was 800 millirems over two hours while working in a Wisconsin nuclear power plant.

The radiation does not worry him. "No," he says. "The Edison people take care of that part. It's just an interesting job." Peterson says he works with a crew of four or five, including his tender and Edison radiation specialists.

The day before, Peterson leaped feetfirst into the refueling cavity pool. "I splashed too much," he says. Edison technicians worried about radioactive water splashing poolside and being tracked elsewhere or becoming airborne. They suggest Peterson use a ladder on his descent into the water this time. He doesn't like the idea.

Diving is like flying an aircraft, Peterson observes, "the accidents happen at takeoff and landing." He doesn't like ladders, citing instances of broken rungs and slipping on them. But he agrees to use a ladder this time.

Now it's time to get that rubber plug in place. Peterson puts the helmet on and gives it a hard crank to fasten it to the metal collar, then lumbers to the pool. He climbs down the ladder and into the water, leaving a trail of bubbles in the clear blue water illuminated by underwater flood lights. His black air hose snakes through the water behind him.

The turbulence of bubbles makes it impossible to see Peterson clearly, so the radiation specialists create a window to the depths. It's a pane of glass in a wooden frame that floats on the surface, flattening the surface and offering a clear view of the underwater scene. One of the technicians dangles a radiation monitoring device near Peterson to keep track of radiation levels near him.

On the intercom, Peterson asks them to lower the plug. Four men in white coveralls carry a rolled length of dark rubber about seven feet long and put it in the water—where it floats. The men stare at it in disbelief.

"Christ!" grasps Smith. "It's floating. This is what is known as a glitch. Apparently nobody had thought it would float." The tube vaguely resembles a shark bobbing in the water.

Peterson rises from the depths, his face visible in the helmet's glass window. He grabs the rubber sleeve and tries to pull it underwater, but it's too buoyant. He tries again, this time wrapping his arms around the sleeve and adding his weight.

Bubbles boiling to the surface, Peterson drags the roll of rubber down, unties it and plugs the pipe opening using a metal brace to keep it in place. A storm of bubbles shows Peterson is breathing hard.

The leaky valve adds 11 days to the outage because the problem was discovered only a week ago and repairs were not planned. "Some leak is okay," explains Smith, "but this is too much."

The valve is called a motor-driven heat removal system isolation valve. It had to be repaired because it was allowing water into the cooling system at higher pressures than the system was designed for, causing a second valve to fail. Two faulty valves could damage the reactor's heat removal system.

Both valves were close to the reactor fuel core. They could be repaired only when the reactor was shut down and all the radioactive fuel removed.

"This is one that is difficult to get to," explains Smith. Making the project even more difficult, Edison discovered the manufacturer no longer makes that type of valve.

"We don't know what we will find when we get into the valve," says outage coordinator Bob Cascarano. "We had a very difficult time finding parts." They were discovered at a Tennessee Valley Authority nuclear plant.

Problems with that kind of valve were not expected. "We were surprised the manufacturer didn't have anything," Cascarano adds. "It's a specialty valve."

The Zion station has hundreds of valves. The technical staff develops skills at learning how to tell when they are failing.

"Most equipment provides clues it needs help," explains Smith, such as rising temperatures or sounds of bearing wear. Vibration analysis and a regular lubrication program are two things that "will catch the vast majority of problems—either prevent or catch them. The rest are caught as they become a problem, as opposed to be maintained away." Smith compares it with putting oil in your car's engine.

Days later, Cascarano reveals that both valves were dismantled and "we didn't find much wrong with them." As for what accounted for the leak, "no way we will ever know for sure. It might have been minor scratches on the valves that caused the leak." They were polished, tested under high pressure, and did not leak. "At this point, we fixed whatever the problem was."

After all that effort and planning, the end result is a mystery. Even the machinery of nuclear power, certified to perform to the highest standards, can be unpredictable.

The valves are reinstalled and Peterson returns to remove the plug from the underwater nozzle.

While watching the diver install the plug, I spent three hours and 55 minutes inside the reactor containment building. My radiation exposure for the day was 8 millirems.

Chapter 24: The Nuclear Crusader

THE ZION STATION AND ROADS LEADING TO IT THIS MORNING ARE COVERED BY dense gray fog that reduces the rising sun over Lake Michigan to a ball of muffled bright light.

One of the pleasures of coming to the plant early in the morning is the refreshing lakefront weather influenced by the "lake effect" that does not always extend far inland. Can't tell what it's going to be until I get there.

Stepping out of my parked car, I hear an unfamiliar sound over the usual turbine hum. It's a sizzling sound, like bacon frying in a pan.

It's "arcing and sparking" from nearby overhead power lines, explains Ed Fuerst, who is waiting for me at the guard house door. The fog's heavy moisture content is causing the double set of power lines going in and out of the station to crackle and snap.

As we enter the station, Fuerst explains that 1.5 to 2 percent of the power produced by the facility is used inside the plant.

"I see my role as a crusader," Fuerst tells me as we settle into his office overlooking the lake. "I fight the good cause and get aggravated. Nuclear power is definitely a hazard to your health—it gives you ulcers."

The assistant superintendent for operation's demeanor is usually so serious and intense, such flashes of humor are surprising. He's a big man, a veteran of the U.S. nuclear navy, source of many nuclear power plant workers, with a degree in physics. His military bearing seems magnified when his temper is aroused. "I butt my head against the politics."

A man enters Fuerst's office and asks him to interpret whether a procedure is being carried out correctly. Fuerst lifts a black ring binder about three inches thick.

"Here's the bible for the Zion station," he says. "This is our license to operate" both nuclear reactors. Edison wrote it and the Nuclear Regulatory Commission rewrote it, spelling out what Edison can and cannot do. "My job is to make sense of NRC requirements." He wants to give me some idea of how stringent those rules and regulations are, and how much they frustrate him.

Five fan coolers and three containment spray pumps combined are 430 percent efficient in reducing steam and pressure inside the reactor containment building in case of an accident.

"It's like having four and a half batteries in your car," says Fuerst. If part of the spray system becomes inoperable and is down to 300 percent efficient, he says, Edison must shut down the reactor immediately.

"So if you are down to three batteries in your car, don't dare leave your driveway," he explains, by way of the car analogy. "We have some [regulations] that are completely asinine. We know it and others know it. NRC won't allow changes."

NRC discovered that a nuclear reactor at the Salem power plant in New Jersey should have tripped off automatically twice, but did not. The agency ordered changes in all Westinghouse Electric Corp. reactors of that type. Reading from an NRC report, Fuerst points out that the problem at the New Jersey plant was dust and dirt, lack of lubrication, and nicking of a circuit breaker surface.

"Why," asks Fuerst, "if they identified the root cause of the failure being lack of maintenance, are we required to modify the automatic reactor trip system? This is a technical industry, which should not be driven by political concerns."

Another irksome issue is fire prevention.

Edison spent $7 million on fire-protection improvements at Zion. "Five million of that was a waste," Fuerst contends, and the cost could rise to $20 million. "Now they say insulate steel beams so they don't catch fire—believe it or not."

I will later call the NRC to verify what Fuerst said. Spokesman Jan Strasma says yes, fireproofing is mandated because structural steel loses strength in fires and causes buildings to collapse.

So the NRC's posture on this issue does not appear so preposterous after all. How do I assess Fuerst's criticism? It appears his zeal in defending nuclear energy matches the zeal of nuclear energy critics. Both sides engage in a certain amount of exaggeration at times. I'm not going to discount what Fuerst says. He has a viewpoint and is entitled to express it based on his 12 years at the Zion station.

Fuerst said he served for a time as the station's fire marshal. Fire prevention goals, he said, are based on controlling a theoretical blaze in a 55-gallon drum full of number two diesel oil anywhere in the plant.

A drum that size would not fit in some places in the plant, he says, but "we have to modify the plant to protect it from the 55-gallon barrel."

Like nuclear plants across the nation, the Zion station was forced to make safety improvements because of concerns raised by the 1979 Three Mile Island accident in Pennsylvania, as well as mishaps at other nuclear stations that revealed safety weaknesses. A 1975 fire at the Browns Ferry nuclear station in Alabama resulted in major NRC additions to standards for firestopping.

"Ninety percent of the modifications since TMI are technically unsound and politically motivated," Fuerst contends, adding that Zion workers often are exposed to radiation while making changes mandated by the NRC.

A Zion radiation protection specialist estimated that 25 to 50 percent of radiation exposures to plant staff were the result of NRC-mandated plant modifications. In the interest of plant safety, plant workers took those exposures.

"Many things the NRC people have done made these plants less safe," Fuerst asserts, "because they don't have expertise and are operating by political considerations."

Three NRC-mandated modifications caused Zion reactors to "trip" the control rods, the reactor braking system, he said, causing sudden shutdowns. Some fire-protection modifications required drilling through concrete walls. Utility officials tend to believe such breaches reduce safety.

"They expect too much of us. They expect us to jump through hoops, but they are not technical hoops. We should be jumping through technical hoops."

Power utilities are required by law to report mistakes or accidents for which they can be fined. "It has turned into a very punitive method of operating," he says. "That's like speeding and stopping for the traffic cop who pulls you off to the side of the road, and he shoots you. Pretty soon you try to get away from the cop. We turn ourselves in. That cop is shooting every time. They slap a fine on you no matter what."

Fuerst believes such citations have lost their effectiveness with utilities. "Maybe it gets them [NRC] publicity, but it doesn't get them change."

"Why do I hate the NRC?" he asks in a way of summary, acknowledging that the agency's actions do not affect him personally. "They grind against me. They piss money away. They make nuclear power uneconomical. They are the enemy."

Before we part, Fuerst walks me outside to a trailer parked between the guard house and the Zion office building to meet ROSA, the acronym for Remotely Operated Service Arm. It's an example of robotics making its way into the nuclear industry.

"It's not an Italian virgin sacrificed to the steam generator," Fuerst jokes, once again coming out of his hard shell.

ROSA is new technology designed to take the place of humans on a job that is highly radioactive. Installed only a month earlier, ROSA "drastically cut down exposure because it's all automated," says Fuerst.

Before ROSA's arrival, workers once known as "jumpers" but now called "entry technicians," did one of the most hazardous jobs in the nuclear industry.

It involves climbing through a hatch only 18 inches in diameter into the bottom of an intensely radioactive, 67-foot-tall steam generator to find and plug leaking tubes. As its name implies, a steam generator is where steam is created to power the turbine.

Two streams of water course through a steam generator in a nest of tubes. One stream comes from the nuclear reactor and is highly radioactive. The other stream of water turns into steam, but is not radioactive. The two streams do not mix.

Each Zion station reactor has four steam generators containing 13,000 tubes each, for a total of 52,000 tubes. The tubes, each three-quarters of an inch in diameter, are packed together in a U-tube configuration so the heat from the reactor water transfers to the other stream of water that turns into steam. Each steam generator contains 45 miles of tubing.

Over time, under intense heat and pressure, some of the tubes crack and leak. That's where the jumpers came in. Before going to work, they train two to four weeks in a steam generator mock-up in a warehouse. The jumper memorizes the tube opening pattern and practices 12 hours a day until his actions are swift and sure.

Only then do they go into the steam generator, manually probing with tools to detect worn tubes and plugging them when found to be leaking. The work lasts two or three hours until they reach exposure limits and must be replaced by another worker. They're sometimes called radiation "sponges."

James Ramage, Zion's ALARA coordinator, says jumpers definitely get the most radiation exposure in a nuclear plant. "We approve a steam generator jumper to 1,500 millirems" of external exposure, he says. "Our policy is we do not give anybody more than 5,000 millirems a year. Legally we could give up to 12,000. You'll never receive more than 5,000 a year at this station."

I did a story about jumpers a few years earlier. At the time, they were paid $60 to $120 a day. They are young, thin, and wiry enough to fit through the 18-inch hatch, and in need of a job.

A jumper might get a radiation exposure of 150 millirems a minute. "You're talking about an awful big dose," Fuerst said. But ROSA doesn't care about radioactivity. Attached to the edge of the hatch opening, it's a mechanical arm six feet long and weighing 140 pounds with rotating devices acting like a shoulder, elbow, and wrist.

"Now ROSA automatically indexes and plugs the tubes," he says. "It's a hell of an advantage over doing it manually." Depending on radiation levels and skill, a jumper might plug 10 tubes before being "burned out" and replaced.

ROSA can work tirelessly and flawlessly by remote control, and it has no radiation exposure limits.

"It's also not rushed, which was the biggest problem with jumpers," says Fuerst. "There's no hurry." He estimates that ROSA has reduced radiation exposure to 100 to 200 people per steam generator.

A trailer parked about 250 feet from the containment building is ROSA's control room, fitted with TV screens monitoring work outside and inside the steam generator. A simple joystick operates ROSA's movements.

Senior engineer Tom Wagner explains that ROSA inserts a tool into steam generator tubes and uses electrical current to detect degradation before the tubes fail.

Robotics like ROSA gradually are phasing out the use of jumpers, but they still are employed in some places.

My next stop for the day is to see Ken Graesser, Zion's superintendent. He calls ROSA "our girlfriend." Zion is among the first power stations to use the robotic arm. He predicts all nuclear power plants will be equipped with ROSAs in the future.

Graesser and I meet periodically to see how things are going. The superintendent says he's heard that some of his employees are not cooperating with me. I tell Graesser I don't mind that, people are entitled to refuse if they choose. It's the hostility that bothers me.

"We've been through a lot here," he says. "There was a lot of news activity. We went through the drug investigation, which was a sad event. It caused a lot of hard feelings. We don't have the luxury of other industries. We are doing something sensitive. It's a fishbowl atmosphere. We have to watch our actions a little closer and be a cut above the rest so we can't be criticized. The people have to have some confidence you are as good as you have to be."

Critics say nuclear power is terribly unforgiving of human error.

"Nuclear power probably is very forgiving of human error," he insists. "Take a look at the design. An inherent characteristic of a nuclear reactor tends to shut itself down as temperature increases—negative reactivity. It's very forgiving. There's nothing exotic about the equipment. It's a big pot with fuel sitting in it."

Graesser describes the nuclear plant as largely a collection of pumps, motors, valves, and piping.

"We've got 10,000 operational procedures in the plant in maintenance and operating," he says. "We constantly evaluate better ways to do it." Safety procedures are revised every two years and all procedures are revised every four years.

Graesser worked in fossil fuel plants before going into the nuclear program. He was a licensed reactor operator for four years at Edison's Quad Cities power station. Like everyone else, he takes radiation protection training once a year and points out that maintenance and mechanical staff members do too. Fire-protection training is another annual event. The industry undergoes constant review.

Training is a major effort at the station, says the superintendent, pointing out that he had just completed four days of management training. "It puts it on a constant businesslike basis."

Graesser has a little softer attitude toward the NRC than Fuerst, but not much.

"I think generally they have good inspectors," he says. "Some are not the best, but that's true in every industry. They're pretty darn tough. I'd like to see improvements there. Their enforcement program is defeating. They are looking for ways to make improvements. It's a very punitive program.

"We want to operate as error-free as possible. Checks and balances prevent us from getting into problems. That's one change I'd like to see—review and upgrade and have a more realistic enforcement program."

NRC recently was checking the Zion station's ALARA program, designed to keep radiation exposure levels as low as reasonably achievable. Graesser agrees that keeping those levels down is necessary for "the survival of the industry."

Talk about boats and fishing, and you see another side to Graesser and people who work for him. The Zion station crowd is surprisingly like their boss.

Graesser goes walleye fishing in Minnesota every summer and boating on the Chain of Lakes near his home in Antioch, Illinois, near the Wisconsin border. "We could darn near have our own navy here," he says, so many Zion workers own boats with Lake Michigan and other waterways nearby in northern Illinois and southern Wisconsin. Plant workers don't go far to go fishing. The state stocks the lake with salmon and other game fish.

A photo of his pontoon boat and his three daughters hangs on his office wall. He's a member of the Antioch Lions Club, which gives him a chance to meet people in town.

"We're all trying to do something we think is important," he says, through community involvement, such as fire departments, rescue squads, church groups, school boards, and sports.

"A power plant group is different from people in other industries," he observes. "Because of the hours they work—with rotating shifts and extended hours—they become a close-knit group. They tend to socialize in their off-hours because of those crazy hours. There's a lot of pride in a place like this."

Shifts end at 7 a.m., 3 p.m., and 11 p.m.

Edison employs 496 workers at the Zion station, 212 in management and 284 members of the International Brotherhood of Electrical Workers. Thirty percent of them live in nearby Wisconsin.

From a top floor window in Graesser's office, I count 20 fishing boats circling over the Zion station's hot water discharge into the lake—identifiable by the glassy appearance of the surface above the point where the discharge wells up.

Small fish are drawn to the warm water, and the big fish move in to eat the small fish. And fishers angle for the big fish. It's a cycle in nature that operates under its own rules and time frame.

It all looks so peaceful, tranquil, and orderly. What could possibly go wrong?

Chapter 25: The Accident

THE EXTRAORDINARY EFFORT TO DISMANTLE ZION'S UNIT ONE REACTOR AND all the systems connected to it and restore them to like-new condition is half finished, a signal turning point.

"We're on the way back out" of the suspension of operations, says outage coordinator Bob Cascarano, with a note of quiet satisfaction.

Uranium fuel is moving back into the reactor core, including the brackets where fuel bundles are stacked. It's the sort of moment that quickens the pulse of a nuclear technician.

"This is new," says my guide, Dave Smith. "We've never had to put in a whole [fuel] core since we did it the first time" more than 10 years ago. That initial operation is described as "smooth as silk. Everyone was impressed."

Assembling the reactor core and all the hardware in the reactor vessel, known as the "lower internals," caused far less radiation exposure than experienced at other nuclear plants doing the same kind of work.

"We almost got nothing at all yesterday," says Carcarano. "That's a real credit to our ALARA [radiation protection] group." It caused an average exposure of half a man-REM, he said, compared with exposures of 3.5 to 15 man-REMs at other plants.

The numbers speak for themselves without getting into details. It's worth noting, though, that the low radiation exposure figure was stated as a matter of pride, and among the calculations monitored in the process of bringing the reactor back into operation. Regulations require it, but it's also a reminder of one of the ways life in a nuclear power plant differs from other workplaces and must be taken seriously.

Water in the refueling cavity turned murky for unknown reasons, and radiation/chemistry specialists are asked to check that out. The water level in the pool is down eight feet, and goes lower when the reactor head settles into place. "You don't want to get the head wet," explains Cascarano. All steam generator tubes were tested and plugged where necessary.

The only surprise, causing an 11-day delay, was the leaking valve, for which a cause could not be identified.

"The bottom of the [reactor] vessel was cleaner than we thought it would be," says George Pliml, an assistant superintendent, except for a few stray washers, a broken strap, and some nuts and bolts.

"We finished a great deal of work; there still is a lot to do," says Smith. To which Cascarano added: "I'll be glad when it's over. It's beginning to wear people down."

It's about 3 p.m., the end of a shift, and Smith and I decide to call it a day after getting updates on the status of the outage. We get in line with a few dozen workers at the guard house exit expecting to walk through a portal radiation monitor, the final checkpoint before leaving the plant.

Men shuffle their feet as the line moves slower than usual. They are getting grumpy at the unexpected delay and start shouting. "What's holding it up? Let's go! Put in some new portals."

The outburst is cut short by the sight of a bewildered-looking man with plastic bags over his shoes going back into the plant.

"Everybody's got it," says the man in the covered shoes. "It" is radioactive contamination, and workers are finding it on their shoes, clothing, hair, and skin.

On their way out, workers go through the portal monitors one at a time. If they trip the alarm, they're told go through it again. Upon a second trip, they're told to strip off clothing in an attempt to identify contaminated items. If problems persist, they're sent to the radiation protection department in the auxiliary building for further testing and evaluation.

Suddenly, the outage that was going smooth as silk takes an unexpected turn. Radiation monitors at the guard house exit flash red lights and shriek electronic warnings like mechanical watchdogs as tainted workers enter them. First one, then another and another.

This is the first warning of what will become the Zion station's worst outbreak of radioactive contamination.

Instead of calling it a day, Smith and I head to the health physics department in the auxiliary building, where the contaminated workers are sent. We're standing in the corridor when a man approaches and asks, "Where do you go when you are crapped up?"

"Crapped up" is slang for being contaminated with radioactivity. Half a dozen workers have come to the radiation protection department, where Len Gesiakowski works.

"Apparently something is happening," he says, bewildered. "If people are actually getting contaminated, we have to check out why because it should not be happening." His remarks show a hesitation to believe it.

As we talk, a man flaps past without his shoes and socks, his bare feet wrapped in clear plastic. It adds a bizarre note to a scene that is becoming chaotic.

The unexpected outbreak is puzzling because all the men were wearing anti-contamination clothing while working in different areas of the reactor building. All of them had changed into their street clothes before the exit radiation monitors stopped them.

"I don't see how we're getting crapped up," says a worker, explaining his job working on a cooling fan seemed harmless enough. Smith and I get permission to enter the radiation protection department where the contaminated workers are sent.

One of them is John O'Neill, a 32-year-old Chicago pipefitter. His shoes were confiscated. Upon arriving in the radiation protection department, he's told he's contaminated. Hearing that, the pipefitter looks shaken and stunned. He's dazed, staring straight ahead, as if trying to comprehend what's going on or what accounts for it.

A frisker finds a heavy concentration of contamination around O'Neill's nose, indicating he might have inhaled it. He blows his nose forcefully, then splashes cold water on his face.

A frisker normally clicks a few times a minute with the passage of stray cosmic rays. When radiation protection specialist Mike Cairns passes a frisker across the back of O'Neill's neck and head, it buzzes furiously.

"Is it still high?" asks O'Neill, and Cairns nods. Wide-eyed, O'Neill looks startled, desperate and seems to be holding his breath. He's been working in the Zion station off and on for four years and this is his first time he was contaminated.

Asked to reconstruct where he was working today, O'Neill said he was on the pump deck in the reactor containment building installing pipe hangers.

"It's mostly my hair," says O'Neill of his contamination. "I had the hood sealed."

He wonders out loud if the interior of the hood was contaminated before he put it on.

Told to take a shower, O'Neill strips and washes in cold water, which closes the skin's pores and is the standard remedy for contamination—washing it away.

While O'Neill showers, Cairns, tall and dark-haired, explains that the frisker found 35,000 counts of radioactivity a minute on the pipefitter's body. "It's got to be less than 100 counts to get out [of the station]. That's a lot to have on your body." It's the most discovered that day.

Dripping from the cold shower, O'Neill steps naked from the shower booth and wraps a white terry cloth towel around his midsection, waiting tensely for Cairns to scan his glistening body. "Is it still high?" he asks. Cairns frisks him and shakes his head, indicating that the radioactivity count is much lower, to O'Neill's obvious relief. He admits to being "a little worried," laughing nervously, and now is "feeling much better."

A technician finds that O'Neill's shoes are not highly contaminated and washes them, before handing them back one at a time. The pipefitter is happy to have them back, shaking his head in dismay and disbelief, after thinking about going home without them.

Four more men are taking showers now. Surrounded now by contaminated workers, I recall Cliff Nehmer's definition of contamination: radioactive material that is someplace we don't want it. Having it on their bodies definitely is a place they don't want it. The theoretical lecture on radioactive contamination suddenly became dramatically real, including how to get rid of it.

"That's a cold shower," comments O'Neill. "Can I go now?" A look of surprise and dread passes over his face, followed by resignation, when a radiation safety technician answers, "You're going for a body count," meaning O'Neill will be scanned by equipment that can detect radioactive contamination inside his body, including his lungs. That scanner resembles a metal coffin. A person lies in it while a radiation scanner passes over from head to toe.

After O'Neill leaves, the technician says, "He had counts around his nose. He might have inhaled something."

The pipefitter is a perplexing case that truly puzzles the radiation specialists. It doesn't make sense.

"It's the guy with contamination in the hair I can't figure out," says Gesiakowski. "If he kept his suit and hood on as he should, I don't see how it could have happened." It's as though a cloud of radioactive gas went through his clothing.

It's a mystery, he admits. "If there were a major problem, everyone would have a problem," says Gesiakowski. Edison estimates that about 100 workers were in the reactor containment building, with another 50 or more working outside in the immediate area. Some workers were contaminated, some were not.

Radiation detection monitors inside the containment building did not sound an alarm.

Gesiakowski doubts it poses a serious health problem. "I don't think so," he says. "Before we let them leave, they must meet our limits. They were working in there and got contamination on their person. But before they leave, we make sure they leave the contamination here."

As for radioactive contamination inhaled in the lungs, he points out that the human body can excrete radioactive material over time, and the process can be hastened with medication.

The day is hellishly confusing as safety experts try to piece together events that caused the radioactive blight. Theories are posed. Maybe solvents are involved. There's talk about a worker struggling with a detached breathing apparatus air hose. And the importance of keeping contaminated hands away from your face.

"It is not unusual to find five people in one day who are contaminated," says Gesiakowski. "It's unusual to have five people turned away from the guard shack. We catch it sooner than that."

That's part of the mystery. With the gauntlet of radiation detection devices workers pass to reach the exit, why did the contamination go undetected until the final checkpoint?

It will be several days before the pieces of the puzzle start coming together.

Chapter 26: Solving a Radioactive Mystery

Of all the rules and regulations the Zion station must follow, Murphy's law—the adage that if anything can go wrong, it will—appears to be another.

In the early hours of the accident, Edison staffers find that 29 workers are contaminated but struggle to discover the source of the contamination.

"We had a significant problem," says Dave Smith. "By examining people, we're trying to find where the contamination is coming from. That's part of the mystery at this point."

Irene Johnson, an Edison spokeswoman, says exact figures on how many workers were contaminated, externally and internally, are not available. "There does not appear to be an imminent health threat, or immediate danger from the levels we've seen," she says.

It's clear that particles of radioactive dust filled the Unit One reactor building.

Another spokesperson, James Toscas, explains, "It is not unusual for people to get contaminated during an outage, but it is unusual for it to happen to a group of people like that. We were not doing anything that should have caused this." Workers were found with external and internal contamination, meaning "they inhaled some of it," he says.

Aside from the hunt for a reason, all proclamations agreed that it was not a major health issue.

Work inside the containment building halted for five or six hours while staff members went through a process of elimination that led to a hapless boilermaker whose air hose broke away from his respirator face mask in a freak accident on a platform he was cleaning between two steam generators.

"We've never had anyone lose it at the mask," says James Ramage, ALARA radiation protection coordinator. "This gentleman's hose came off at the mask for some reason at 25 psig [pounds per square inch gauge]," causing a rush of air from the hose. "In 10 years, I've never heard or seen anyone lose supply of air at the respirator."

Hoses with a lock fitting also connect at the belt and automatically shut off air flow when they disconnect. Hoses screw on to masks and do not have the automatic shutoff safety feature. The boilermaker struggled for 20 to 30 seconds to reconnect the hose to his mask, before radiation safety specialists told him to get out of the area immediately.

"It was like blowing leaves off your driveway," said Smith later when reasons for the widespread contamination became clearer. "The hose was whipping around a contaminated area," kicking up a storm of radioactive dust and debris.

Actually, it was not that simple. When things go wrong in a nuclear plant, something called a cascade effect sometimes happens. One bad thing leads to another.

Ten feet from the stricken boilermaker was a portable ventilation unit blowing filtered air into the containment building.

"By him blowing stuff off the platform," says Ramage, "it blew it in the discharge of the ventilation unit, and blew it straight up into the air and caused it to go airborne in the containment."

It was two mishaps, one following the other, causing a chain reaction of contamination where 146 women and men were working.

The portable ventilation unit "was not designed to cope with the air hose flailing around," says Smith. "While this is unusual, we don't experience this very often, nobody is in any danger at all. There is no health risk. We don't have numbers. The majority of people with body counts show it is an external contamination, rather than internal. They know that by the way it is decaying off."

Here was another chance to understand in real terms how radioactivity works in the human world, not theoretically. Radioactivity is like a fire that eventually burns itself out, depending on the radioactive material. Some burn out in seconds, others in minutes, hours, days, weeks, decades, or centuries. Scientists measure that burnout rate by half-life, the time it takes for half of it to burn away, or decay and disappear.

While work in the containment stopped, Zion technicians took air samples and hunted for the source of the contamination. Ramage and his crew, meanwhile, examined the contaminated workers, looking for the "body burden" of radioactive material on or in their bodies.

Of the 29 workers reported contaminated, said Ramage, their levels of contamination ranged from 1 to 5 percent of NRC allowable body burden levels. Within 24 hours, body burden levels dropped 500 percent for 99 percent of those examined. Levels did not drop that far for one or two individuals.

One thing particularly puzzled me about the accident and its late detection of widespread contamination at the guard house exit. The Zion station is loaded with radiation monitors of all kinds. I asked Ramage why the contaminated workers were not detected in the auxiliary building, where they changed clothes after leaving the reactor building. I remember the friskers, and Nehmer's warning to scan my body with it for a full two minutes before leaving the step-off pad area. What happened? I assume at least some of the workers used the friskers before leaving the auxiliary building.

The sensitivity of auxiliary building radiation monitors "can't be set as sensitive as the ones at the guard building," he answers, because the auxiliary building is radioactive with higher background levels of radioactivity than the guard building. If set too sensitive, alarms would be sounding all the time in the auxiliary building. Monitors in the guard house, he says, are "set at the lowest limit of detection."

It is a reminder of the complexity of the nuclear world. Those monitors so important to personal health and safety can be set to detect certain levels of radioactivity. It strikes me like programming a watchdog to bark at some intruders, but not at others. It's complicated.

Of the 29 contaminated workers Edison mentioned initially, 24 were caught at the guard house. The rest were caught elsewhere.

Another factor, says Ramage, was that the radioactivity levels "were within a hundredth of what they could receive legally. With the limits where they are, they've got no problem."

Immediately after the accident, Edison reports that 29 workers were contaminated. A month later, the Chicago utility disclosed that 61 workers were contaminated. The 29 mentioned inhaled radioactive material and suffered internal contamination.

For some reason, Edison did not mention workers who were contaminated on the outside of their bodies in its contamination count at the time of the accident. Maybe the company considered that routine. Edison did not explain why it was counting only those with internal contamination at first.

Edison calls this its largest contamination event in terms of the number of people involved, but not in severity. Most exposures were in the 1 to 2 millirem range.

They were 1 to 5 percent of allowable NRC limits.

Edison's worst case of exposure at the time occurred in 1981, when a carpenter at the Dresden station got 22,000 millirems or 22 REMs. NRC fined the utility $50,000 for that event.

I asked Ramage if the containment building contamination incident was a good example of ALARA, the practice of keeping radiation levels as low as reasonably achievable.

"It's a very poor example," he answers. "Generally, we don't have this problem. Generally, we keep a very good program. We review all the jobs before they start. We review any job where there is a chance of radiation exposure or contamination. We talk with the job superintendent. We write up the event. We sit down with people. We will recommend temporary lead shielding," as long as installing the shielding does not result in high exposure rates for those installing it.

"We follow the job as it goes along. All jobs in the auxiliary building are done with radioactive work permits [RWP]. RWPs will tell the individual what the dose rates are on the job. What the protective clothing requirements are. Also what the exposure levels for the job are. How we can reduce the exposures and we document them."

That's about as concise a description of how the ALARA program works in a nuclear power plant as anyone would want. By those strict standards and procedures, the contamination outbreak slipped past and took everyone by surprise, Ramage admits.

Surprise is the last thing anyone wants in a nuclear power station. The contamination episode was a *big* surprise.

A month later, John O'Neill is back in the Zion station and I ask him if he is recovered from his encounter with contamination. "Oh, yeah, the inside, too," tapping his chest. "It was gone in two days, not too long." Tests showed he did inhale contamination. Doctors say the lungs can flush themselves out.

"Most of the stuff we breathe, we clear out," said a doctor at Northwestern University at the time. "Fortunately, the body is built that way. You also cough and get rid of it. If not, we would drown in our own dirt in the city."

As for the *Chicago Tribune's* coverage of the accident, there were four short stories tracking developments in the contamination. Two appeared on November 4, a third on November 5, and the fourth on November 8.

The coverage was calm and in keeping with *Tribune* professional standards. The event was noted and explained to the degree necessary to describe radioactive contamination, how serious it was considered, and potential health consequences.

Reporting the event posed a dilemma for me. As a reporter, I knew I could not ignore it. But I suspected any rapport I had gained with the Zion station staff was shattered by reporting the accident. Nuclear workers hate unfavorable publicity. I waited to see how the Zion staff would react.

Chapter 27: Zion Reaction

DAVE SMITH GREETS ME THE NEXT TIME I GO THE ZION STATION AND TELLS ME the superintendent wants to see me "about the stories" on the contamination accident.

Climbing the steps to the third-floor office, I have the feeling akin to being summoned to the school principal's office. Maybe it's my imagination, but people I pass in the station seem less friendly, aloof. We enter Ken Graesser's office as a meeting is breaking up.

A copy of the *Chicago Tribune's* first day story on the accident is on his desk. It's a column wide and 11 paragraphs long, with the headline reading: "30 in nuclear plant at Zion contaminated." After inviting me to sit down, Graesser gets to the point immediately, no small talk.

"The only problem I have [with the story] is saying low-level radiation gives you cancer," says Graesser. He insists the word cancer "is put in [stories] to scare people."

That first story, written by me, said: "Low-level radiation exposures normally do not produce immediate adverse health effects, but they could show up years or decades later as cancer, according to health authorities."

I know the health effects of exposure to low levels of radiation is controversial, I tell Graesser. But to omit references to potential consequences of radiation exposure would have invited criticism. Also, I pointed out that Cliff Nehmer, my health physics trainer, mentioned the possibility of health effects, including cancer, and I could not do less.

Graesser says cell damage is assumed in any radiation exposure, but the body repairs itself, while eliminating radioactive material that loses strength over time.

I spread copies of all four *Tribune* stories on Graesser's desk. The second story, also written by me and 11 paragraphs, had this headline: "Edison seeks leak source in Zion plant." And it contained this paragraph: "Radiation from radioactive material can weaken over time and it can be flushed from the body," the points Graesser made.

The third story, just three paragraphs long, was headlined: "Radiation levels of 29 Zion workers are low." The fourth story, six paragraphs, written by me, was headlined: "Hose break kicked off Zion mishap."

Graesser turns to the fourth article, explaining how the hose uncoupling caused the accident. He calls it "the most accurate story ever to come out of Zion station," later adding, "generally I was happy with all the articles."

My fears about getting a spanking from the superintendent subsided. I was curious, though, about how Graesser would explain the health hazards of radiation exposures. So I asked him.

"Some people feel there are some health effects, but specifically what they are is not known," he says. "I see nothing in 30 to 40 years at Hanford," he says, referring to the site in Hanford, Washington, that produced plutonium for nuclear weapons, including the atomic bomb dropped on Nagasaki, Japan.

Graesser is saying that the Hanford atomic weapons facility produced material far more dangerous than those found in nuclear power plants, and "nobody has come up with substantiated effects of cancer. People say there may be."

(The passage of 30 years tells a different story. A mass tort lawsuit filed against the federal government by 2,000 residents living downwind of the Hanford works charged that they suffered cancer and other diseases because of radioactivity released from Hanford. In 2015, the U.S. Department of Energy resolved the final cases, paying more than $60 million in legal fees and $7 million in damages. For historical reasons, though, it's important to understand views that existed in their time, which in this case is 1983. Graesser's views on the hazards of radioactivity are not totally different from those held widely in the nuclear power industry even today.)

Cell damage happens with any radiation, Graesser contends, including X-ray. "Genetic effect is a potential. We won't know that until two, three generations down the road. Concern for low-level radiation, I haven't seen anything to substantiate it."

Turning from the abstract and theoretical, the superintendent mentions something totally unexpected when discussing the continuing examination of the contaminated workers. Sixteen were contaminated with cobalt-60, a powerful and long-lasting byproduct produced artificially in nuclear reactors. It has a half-life of 5.2 years.

"Cobalt-60 is the thing they are concerned about here," says Graesser. "There's only 16 with a body burden of cobalt. Pores gather it. It might take two or three showers to get rid of it. We're talking about a speck on the hair that can give you what looks like a burden" because of its intense and long-lasting radioactivity.

In all the discussions about radioactivity I've heard at Zion, this is the first time anyone mentions cobalt-60. Research showed that it is a byproduct of typical nuclear power plant operations.

"The worst is an itchy nose," says Graesser. "He rubs it, and then you have a contaminated nose. That's a tough one." The answer is scrubbing nostrils with a Q-tip.

Airborne contamination is not unusual, says the superintendent, "but we can usually find the reason very early. We were doing nothing in this workwise to cause that accident. It took a long time to find that hose came off."

These periodic meetings with the superintendent give me a chance to catch up on issues from the past. I ask Graesser if he has anything to say about the NRC investigation into allegations of drugs, alcohol, and sex in the station.

"There were allegations made by individuals," he responds. "The allegations were investigated by NRC. It was one of the first investigations of this type performed. The report showing there was not a drug problem at Zion speaks for itself."

Edison has a drug policy, he says, and "anyone found taking drugs on plant property will be dismissed." Drug use off plant property could result in disciplinary action. "If somebody comes and says he has a drug and alcohol problem, we aren't out to fire people but to help them. We have too much invested in them in training."

Graesser returns to an issue that is still troubling him. Unfavorable publicity is on his mind, just when I thought we were done with that. He exclaims:

"People in the [nuclear] industry see things in the paper. I wonder why we get beat over the head and others get away with all sorts of things. I hope people don't think I'm suicidal going to work in this industry. I have a wife and kids. I wouldn't do anything for an hour to endanger their lives."

I've heard this before. From the nuclear power folks and others in industries seen as dangerous to public health, including chemicals and petroleum. With the exception of the Three Mile Island accident, which caused no known fatalities, the record of nuclear safety in the United States tends to prove them right.

If popularity and acceptance are their goals, I tell Graesser, it's an uphill battle. Many American institutions believe they are misunderstood or oppressed—doctors, lawyers, politicians, media. All are regarded with suspicion, and electric power utilities are not alone.

About a month later, Dave Smith weighs in on the contamination event when we meet in our makeshift office in the parked trailer. He's brooding and frustrated by the disorganized way it unfolded, and argues that reporters should act like engineers and write their stories when all the facts are known, rather than "rush to print."

On the first day of the accident, "nobody knew what the problem was. We were forced to give information when we were not sure at the time of the accident. We wanted to get a better idea of how extensive the problem was—the cause of the problem—before we had to tell our story to people who didn't know what we were doing." Later, "we knew the extent of the problem and knew it was significant." He was against speaking about the accident while it was happening.

Our discussion, which sometimes turned heated, was a chance to clarify a few things about our working relationship. Smith was annoyed that I went to James Toscas, an Edison spokesperson, for information during the accident. I reminded Smith that I first went to him with a question, but he said he could not give me an answer until the next day although he had confirmed that the accident was "significant."

That, plus information I already had that something like 10 workers were contaminated, was enough for a story in the *Chicago Tribune*. Since Smith told me he would be unavailable until the next day, I turned to Toscas for further information. And, I did not think that Smith's role as my escort in the Zion plant meant I was forbidden to speak to anyone else at Edison.

Somehow, our talk turned into our worldviews, his as an engineer in a nuclear plant, mine as a reporter. They are very different. He tries to keep things organized and tidy. I see the world as a messy place, like the contamination accident, full of surprises that must be understood and explained.

"I'm mad as hell," Smith says, pushing a chair on wheels across the room when I say power utilities are partly responsible for information about themselves, and those sources can be inaccurate, not only reporters. Smith tells me he believes Edison is taking a risk by allowing me inside the Zion station, but did so despite the risks.

Smith feels misunderstood by the public. I point out that his industry missed a chance to gain public approval by refusing to answer questions about its impact on health and the environment.

In earlier days, Atomic Energy Commission safety and licensing boards refused to hear testimony on environmental concerns when licensing nuclear power plants, saying hearings were limited to questions about plant safety.

Residents living near the proposed sites came to the hearings with questions about thermal pollution, the hot water discharged from power plants into lakes and streams. This was a starting point of early environmental concerns. People were told that was outside the scope of the safety hearings, turning curious onlookers with reasonable questions into bitter opponents. While covering such hearings, I saw attitudes harden because the nuclear power organization, the commercial industry and government agencies, was unfriendly toward people with questions about how a nuclear station in their neighborhood might affect their lives.

I tell Smith there were essentially two groups of people back then: those who opposed nuclear power and those who simply had questions they wanted answered. Some answers might have paved the way toward a better relationship with the public.

A 1971 landmark court case involving the Calvert Cliffs nuclear plant in Maryland broke that impasse, ruling that the AEC must consider the environmental impact of building and operating an atomic power plant.

The AEC was abolished in 1974 and other federal agencies were created to take its place.

Smith says people in the nuclear power industry simply want to preserve their jobs, and "I just want to be left alone to do my job." I seem to touch a sensitive nerve when I point out that nuclear power people want to be left alone, but they also want the public to like and understand them. They can't have both, I say, since one involves isolation and the other engagement. "Now I'm mad," says Smith, clearly distraught. We are like beings from different planets trying to communicate.

Chapter 28: Nuclear Women

IT COULD BE ARGUED THAT THE ENTIRE NUCLEAR POWER INDUSTRY RESTS ON A foundation built by a woman—Marie Curie, a nuclear energy pioneer, the first person to claim two Nobel Prize honors for her work on radioactive materials in 1903 and 1911.

She died as a result of radiation exposure.

Another woman, Queen Frederika of Greece, was known in the 1950s as "The Atomic Queen" for building a network of scientists, engineers, and politicians to support nuclear physics research in Europe after World War II.

Women have a long and honored place in nuclear history. The International Atomic Energy Agency reports that women account for 22.4 percent of global nuclear industry employees.

A woman's place definitely is in a nuclear power plant, judging from the number I saw working in the Zion nuclear station, many of them as radiation safety specialists.

Forgive me if I seem lecherous, but my attention was drawn to Zion station women the day the top of reactor Unit One came off and I saw what appeared to be a double standard between men and women.

Once the delicate head lift operation was finished, workers in baggy canvas coveralls and hoods trooped to the step-off pads in the auxiliary building to remove their anti-contamination togs. At first, everyone looks alike; it's impossible to make gender distinctions with all that clothing.

The men stripped to their underwear and T-shirts, or were bare-chested. One woman was in the group, which became apparent when she removed her hood and coveralls. She was wearing a one-piece bathing suit with shorts over the bathing suit—more covering than men were allowed.

Sandy Holven, a uniformed guard, handed the almost-naked men their identity tags as they walked past. I asked her what it's like to see men like that all day. "It's not fun," she answers.

This episode suggests another way that the nuclear industry differs from most others: workers routinely strip in front of each other. I could not help wondering how the Zion staff reacted to this semi-nudity. This, after all, is the only nuclear station in America that was investigated by the NRC for alleged sex, drug, and alcohol violations. So I began asking the staff about their reactions to this odd dress code.

In the auxiliary building, a woman is wearing a T-shirt with a slogan on the front: "A little nukie never hurt anyone," suggesting a sexual informality in the workplace, and a play on words involving nuclear energy.

Jeannie White, a blue-eyed blonde, is a radiation protection specialist and married to one of the Zion station's foremen. I ask her reaction to seeing men without their clothes.

"Most of the time it grosses me out," she says. "These guys work so much, they don't have good physiques." But, I press on, what about the double standard? Women don't strip down to their underwear as men do.

"I'm in a body suit," she says forcefully. "They don't pay me enough to strip down to my bra and panties. This is not burlesque. This is nuclear power. I strip down to the level that is professionally acceptable for my job. It is within radiation standards."

White compares herself to a nurse, saying she is a radiation-chemistry technician. "Have you ever been in a hospital?" she asks. "Ever been washed down by a nurse? If it was in a bar, they would turn red in the face."

Undressing in front of women, like working in radioactivity, says assistant superintendent Ed Fuerst, "you get used to it. You come to appreciate it and get used to it."

My informal survey continues for a short time. Ernest Abbott, a fuel handler, says, "I'm not looking at women. I'm performing a job. I never think about it." Tim Kresal, a mechanic, says, "You get used to it." Dale Anderson, a mechanic, says, "You get immune to it. It's like the beach." Ken Bohning, a station man, says, "Are you serious? It doesn't bother me." Dale Gibson, a mechanic, says, "If it doesn't bother them, it doesn't bother me."

The survey reveals the Zion staff as blasé and adaptable, not prudes.

My attention turns to something more in keeping with the line of work started by Marie Curie and those who followed her in the business of nuclear energy. That would be Leslie Holden, 26, a blonde with a finely chiseled face and wearing jeans and a burgundy and gray checked shirt, informal like the men around her.

"Scree—Holden," she says, answering a ringing telephone in the Zion nuclear power plant control room, where she is one of those in charge. In answering the phone, Holden announces her name, but also her title: Scree. It's short for shift control room engineer (SCRE), a job that requires her to be coolheaded, calm, and focused on the big picture while others around her might be losing their wits in a major accident. It tells the person on the other end of the line that they are speaking to one of the top people in the control room. She is a nuclear troubleshooter with a title that appeared only recently.

Nuclear energy is a male-dominated world, but Holden is living evidence that is changing, prompted by the 1979 Three Mile Accident in Pennsylvania. The Nuclear Regulatory Commission mandated shift control room engineers after that accident revealed that reactor operators were so busy answering hundreds of alarms, they didn't have time to think.

"It all goes back to Three Mile Island," observes Holden. "They didn't have an engineer [in the control room]. The people were too involved. Nobody put it together and said: 'This is what we've got.'" Reactor operators made matters worse by turning off reactor cooling systems, causing the fuel core damage.

During normal operating conditions, Holden is a supervisor in the control room responsible for making sure the reactor fuel core is covered with cooling water. And she's responsible for keeping track of plant safety as equipment is taken out of service.

"During an emergency," she says, "I remove myself from that job and look at the job in an objective manner. I look to be sure they are following procedures at the outset of an emergency, make all the phone calls."

Federal regulations require Edison to notify a long list of federal, state, and local disaster management agencies of an emergency.

In an emergency, she must be detached in the worst of times, "rather than being carried away in the process of equipment malfunctioning and opening the valves," she says. "Keep it safe as possible and manage off-site [radioactivity] releases."

Holden is one of six SCREs at Zion, but the only woman. She is a licensed senior reactor operator, with a mechanical engineering degree, specializing in energy conversion systems, from the University of Illinois in Chicago.

"I'm responsible for everything that goes on within these walls," she says of the control room for both of the Zion station reactors. "It's a challenge, a chance to put my education to use. I work with a lot of people."

While Holden attends to those duties, I ask her to think beyond her role at Zion to her role in the world of nuclear energy. It's a man's world, isn't it?

"I think it takes a certain type of person to want to be a nuclear person, or a doctor," she answers. "It is a highly specialized field. I don't know how many say, 'Gee, I want to be a nuclear power person.' I knew I wanted to be a nuclear power person since I was a junior in high school. I was fascinated by it.

"There was a lot to be learned. A lot of people don't understand nuclear power. You'd go home and ask a question and they'd say, 'I don't know.' Maybe because I got a lot of 'I don't know' answers that I decided to find out."

Holden grew up in Chicago and attended Lane Technical High School, long a high school for boys before turning coed. "I've been in a man's world a while," she smiled. While still in high school, Holden went to work for Commonwealth Edison as an intern. After graduation, she joined the company full-time and "I went nuclear."

She worked on equipment procurement and evaluation during construction of Edison's Byron and Braidwood nuclear plants. During that time, she met Leslie Bowen, a senior reactor operator.

"She encouraged me," says Holden. "It was interesting to see a woman who knew so much about a nuclear power plant." Holden joined the Zion technical staff for a year, then took a year to become a licensed senior reactor operator, a management position.

Would she encourage women to make nuclear energy a career? "Yes," she says. "I wouldn't encourage them any more than a man. Being a woman or a man doesn't mean as much as a person wanting to be involved." Because of rotating shifts, she says, it is not a job for everyone.

How do the men treat you? I ask. "Wonderful," she answers dryly. Is there any difference in the way men treat her? "If there is," she says, "I'm oblivious to it. I run into confrontation, but it has more to do with my lack of experience than that I am a woman, at least I take it that way." She is not treated like an outsider.

Does Holden consider nuclear power a man's world?

"It must be," she answers. "There are more men in it, but that's because there are not enough women who want to be in it. The fact that I have four years with the company full-time, I don't know many women who want to be here."

Holden is married to a Zion nuclear engineer. "Most women my age have families and husbands. It took a lot of time to get the [reactor operator] license to be here. My husband also being involved is understanding of the circumstances." He works days, but Holden works rotating shifts. "I don't see him for seven days," she says. She works six straight days with two days off.

My radiation safety training instructor made a point of cautioning women of childbearing age to be careful. I asked if radiation worries her.

"It is something to be aware of and to care about yourself," she says. "If I were pregnant, I wouldn't want to go into the steam tunnel. It's so hot I'd faint. We are not allowed in the auxiliary building once we know we're pregnant. My job is not considered a high-exposure job. I don't plan on having children for at least five years. It is something you should be aware of, but not concerned about."

Holden estimates her monthly radiation exposure is "definitely under 50 millirems."

Woman interested in a career in nuclear energy might consider how Holden did it, starting in high school and driven by determination and goals, even after she got a job at Edison.

The Zion station was looking for employees, she pointed out, "so that's why I volunteered. I asked to come up to the station, and I wanted to be a Scree and was interested to be a Scree."

Because the Zion station is close to the Chicago metropolitan area, the NRC requires special safety requirements. Holden was under observation for a year to be sure she was technically and psychologically suited for the job.

The toughest part of the job so far, she says, is the outage. "I don't have a whole lot of experience with outages, and training doesn't cover outages. There is really no way you can train for outages. This is a bad outage to start on. This is not a normal outage. There is a lot going on."

But her outlook is positive.

"I'm here because I like it," she says. "I'm adequately trained to do what I'm supposed to do. It does not involve the fact that I am a man or a woman. It takes a certain type of person to do this. Everybody is here not just because it is a job. They picked the industry. Everybody cares. We don't find a lot of indifference. When something happens, and you say I want you to do this, they are doing it."

Generally, Holden believes people at Zion are handpicked by Edison.

"The people are very smart," she says. "They're sharp. You are either sharp or you don't make it. People will detect if you are smart." In other professions, she believes, people are less aggressive and can't think on their feet.

"Everyone respects everyone here," she says. "There are levels of respect, some are respected more highly than others." Is she respected?

"I hope so. I don't get a lot of talk back," she answers. "Usually, when I tell people to do something, they do it. Most questions are legitimate questions. I have not been explicit enough, or there are changes I wasn't aware of. We work together."

At Zion, the control room where Holden works is at the top of the station's operations chain. She went there for a challenge, and to work with some of the brightest, most alert people. They get respect.

Down below, on the ground level, is another group of people who get the least respect. They are maligned and belittled. They are part of the nuclear energy world too.

Chapter 29: Nuclear Guards

THE RESPECT, TRUST, AND ADMIRATION LESLIE HOLDEN EXPRESSES FOR FELLOW workers does not always extend to one group at the Zion station: security guards.

It is a paradox. The nuclear power plant security force basically is a private army whose primary duty is to repel terrorist invaders, either by fighting them off or preventing them from reaching sensitive areas until help arrives. Plant safety depends on them and their willingness to sacrifice themselves if necessary.

They are expected to hold off a blistering attack by at least five armed invaders firing machine guns, although everyone agrees their chief on-the-job enemy is withering *boredom*.

Yet the Nuclear Regulatory Commission investigation into sex, drugs, and alcohol at the Zion station focused heavily on its security force, including an accusation that security guards would abandon their posts and run if attacked. It should be pointed out that the abandonment accusation came from one person appearing on a televised newscast, who did not mention the Zion station specifically. NRC pursued the matter as though the allegation related to Zion and found no evidence to support it.

The rest of the allegations against Zion guards came from co-workers telling NRC what they had seen, heard, or smelled. That included guards coming to work intoxicated, engaging in sexual activities, using marijuana or drugs, having inadequate training or drills to withstand attacks, and failing to detect alcoholic beverages while screening employees coming to work. You could say their friends blew the whistle on them.

Six of the Zion guards admitted using marijuana while off-duty and two guards were fired for refusing to cooperate with the investigation. NRC largely exonerated the Zion guard staff, but the attack on their reliability, based on allegations from people working with them, left the guards feeling their reputations were sullied and gave the impression they were inept or addled by drugs or alcohol.

"People think of guards as those they see at a fair or at K-Mart," says plant superintendent Ken Graesser. "Our guards are highly trained. They take special firearms training, and they are paid enough to support a family."

Nuclear plant security swiftly became a global issue. I mention to Graesser that only a year or two earlier I visited the Zion station manager's office without any intervention by guards.

"A lot of our security came about because of terrorist activities in other countries, plus demonstrations in this country," explains Graesser. "It is a multifaceted system and very sensitive. It is somewhat an overreaction. It developed from fuel-reprocessing plants, and we are not a fuel-reprocessing plant."

Such plants produce plutonium, a radioactive material that can be used to make nuclear weapons. Authorities feared that terrorists might invade them and steal plutonium. In 1977, President Jimmy Carter banned reprocessing commercial reactor spent fuel for that reason.

"There was a worry that somebody would get plutonium out of here," Graesser admits. "There is no way anybody can get plutonium from this plant."

But that's how Zion station became a veritable fortress, surrounded by two chain-link fences six feet high and topped with three strands of barbed wire. Guards appear to face challenges far less formidable than terrorists, so I try to gain an understanding of their typical daily routines.

About two dozen guards serve on a shift.

As my escort, Dave Smith, and I head toward the onsite security office, he warns me, "we will be dealing with information that is protected by law. You can't have it. The point is to protect the security system installed. If we're getting into sensitive areas, I'll stop it."

Zion station guards are employees of a contractor, Burns International Security Services. Gregory Gilliland is the Burns Security site supervisor. He points out that many layers of security exist at Zion, beginning with the fences and bomb detectors at the portal entry monitors. Employees insert their identity badges into card readers to gain entry into individual station zones.

"We have control of necessary people in necessary areas," says Gilliland, explaining that more than 100 card readers around the plant are programmed to allow employees into specific areas.

In a way, the Zion station is booby-trapped to detect intruders by setting off an alarm if anyone tries to open a door without an identity badge or one not programmed for that zone. Card readers are a fairly old technology, but reliable.

What about the guards? Who are they? How are they recruited?

"It seems to be a lot of ex-military, national guard, part-time or ex-policemen," replies Gilliland. "We've had a few ex-military women and a slim few policewomen. The ones who end up staying the longest are males, 21 to 30. The ones who are older or retired tend to drift away." Turnover is fairly low; 90 to 95 percent of the guards remain.

Gilliland boils down their duties to access control, screening visitors and Edison personnel, verifying identity and citizenship, and patrolling the building and protected areas.

They are trained to respond to emergency situations and search areas where an alarm sounds. Firearms training improved since hiring an instructor and practice sessions several times a year at a shooting range, including women guards.

"They do everything the men do here," says Gilliland. "My top two shooters, one is female."

Training for invaders, "we think of everything possible that could happen and plan for that," says the Burns supervisor, including surprise drills, which the guards highly favor.

"I know we can hold an attack off until outside assistance arrives," says Gilliland. "We can defend ourselves until we get outside assistance."

Guards work eight-hour shifts, although that can go longer with overtime during an outage. "I make $22,000 a year and there are guards who make more than me," says the supervisor. "I won't allow anyone to work more than 60 hours a week. After that, your efficiency goes down." The salary, he says, is comparable to what any factory in the area is paying. (Adjusted for inflation, $22,000 in 1983 is equal to $62,845 in 2022.)

The hardest part of being a nuclear power station guard?

"They'll tell you the sitting and nothing to do—the boredom," he answers. "Most of it is monitoring the badges. We try to rotate the posts to relieve the boredom. There are no tough jobs, no stressful manual labor type of jobs. The only thing is boredom. At night, there is no flow of workers and nothing to monitor at some points.

The best part of being a guard? "They are all going to tell you the paycheck," replies Gilliland. "They will tell you they are satisfied. They all know when payday is."

It's lunchtime. Smith and I head for the guards' lunchroom to talk to them about their work while on lunch break. On the way, Smith gives me an update on the outage. All the fuel is installed in the reactor core, and the top of the reactor is back in place.

The lunchroom is constructed of cinder block. It has a sink, a refrigerator, a table, a coffee maker, and a bulletin board. A cluster of off-duty guards sits at the table.

Sue Haskins of Beach Park, Illinois, has been a guard for four years, getting the job after she saw an ad in a newspaper and applied. She's married and the mother of seven children, and she's pretty good with a .38 pistol.

"I shot a 250 out of 300" during firearms practice, she says. "I didn't have any problem with it. Some people don't make it."

But is she prepared to repel an invader firing a machine gun at her?

"Machine gun," she says, her face expressionless, as though turning the idea over in her mind. "The first thing I'd do is try to get out of his line of fire. We're trained to hold them off." She has no doubts that she could do that. "Yes, I could," she says, her voice rising in determination.

The hardest part of her job, she says, is the continuous training. "There's always something new and different," she says. "You get it down pat one way and it is changed. Something different."

Don Collings of Waukegan is a handsome man with blond hair and a mustache. He became a guard because he needed a job and was familiar with the Zion station through an earlier job. Collings served as a jumper, darting in and out of the bottom of a steam generator to repair tubes. I recognized Collings because I interviewed him a year earlier for an article I wrote on jumpers.

Out of curiosity, I ask how much radiation exposure he took as a jumper. "I pulled 2,400 millirems in seven, eight, 10 times in the steam generator," he answers, over 20 days. Shortly afterward, he applied for the guard job.

"I like working with people," says Collings. "This is a great opportunity to do so." He started on the night shift, which he disliked. "I don't function well after dark," he says. But now he's on the day shift, which is "the greatest."

Collings does not hesitate when asked if he could repel an invader. "Yes, no question about it," he replies. "I've been shot at before. I spent a tour in Vietnam.

I'm still a naval reserve chief hospital corpsman and spent a tour with the marines [in combat]." While still on active duty, Collings was chief of security at the Great Lakes Naval Training Center in North Chicago, Illinois. "I had some security background," he notes.

"Most visitors who were never here before are awed by the security measures," says Rick Hiser of Waukegan, who became a guard after 20 years in the navy. "If anything, they are more submissive. People who have never been in the plant are easy to deal with because they are totally lost."

Collings says: "When they step in and see the monitors and see three guards with .38s, they are very humble in most cases."

Hiser adds: "And when we give them a pat search, they know this is for real."

To which Collings adds: "We have to relax them because I think they walk in here a little bit fearful, which ain't all bad. We're in a high-visibility type situation. Anytime something minor happens, it is blown completely out of proportion. It goes with the territory. If it happened in a coal plant, you'd never hear about it. It's the term 'nuclear.' It's the connotation in peoples' minds. Our challenge is to educate people."

I ask them to describe any surprises, and Gilliland answers. "We had a drunken sailor going to the Great Lakes naval center and we caught him climbing over the fence. He thought it was the naval base."

Hiser answers this way: "We don't have many surprises. Our job is to make sure we don't have many surprises. Drivers coming here comment, 'What is this, Fort Knox?' Many go through security and don't want to come back again." He also has come to terms with the idea of confronting invaders.

"After 20 years in the military, I realize that is what I'm hired here for. If that situation arises, it is part of the job. If I don't like it, I'll find a different job. After being in Vietnam, I don't like anybody shooting at me. But it is a hazard of the job. There is a risk involved."

All the guards say their favorite duty is escorting visitors to wherever they are supposed to go. It is a job that offers variety and a chance to meet people. "I'd rather be on escort five days a week," says Hiser.

All agreed that boredom is the worst part of the job. "The answer is your supervisors switch you around [to other duties] so you can't be bored," says Hiser. Before that policy went into effect, guards stayed at one post for eight hours.

"If you've been at a post for eight hours," says Collings, "half of us would be asleep." If you come to work tired, adds Hiser, the boredom level rises.

The lunchroom roundtable discussion becomes highly animated and rapid-fire when the NRC investigation comes up.

Sharon Garza of Kenosha, a guard for five and a half years, contends "it was all blown out of shape. Two security guards were fired. I don't think they should have lost their jobs over it."

"That was a shame," says Haskins of the investigation.

Upon hearing of the charges against the Zion station staff, Hiser's nine-year-old son asked, "Dad, do you work in a place like that?" To which Hiser answered: "Son, do you believe your dad would work in a place like that?"

"We were degraded by it," Hiser says. "That shattered my character."

Haskins again: "It was ridiculous. Disgruntled people. We have a lot of good, competent people. These people are super people. You always have a couple of jerks. Almost everyone drinks, but it's normal. We stop and have a few after the job. I have never seen anybody get out of line. You take a lot of flack from people—Edison and contractors—who resent security. They remember when they were building the station [and had free run of the place]. Now they can't go anywhere."

Hiser: "You have to be polite. Most of them are nice, but a small percentage lose their tempers. When the story broke, they said: 'We were right about you guards.'"

Haskins: "If there was all the sex and drugs going on here, I was missing something."

Hiser: "The same lie is reported so many times, they believed it."

Haskins: "It came to the point where you didn't want to say where you work. Truck drivers from way out of state were asking me about it. It was really tacky and presented us in a bad light. We have to take blood tests and urine analysis. If anybody takes drugs, they won't last long. It still pops up. It always will. A lot of people have nothing to do. The plant. The guards. The drugs. That's what people remember."

Many guards are family people with kids, says Hiser. "The kind of people who want to make a living."

The discussion seems to burn itself out, and the guards return to work.

When on patrol duty, guards either walk or ride a vehicle on the grounds inside the double fence. I meet Allan Griffiths of Kenosha, a guard for one year. He is on what is known as IRP—inside rover patrol. I get inside the truck with him. An asphalt roadway circles the station.

"We're watching the fence zone, the buildings, anything unusual," he says, while traveling five miles an hour. It takes five to 10 minutes to circle the plant. He can travel for an hour or two and switch duties with another guard, or he can work seven or eight hours.

"Rats and birds, that's all I've seen here," says Griffiths. The vehicle passes white-capped waves on the Lake Michigan side of the plant with seagulls wheeling in the sky. He averages 90 miles a day.

He's looking for "lights, anybody walking around without a picture badge, making sure it is secure. That's it." The truck makes a low hum in low gear. He puts his badge in card readers keeping track of his travels like a time punch.

The route? "It doesn't make any difference. You get tired of going around in a circle one way, you go another way." There's only two ways to do this job, clockwise or counterclockwise. He might be called to check door alarms or fence alarms, or to help someone whose key card does not work. The foot patrol takes a different route.

"If I'm around people, it's not bad," says Griffiths. "We have a radio here to listen to," referring to the radio dispatcher and radio messages. "That makes it a favorite post because it's something to listen to," he says. "In the building, I'll just wander around."

We finish a cycle, four-tenths of a mile.

If on vehicle patrol duty, Griffiths would be among the first to encounter invaders "if they came through the fence," he says. Television screens also monitor the fences, he points out, which would provide quick evidence of intruders.

He agrees he could be assigned to fend them off. "You could see before you get there what you are getting into," he says. "You've got a radio and can call for help. The vehicle offers some protection."

Dave Smith warns me against discussing the response of guards in case of intruders. "It would be safe to say Allan would not be the only one to respond to intruders," he says.

Griffiths admits that much of his attention while circling outside is focused on those two chain-link fences, spaced about 12 feet apart, and the gates.

Later, I ask Gilliland how he handles tired guards on rover patrol. "You don't want to run into someone or something. If you are starting to doze off, you tell the shift supervisor and he'll find something else for you to do."

Gilliland also comments, "you can tell the outage is winding down. All the trailers are gone," referring to the makeshift offices and equipment carriers parked outside the plant.

After thinking about comments made by the guards at the lunchroom round-table, it appears the guards are more compassionate and sympathetic toward their co-workers than the co-workers are toward the guards. Maybe it's resentment of their authority or they are seen as less professional than the rest of them. There were lapses in the way they checked food and beverages coming into the station that the NRC ordered corrected. With some exceptions, the allegations against the guards were not upheld any more than the allegations against the rest of the staff. Call it a draw.

Chapter 30: Simulating Stimulations

Vigilance is expected from the guards prowling the fences in anticipation of someone storming the power station with destructive intentions.

Another kind of vigilance is expected at the reactor control panels, one that is not as visible as armed guards in dark uniforms but more urgent. An operating nuclear reactor requires constant attention and—here's another paradox—reactor operators face the same problem with sleep-inducing boredom as the guards when a reactor hums along quietly doing its job.

Here's the difference: there are no reports of hordes of invaders storming a nuclear plant, but several notable examples of reactor operators faced with nuclear reactors going haywire in the United States, the Soviet Union, and Japan.

Reactor operators must act quickly in an emergency and the training for that is unrelenting.

One of the tools is the Westinghouse Nuclear Training Center on Shiloh Boulevard, walking distance down the road from the Zion station, which is an industrial island.

Shiloh, of Hebrew origin and meaning "his gift," is the road running roughly a mile and a half from the power plant to the center of the city, marked by the circular Dowie Memorial Drive. In this way, the nuclear plant is connected directly with the city's founder.

A cattail marsh borders the power plant on the west. It's open space until arriving a quarter of a mile further to the Westinghouse facility.

Beyond that, it's another quarter of a mile to a cluster of commercial buildings at Deborah Avenue, then another quarter of a mile before arriving at the residential area closest to the power plant marked by Edina Boulevard, Elizabeth Avenue, Elim Avenue, and Elijah Avenue—more Biblical holdovers from Dowie's time.

At the Westinghouse Nuclear Training Center, a school for nuclear power reactor operators, the goal is "to provide safe operating-based training for nuclear power plant operators throughout the world," says Jerry Scholand, manager of the Westinghouse facility. It's visited by 1,500 trainees from around the world each year—almost all reactor operators.

Behind the Westinghouse control panels are two computer-operated simulators designed to test the limits of a reactor operator's nerves and knowledge. One is an exact duplicate of the control board for Zion's Unit One reactor, complete with beige panels with black handles, red and green lights, digital displays, and gauges outlined in black.

The physical control board is the same, explains Pete LeBlond, Zion's training supervisor. "With the exception that valves never leak, it is quite close." Commonwealth Edison accounts for most of the trainees sent to the facility.

The other simulator features the standardized Westinghouse nuclear power plant design. Training applies to any nuclear plant to some degree. Basics involve startup, shutdown, and cooldown. Then it gets more complicated.

"Our malfunction list is 400 long," explains LeBlond. "With a good instructor, you can put combinations of those together in a limitless way. There's no limit. We've been operating 12 years and we haven't hit them all." The facility has 80 instructors, each of whom train three students at a time.

"Our instructors are pretty clever in covering up an incident, so it is difficult to say what went wrong," says LeBlond. A pump fails and the student "misses that fact until the world comes down around him. A program might fail but signals are not clear, and he might be looking in the wrong place."

Years ago, instructors tested trainees with one or two malfunctions, but that has evolved into training for more complicated "rolling malfunctions" that pile up on you. Using an automobile as an example, that would be like having your brakes, windshield wipers, headlights and hydraulic steering failing one after the other.

"If you fix the first one," says LeBlond, "then you have little trouble when the second one hits. If they don't fix the first problem by the time the second one hits, it's fairly serious."

Students are videotaped during these exercises. Sometimes one might protest,

"I didn't do that!" when faulted for an error. They replay the videotape to show what he did. The simulator also runs forward and backward, telescoping time so the results of an exercise are seen swiftly.

"People fall into a different mode in an emergency and don't realize what is going on around them," explains LeBlond. Flustered or paralyzed with anxiety, an operator might forget what to do next. This is where the training kicks in.

"You can't let the world fall down around them," says LeBlond. "You say, here is the information available and here's how you interpret it."

Scholand adds: "The purpose [of the training] is to keep them in a thinking mode, rather than in a myopic mode" fixed on a single problem.

"It's interesting," says a trainee. "Those guys are dreaming different ways to fool us." Instructor John Garber says, "It's challenging on the part of the instructor to try to put something in there [the simulator] that will fool them." Does he fool them? "Not too often," he says, "especially if they are Zion people. They are on their own machine. If I have a different class not familiar with the Zion plant, it is easier."

Every trainee must know emergency operating procedures (EOP), consisting of six volumes of 2,265 pages. Copies are in the simulator control room and in the Zion reactor control room. Thirty-three volumes describe all operations in the plant.

When problems strike, an operator must flip to the right EOP section and follow steps spelled out there. The operator must diagnose the problem, take immediate action to control it, and end it.

Accident diagnostics uses a flow chart or checklist beginning with a description offering "yes" or "no" choices. If "yes," turn to a set of instructions. If "no," turn to another. The operator follows descriptions and instructions until the problem is solved. The flow chart reminded me of income tax forms that guide taxpayers to relevant choices, while skipping others.

All of this takes place in a theater-like setting divided into two parts. The reactor control board with an aisle for students is in front. Immediately behind them and about three feet higher is a glass-enclosed control room overlooking students at the simulator control board.

Carl Schultz is an Edison assistant staff supervisor re-qualifying as a reactor operator. A husky, white-haired man, he says, "You don't think this is a simulator." He's one of 25 trainees in the building. "As far as you are concerned, it is a real plant out there. You are trying to mitigate the situation and bring the plant back to normal. Some things you know instinctively. But you always go to the procedures to be sure you did not skip a step."

It's an important point. The nuclear safety world is always changing. The Nuclear Regulatory Commission might order plant modifications and new procedures. As part of his training, Schultz must understand how 57 key systems function in a nuclear power plant.

"We are required to come here three days year," says Schultz. It's his 10th time to be simulator-tested. "Once you have been trained in something, it's like driving your car. When you first started driving, it was stressful. But not after you drive for a while. Trying to analyze problems can be stressful, but the plant tends to control itself."

Schultz points out that a major difference between the simulator and a real reactor is that problems on a simulator are time-compressed.

"In a typical plant, what we went through in safety systems would happen in hours," he says. "We did it [on the simulator] in 15 minutes. You work faster than in a real plant. You work on multiple problems. You don't usually have multiple problems [in a real plant]."

All the training exercises, no matter how complicated, end in putting the simulated power plant in a safe condition.

"I don't think it's stressful, compared with operating a plant," says trainee Pat Allen, a shift foreman at Edison's Byron nuclear plant. "If you make a mistake here, you start again. That is the satisfaction of simulator training. You can get a multiplicity of events. You may not resolve all of the items, but you put the plant in a safe condition no matter what it is. I like it because it gives you a chance to practice things you don't do in operation. It reinforces your confidence in procedures."

Ninety percent of the training is for licensed reactor operators and senior reactor operators. Some are training to become reactor operators while others are re-qualifying, like Schultz. The rest are technical engineers and management officials in need of background training.

Training can last up to 14 months, including classroom studies and time on the simulator. It breaks down into five phases ending with reviews and oral, written, and simulator exams. The first stage is devoted to 100 percent normal operations. Operators first must understand what is normal. The first phase also is devoted to nuclear theory, thermodynamics, radiation protection, and materials science.

The second phase covers plant systems and instrument malfunctions. The third is spent on the simulator and half the time devoted to malfunctions like steam leaks and loss of reactor coolant. The fourth, still on the simulator, involves major and minor malfunctions such as pump failures, pipe breaks, instrument failures, electric power loss, and human error. The fifth phase covers simulator overviews, followed by written and oral exams with one more day on the simulator.

Training supervisor LeBlond points out that "the training is not cheap. We're selective [about] who we send here. For us to take a degreed engineer and make a control room supervisor of that man currently costs $140,000 to $150,000 for one man over a period of 14 months, including salary." The training lasts days or months, depending on how much training a person needs. The failure rate is about 10 percent, trainees who wash out.

"You have people great with their hands," says Scholand, "but not great with book learning. Other people are good with books, but can't operate." Six continuous hours on the simulator is the normal class duration. Eight hours is too long, says Scholand, "and passes beyond learning."

Edison requires an evaluation of reactor operators every year.

Training involves a certain amount of play-acting to evoke realism. Tom Fueston, an instructor acting as a load dispatcher, calls the senior reactor operator and tells him to reduce power from 100 percent to 75 percent, using a combination of boric acid in the cooling water and control rod insertion. A lighted digital display drops gradually from 100 to 75. Reactor operators must learn to make such adjustments slowly and incrementally.

Students learn that a reactor will attempt to automatically reduce power if false or incorrect signals are entered in the control room.

Three men are standing in front of the reactor control panel. Whatever they did causes a horn to sound: "Whoop, whoop!" Lights flash, signaling instrument failure in the control room. "We got a problem with 505 impulse pressure," says one of the students. He calls for an instrument mechanic to inspect the instrument circuitry feeds. Playing a balky mechanic, Tom Fueston says he's busy. Reactor operators must be assertive. One of the students, acting as a supervisor, tells the mechanic that the instrument failure is his top priority, and to check it now.

The flashing lights remind me of something I heard while covering the Three Mile Island accident near Middletown, Pennsylvania. Federal investigators said the TMI reactor operators were confused and overwhelmed by hundreds of flashing warning lights and sirens during the accident.

Because of a minor malfunction at 4 a.m., March 28, 1979, the Unit Two reactor automatically tripped off. A pressurizer relief valve on the reactor cooling system popped open, and should have closed in 13 seconds. But it stuck in the open position, causing a major loss of reactor cooling water. Reactor safety systems automatically flushed more cooling water into the reactor, but reactor operators (unaware the relief valve was stuck open) mistakenly cut back the flow of vitally needed cooling water, overheating and damaging the reactor fuel rods.

"At Three Mile Island, if everyone left the control room, everything would have been fine," James Toscas, an Edison spokesperson, once told me, meaning the stricken reactor was trying to correct itself automatically, but the reactor operators prevented that from happening. It tends to show that reactor safety is only as good as the people running the plant.

The accident was chalked up to a combination of human error, poor training, mechanical failures of several kinds, and confusing signals. In nukespeak, it was called a man-machine interface failure. The nuclear industry learned from the experience and upgraded nuclear power plants accordingly, including a new indicator to show whether the pressurizer relief valve was open or closed.

I ask Scholand if he could reproduce the Three Mile Island scenario. I want to see the control panel light up with a blizzard of lights, as the Three Mile Island reactor operators might have seen it at the height of the accident.

"The operator at TMI could have ended the problem by closing the valve, but didn't do that," says Scholand. But he sets into motion the kind of accident I asked for, and I time it.

A siren sounds. "We've got a radiation alarm," he says. I count four red lights on the control panel, showing a valve indicator in full open position. The person playing the role of a technician, called "Charlie," is told to check the valve. It is checked and closed. The exercise lasts one minute and 44 seconds before the problem is snuffed out. Clearly, lessons were learned from the TMI accident.

Scholand suggests trying something harder, a major loss of coolant. They program a 32-inch cooling water pipe breaking in half. Reactor pressure drops from 2,200 pounds to zero. An auxiliary feed water pump won't start, a major complication. It would be difficult to find a more punishing event for a nuclear reactor.

"It looks like we got a break—a big one," the senior shift supervisor says calmly. Another instructor says, "no auxiliary feed water and will not start."

Ninety-three lights flash on the control panel.

The shift supervisor gives a student instructions and adds, "pronto!" He is calling instructions like a quarterback. He reads through the "general emergency" pages in the emergency operating procedures.

"It's sure a mess," says the operator. The scenario causes radioactive contamination in the power plant and Fueston walks through the control room with a radiation scanner, as he would in a real accident. He plays the role of a confused or stubborn worker to add to the realism.

Control rods drop into the simulator reactor in 10 seconds and the reactor shuts down within a minute. An emergency reactor shutdown is known as a SCRAM or a trip. Even after a reactor is shut down, temperatures in the fuel core are extremely high and the core needs to be bathed in a constant stream of cooling water.

The simulated accident continues for 16 minutes before Scholand freezes the control panel so I can count the white, red, blue, and yellow warning lights and indicators that flash on as the emergency scenario progresses. There are 403 of them, some scattered across the control board and others grouped together.

Those in the control room call it a "casualty," not an accident. Fueston cautions that the scenario fed into the simulator is far worse than what happened at Three Mile Island, which played out over hours before a new group of reactor operators arrived at a shift change and recognized what was happening more clearly than those on duty when the accident started. With fresh eyes, the new crew of reactor operators brought the accident under control.

What we're doing at the Westinghouse training center is an exercise. Among lessons to be learned, says Scholand, is "who's got control?" In a bad situation, he explains, "somebody might seize control from the shift supervisor."

If the scenario Scholand fed into the simulator were real, he explains, consequences would last for days. The first goal would be to be sure the hot reactor got enough cooling water from the emergency core cooling system and lend support to the power station to return to normal conditions.

I leave the training center with an appreciation of the training that reactor operators go through, including requirements that the training is repeated periodically. It was a bonus, I think, to get some idea of what the Three Mile Island operators saw and heard during the accident, including a control board that lit up like a Christmas tree, accompanied by sirens and horns. Federal investigators said that could be overwhelming, and I got an idea of what that meant.

Days later, training supervisor LeBlond tells me that, after giving me a good explanation of how the reactor simulator works, I followed with a question about how many lights went on at the TMI accident. He thought that was "disappointing and trivial." He adds: "You had your mind made up when you came into the control room."

I ask him what he meant by that, since I had no prior knowledge of what to expect from the simulator. He answers that when I mentioned TMI and confusion in the control room, I had already decided control rooms were confusing. As a parting shot, LeBlond says, "You don't operate in real time," meaning the printed *Chicago Tribune* reports were slow and outdated compared with radio and television. Internet services were beginning to emerge.

I've pondered that exchange. I began the Zion power plant assignment hoping to reach some mutual understanding between me and nuclear workers and a better appreciation for what both of us do. Judging from my exchange with LeBlond, we were not gaining much traction.

Nuclear workers tend to believe reporters are a bunch of horses' asses, not to be trusted. LeBlond's remarks tended to reflect that. He was hostile and suspicious, questioning my motives. They expect me to get it wrong, or ask stupid questions.

I still think the question about the TMI control board was not too outlandish, since federal investigators said the massive burst of warnings on the control board contributed to the confusion of the accident. They said it was overwhelming, and I wanted to see what overwhelming looked like.

In retrospect, since the simulator did not exactly recreate the TMI accident, the experiment fell flat. And I looked like a horse's ass.

My next visit to the Zion plant involves an encounter with "the enemy," another group that gets a fair amount of criticism from the nuclear folks.

Chapter 31: NRC Regulators

THE U.S. NUCLEAR REGULATORY COMMISSION IS "THE ENEMY," AS ZION STAtion's assistant superintendent Ed Fuerst sees it.

It's possible that the nuclear folks despise federal regulators as much as reporters, if that is possible. It might be fair to say nobody really likes to be regulated. But that is NRC's job, one that is complicated by public and political pressures.

Some nuclear critics go so far as to wish the genie of atomic power could be put back in its bottle, but that would be like trying to un-ring a bell. That's not one of the NRC's goals.

Nuclear power is a highly emotional issue, and NRC gets caught up in emotional conflict, including the circumstances of its birth.

The United States Atomic Energy Commission governed the nation's nuclear matters from 1946 to 1974, promoting and regulating peacetime development of atomic science and technology until it was strongly attacked for favoring the nuclear electric power industry while going easy on nuclear safety and environmental protection.

Expecting one agency to foster the nuclear enterprise while controlling it proved to be a conflict of interests over special interests.

The Energy Reorganization Act of 1974 abolished the AEC and gave its authority to two new agencies. The Nuclear Regulatory Commission got its enforcement powers. The second agency, The Energy Research and Development Administration, in 1977 became the U.S. Department of Energy in another reorganization.

The Nuclear Regulatory Commission began operating on January 19, 1975, as an independent agency that regulates commercial nuclear power plants and other uses of nuclear materials, focusing on protecting public health and safety. Only 12 years after its launch, a congressional report found the NRC getting "cozy" with the commercial nuclear power industry and handing off some of its regulatory authority to the industry.

The motto on NRC stationery is "Protecting People and the Environment."

Judging from comments by Ed Fuerst and Ken Graesser, the relationship between Commonwealth Edison and the NRC can't be very cozy.

Watching how the NRC relates to the Zion station offers a new perspective when seen from inside the station. There, you'd be likely to see Frank Dunaway, curly-haired with a red beard, strolling through the station's control room like a cop walking a beat, looking for trouble.

Dunaway is an NRC resident inspector assigned to the station.

"The primary function of the resident inspector program is to provide a daily presence in the plant," he explains, "making sure that they are in compliance with codes and safe operation of a nuclear power plant. Health and safety of the public is what I mean by safe. We are chartered to protect the health and safety of the public."

Doing that involves more than a casual look around.

"One of the requirements of the plant is to provide me with full access to the plant" to perform several monthly inspections, which Dunaway calls "modules." These are sets of procedures that cover hundreds of items involving maintenance, operating safety, surveillance, and verification that requirements are met. An inspection procedure might take a week or more.

Because the Zion station is so close to Chicago, NRC performs more inspections there than it does at other nuclear plants, though the agency has imposed no power restrictions on the station.

"We perform independent inspections in areas we feel are warranted," says Dunaway. "I use my engineering training and judgment to make that decision."

I'm talking to Dunaway in a conference room in the Zion station in a roundtable discussion with several regional NRC officials who responded to my request to get their side of the story on how the Edison staff gets along with the NRC, and to respond to complaints by Ed Fuerst.

The others at that session were Charles (Chuck) Norelius, director of the division of reactor projects; Mark Holzmer, senior resident inspector; and Dwane Boyd, section chief.

Boyd makes a point that is important to what Dunaway said about his training and judgment. "We have our own [reactor] simulator," he says. "We train them to look for things that are not normal. We are trained to see if the plant is not operating correctly." Dunaway is a licensed reactor inspector.

Picking up on that, Dunaway describes a typical day at the Zion station, beginning at 7 a.m. "I walk around the plant," he says, before going to the superintendent's office for a briefing on what is happening in the plant that day. He will review records and test procedures for completeness and accuracy before going to the auxiliary building and the containment building, if it is open.

In the control room, he walks down the room-length control board observing the position of switches, gauges, and panel enunciator lights before reading the reactor operators' log and having a discussion with reactor operators about activities of the day.

"I go in and say that enunciator light wasn't illuminated yesterday," says Dunaway. "I read the log and talk to the operator to be sure he knows it is there and the cause will not put them in violation of their [operating] license." He has been inspecting the Zion station for a year and a half.

After the first couple of hours, says Dunaway, "it becomes unstructured." He's guided by his training and instincts. "I look at what is going on."

Norelius explains: "At least 20 percent of their time, inspectors follow their noses," making independent inspections.

These inspections can result in citations for violating regulations or operating codes. The NRC officials give me a month-by-month rundown of reasons the station was cited. They included inadequate procedures for instrument calibration, failure to keep an entrance to a high-radiation area locked, and missing required inspections of a diesel generator and a fire pump engine. The station was cited about half a dozen times in the last year and a half, but not fined because the violations were not severe.

"I don't interact with these people offsite," says Dunaway. "Onsite, I characterize our relationship as professional respect. I don't consider myself buddy-buddy with these people, or the opposite of buddy-buddy."

Any antagonism or hostility?

"I haven't encountered any," says Dunaway.

Has anyone tried to punch him?

"No, there are federal regulations against that," he says straight-faced.

"Objectivity is one of the concerns we have," interjects Norelius. "We try to be alert to lack of objectivity. We have a program of rotating inspectors. The minimum amount [of stay] is three years; it can go to five years. We try to stay neutral. We have regulations to enforce and are not shy about enforcing them. We also don't want our people to become antagonistic. Our objective is to stay neutral and do our job."

NRC has been accused of being both a policeman and a promoter of nuclear power. I ask them if they feel they are promoting nuclear energy.

"Talk to the people who are regulated," says Boyd dryly. I have, and they say you are too strict, I say.

"We have rules and regulations," says Boyd. "We see they adhere to them. Our function is to enforce. We don't vary from that."

The NRC regulators begin to match Ed Fuerst's passion when I ask them about the assistant superintendent's accusations that the NRC's requirements are costly, even dangerous, and don't make sense. That about half of the plant modifications are done at NRC's request, exposing workers to radiation, and are not productive.

"We don't care if they produce electricity or not," replies Boyd sternly. "We care about safety. They operate safe or not at all. These things imposed on them have been carefully thought out. How do you prevent another TMI? They are costly, do assume exposure to get accomplished, but are considered by regulators to be necessary. We don't care about how expensive they are, but we do care about exposure. We do not know of anything being dangerous."

Norelius admits that after the TMI accident, NRC issued so many mandates for nuclear plant changes "they overwhelmed plant staff." The new requirements were filtered through a group assessing the need for those changes, "but once issued, we feel it is required."

A fire in the Browns Ferry nuclear power plant in Alabama resulted in mandates to separate electrical cables, which required breaking through concrete walls in power plants. Norelius says separating cables was done in the interest of safety.

I mention Fuerst's complaint about guarding against a fire like one from a barrel of burning diesel fuel.

"You've got to start somewhere," answers Boyd. "You can't have nothing. You've got to have something."

Utility people say you are picking on them, whose goal is to provide electric power, I say.

"The requirements are intended to make the plants safe," answers Norelius.

"We aren't picking on them. We are enforcing the requirements. We pick on them all equally."

All the NRC officials agree that Edison has the main responsibility for safety at the Zion station. Beyond that, "to inspect and enforce; that's what we do," comments Holzmer.

NRC is described in a lot of ways. How do they prefer to be described?

"We are regulators and make certain that the rules and laws are adhered to," says Boyd.

The rules inspectors enforce come from commission headquarters in Washington, DC. "We don't grant exceptions to the rules," says Norelius. "If somebody complains, we say fine. We'll get you in touch with somebody at headquarters."

I ask them to explain the toughest part of what they do, and hit upon a topic I did not expect and find especially intriguing: communicating.

"One of the most difficult things is to communicate accurately with the licensee," answers Dunaway. "Sometimes you ask the wrong question. Sometimes you get the right answer to the wrong question. That role of communications is probably most difficult. The most successful inspectors are those who communicate the best—who convey their concerns accurately to the licensee."

I think back to the times Dave Smith and I clashed as we tried to communicate. It turns out that even nuclear power people have trouble communicating among themselves.

"We're human beings," says Dunaway, to which Norelius adds, "the complexity of the machines."

That's another revelation. Nuclear power technology is so complicated, even the experts stumble when trying to communicate.

What don't the regulators like to do?

"I don't enjoy being a tough guy on a regular basis," answers Norelius. "I don't enjoy telling somebody we will fine you $100,000 for a problem. I do that, but I don't enjoy it."

What do they like to do?

"I do like my job," answers Dunaway. "It is a technical challenge. Something I feel positive about. At the end of the day, I feel I have done something constructive. I go home and feel very happy with the job I've got." Dunaway came to NRC from the nuclear navy. He was a nuclear power training officer for seven years.

Like me, the NRC inspector is grateful for the chance the extraordinary 10-year maintenance outage gives him to explore areas normally not open for inspection.

"This is my second outage," says Dunaway. "I have the opportunity to inspect areas I haven't looked at before. I had not been in Unit One containment. It changes day to day. I get to watch them moving fuel. It puts variety into the job. You get to see different things."

What does he think of the people working in the Zion station?

"People who work here are a competent group of technicians and supervisors," answers Dunaway. "I really feel they work hard to comply with regulations and operate safely. I have a professional respect for them."

Before parting, I ask for their assessment of the Zion station's safety performance.

"Zion has improved in their performance over the years," replies Norelius. "They had a fair amount of problems since the late 70s. Today, we don't have any major problems. We've had findings in the radiological area that needs some improvement."

NRC rates power plants on a three-category system. Category one indicates less NRC attention is needed and the licensee is performing at a high level. Category two indicates a normal level of NRC attention is needed and licensee performance is satisfactory. Category three indicates more NRC attention is needed and the licensee's performance is acceptable but weaknesses are evident.

"All areas at Zion were rated category two," says Holzmer.

My takeaway from the discussion is that the NRC regulators are kinder toward Zion than the Zion folks are toward the regulators, similar to what I found with the security guards.

Kindness aside, the regulators are dealing with technical issues far more complicated than keeping intruders away from the fences. They are a more brainy group with the brainy challenges of splitting atoms safely. They can't put the atomic genie back in the bottle, but they can try to keep that genie in a safe place while performing its work for modern society.

Chapter 32: Farewell Reactor

It seemed like a good idea to enter the reactor building one more time, after the buzzing beehive of activity quieted down to a lull and before it is locked up and becomes unfit for human survival.

I watched Phillip "Ski" Stachelski, Larry Thorsen, and others dismantle the reactor system and saw major parts lying open, their pieces scattered across the decks like items at a giant yard sale. I want to see how it looks now that all the parts are put back together and ready to operate after months of effort.

Safety coordinator Dick Principe agrees to escort me there. In the auxiliary building, we get entry permits allowing us to have radiation exposures up to 100 millirems and an instruction sheet saying, "Don't go into any roped-off areas" and "take in only what you need." It is a warning against bringing in trash that might become radioactive.

The auxiliary building has the familiar warnings of radioactive areas, including black and magenta tape on the floor. We sign the necessary paperwork and pass a steel mesh cage containing contaminated tools. Then to the room with the anti-contamination clothing in bins and barrels.

It's been weeks since the last time I suited up before going into the reactor containment building and I'm feeling a little rusty. I collect all the items needed and go to the area where workers undress. I strip to my T-shirt, shorts, socks, and shoes.

The yellow plastic liners go over my shoes. Then the white canvas coveralls, which have cloth straps at the bottom of the legs to hook under the yellow foot liners. Then yellow rubbers, the hood, soft white cotton glove liners, a film badge, and a dosimeter in the breast pocket. Then orange rubber gloves, which are taped shut with duct tape to the coverall sleeves. I skip putting a notebook and a pen in a plastic bag this time.

A guard takes our identification badges and puts them in a rack until we return. We walk past the step-off pads that we will use upon our return from radioactive areas, then into a place known as the "cross-town area." This is a connecting area between reactor Units One and Two.

My nose itches. Like a pro, I rub the inside of my hood against my nose.

Machinery and boxes of tools are stacked at the doorway for removal. "It's all coming out," says Principe. Tool chests are labeled "Caution. Radioactive material." They are taken to a decontamination pad on the floor below and placed in a shed for future use.

The hatch to the containment building, which reminds me of a giant bank vault door, yawns wide open. It seems less forbidding this time, because I know what's beyond that door. It's not exactly a safe place. I remember John Miller saying in his booming baritone voice, "If I jump up and down and scream and holler—run!"

But I've been there before and survived. Subsequent attempts seem routine now. I'm learning how it feels, as nuclear workers do, to accept it. So far, so good.

We enter the reactor building. Unexpectedly, I hear a symphony. I was startled once before by the musical sounds of a waterfall in the refueling cavity. This is different. Without the clang and clatter of activity, the sounds of radiation monitors are singing their siren songs.

"Beep," goes one. Principe notices it too and says it will sound faster if it detects more radiation. There's a steady "boop-boop" every two to six seconds. And "bingg-bingg," resembling sonar. A busy "boing" sounds lonesome. It is a euphonious chorus of warnings, although the slow tempo now indicates there is not too much to worry about.

A radiation protection technician confirms that, telling us that the radiation field just inside the containment building, where we are standing, is 2 millirems. But that will climb, he says, to between 40 millirems and 150 millirems the closer we go to the reactor head, which is now solidly fastened to the reactor. Black cables hanging from the sides of the reactor head power the control rod drives.

"It makes a big difference," remarks Dan McCormick, a technician, now that the top of the reactor is in place, offering some protection from the intense radioactivity of the reactor fuel core.

The radiation monitor music sounds through the vast enclosed containment building. It heightens the impression that the reactor is waiting, waiting for the day it will come to life again. Its pulse is beating slowly. But it's beating.

The refueling cavity is dry. Workers attend to final cleanup duties, like removing yellow sheets of protective plastic from the floor.

For a bird's eye view, Principe and I climb about 80 feet up a ladder to the control seat of the overhead polar crane. From there, we can see some of the centers of outage activity, the four steam generators where ROSA did her work and the four big cooling water pumps that Stachalski worked on like a surgeon.

Seeing those interconnected parts from above makes the reactor system look understandable, better than a schematic outline on paper. Many levels of steel grating platforms are connected by a series of ladders.

Principe and I climb down from the polar crane, then down to the pump deck. He points through the grating to the base of a steam generator, where the boilermaker lost the air hose to his face mask.

"The man was working on the grating down there," says Principe. "There was a filter unit on the floor and it blew it all up through here." It's easy to see how the radioactive debris spread so fast, blasted upwards through the open steel grating flooring by the ventilation unit.

On our way out of the containment building, we chat briefly. Principe is the father of a 13-year-old boy and is from Kenosha, Wisconsin. He was a naval engineer before joining Commonwealth Edison.

"I enjoy fixing things," he says, "repairing things and making things." That is typical of his co-workers at Zion, who tend to be mechanically gifted. Some of them assist in building their own homes, or remodel their own homes. Some race stock cars on weekends, he says, and many are paramedics who volunteer with local fire departments.

When they started the Zion station, says Principe, it was staffed with older mechanics from other Edison plants. "In 10 years, we've had a turnover," explains Principe. "At 37, I'm an old-timer here." He's been working at the Zion station for 12 years, as it was being constructed.

In the auxiliary building, we slowly remove our anti-contamination clothing on the step-off pads, a step at a time, never stepping back into an area that might be contaminated. I make a rookie mistake.

Dumping my coveralls in a barrel, I forget to remove my dosimeter and film badge from the breast pocket. Standing there in my shorts, I tell Principe of my blunder. He gets a pair of cotton glove liners and digs through the pile of discarded—and possibly contaminated—coveralls, until he finds my missing radiation protection devices. He passes them in front of a frisker to see if they are contaminated, then hands them to me.

Upon leaving, we fill out our entry cards. We were in containment for an hour and 25 minutes. The dosimeter tells me my exposure was 9 millirems.

Allow me a moment of reflection. I wrote about the errors of others in my time at the Zion station. In fairness, I should say more about mine, forgetting to remove my radiation protection devices from my coveralls.

The incident reminds me of the importance of constant attention to safety in a nuclear power plant, being prepared and knowing what to do in advance. Planning ahead. And not only for my own safety, but for the safety of others.

Principe had to dig into the barrel of discarded clothing to retrieve my dosimeter and film badge, exposing him to potential radioactive contamination, an exposure resulting from my carelessness and lack of attention. I'm thankful to Principe for that and sorry it happened. It also taught me something about life in a nuclear power plant.

Stay alert.

Chapter 33: The Control Room

"WHOOP-WHOOP-WHOOP."

The birdlike call sounds an alarm in the Zion station control room, the nerve center of the nuclear station.

From floor to ceiling and spanning a distance of 75 feet, the room is jammed with gauges, computer screens, dials, handles, buttons, controls, and indicator lights in red, green, blue, and white.

Twin control panels for the two Zion reactors stand side by side. Clusters of green lights on the Unit One panel show it is not operating. Clusters of red lights on the Unit Two panel show it is running at full power.

George Keene, 33, wearing a purple-and-white Northwestern University baseball cap, strolls to the massive Unit One instrument panel and touches a button that silences the whooping call.

"I'm acknowledging the alarm condition," Keene explains. "You have a couple of hundred alarms. We are in a 10-year refueling condition. There are all types of systems inoperable and de-energized. Different people are doing testing out in the plant, calibrating instruments. I know they are working on this system. They are anticipated alarms. Some procedures call for verifying the alarms and some don't."

A high school graduate with one year of college, Keene climbed the rungs of the technological ladder to become a reactor operator licensed by the NRC to handle the controls of a nuclear reactor.

In my naïveté, I once thought that only those with degrees in nuclear physics were allowed to touch the controls of a nuclear reactor. Keene is living proof of how it really works. After graduating from high school in Oak Lawn, Illinois, he studied business administration for a year, then worked a year as a printer's apprentice.

He applied for a job at Commonwealth Edison, where his father worked. Equipment attendant, an entry-level position at Edison's coal-burning Ridgeland power station in Chicago, was his first job. He bid for an equipment attendant job at the Zion station and got that.

"When I came out here, the station was under construction," says Keene. "We were on our own and roamed the plant with blueprints. Within a few months, we started formal training."

Keene was selected to attend the Westinghouse reactor simulator course for a year. In that time, he was promoted to equipment operator to manipulate equipment at somebody's instructions.

"It was at that position as equipment operator that I got most of my license training for a nuclear station operator," says Keene. "I was in nothing but training for two years. You do not work in the plant, but are strictly a student."

Keene is an example of a tradition in nuclear training. By working as an equipment attendant and as an equipment operator, he learned the nuts-and-bolts end of a nuclear station. When he became a reactor operator, he knew what machinery was linked to those dials and gauges on the control room panels.

"That's the idea of the program," he says, "as you come up with the in-plant experience. When you come into the classroom, you tie it all together."

"We grow our own," comments Dave Smith, referring to the in-house educated reactor operators, calling them "the most highly trained and best qualified people we have in the plant."

In recent years, Edison has been recruiting college-educated people, says Keene, "but for my job, control room operators license, it's usually people out of high school. They will be on shift for three years as equipment attendant, then they send them to school for 18 months and they get their RO [reactor operator] license."

But it's a grueling job in which the training and testing never stops. Keene is thinking about moving to another position.

"At this time, I'd like to get into something that doesn't require the [reactor operator] license," he says. "To maintain the license every year, the NRC is putting more and more guidelines for maintaining a license. Once a year, you have to take a requalification test, which is like taking your license test over. The requirements get stiffer and stiffer.

"It puts a lot of pressure on the guys to pass requalification every year. As for myself, it gets harder to pass every year you are further away from your initial training. It seems like once a month, you are getting a test, or a quiz, or taking a physical. It is a lot of hassle."

Keene also is getting weary of shift work. "Shift work wears you down," he says. "You never have the same days off." It leads to burnout. "One term they use is working in a control room is like working in a fishbowl—everyone looks at you from the outside." He wants to get out from under the pressure. And speaking of pressure, I ask Keene if he was involved in the NRC investigation into the Zion station.

They all were, he answers. "I thought it was disgusting. It was terrible. They had people around here who were nervous wrecks about the whole thing. You go into a little room and are questioned by government officials. For a while they went back into your whole life. Just the whole atmosphere—lawyers, FBI—it created a lot of tension."

I ask Keene if he used alcohol or drugs.

"I don't care to answer that question, to tell you the truth," he replies. "The whole atmosphere was really bad."

I ask Keene if a college background might have made it easier to keep up with NRC requirements?

"It would make it somewhat easier, but not that much significant," he says. "Ninety-five percent of the tests is acquired knowledge at the station and procedures. One thing that might help in college is familiarity with tests, that helps."

While talking to Keene, I'm told the shift engineer, a top boss in the control room, wants to know who brought me into the control room without telling him first.

Dave Smith says he forgot to mention me.

"If you are not a familiar face, you need permission to come to the control room," Tom Flowers, an ex-navy man, tells me. At the age of 38, he is "the most experienced man on duty."

This is in 1983, when the nuclear industry is young and a major source of people with nuclear experience is the U.S. Navy.

"I knew the admiral," Flowers says proudly, referring to Hyman G. Rickover, who directed the development of naval nuclear propulsion and controlled its operations for three decades as director of the U.S. Naval Reactors office.

A graduate of Chicago's Fenwick High School, Flowers has been a licensed reactor operator for 11 years and has 20 years of experience with nuclear energy.

He joined the navy after high school, went to nuclear power school, and learned to start naval reactors. He joined Edison as an equipment attendant at the Dresden nuclear power plant. He's working toward a mechanical engineering degree.

Sporting a blond mustache, Flowers wears a Florida football jersey with the number 83 and a baseball cap.

The toughest part of his job?

"The fast pace or boredom," he answers, echoing a common sentiment at the plant. "We have a lot of boredom. Sometimes, when you have both units going, you sit and wait for something to happen."

"One of my worst experiences was a small fire in one of the diesels," lasting about 30 seconds, says the shift engineer. He was required to notify 15 or 20 Illinois and Wisconsin emergency management agencies of a low-class emergency.

"I get the impression they wanted us to make phone calls before we put out the fire," says Flowers. "Everybody wants to be notified first. It's ridiculous. We took a bum rap from a lot of people on the phone calls."

Not that fires are out of the ordinary, he adds. "You have a million miles of wire, you have fires. Relays burn out. Pumps fail. Anytime you have smoke, we call it a fire."

Flowers believes nuclear energy is safe. "I've been in the business 20 years and I haven't seen anybody killed, aside from construction accidents," he says. This is before the accidents at Chernobyl in the Soviet Union and Fukushima in Japan. "It's probably one of the safest industries because people are aware of safety."

Fuel cores of commercial nuclear power plants are "slow acting," he explains. "If you have a major mistake, you might have a major problem like TMI. Things can happen fast around here, but not of consequence that endangers the public. It's mainly the danger of losing your production of electricity."

Flowers says he is in charge of operating the plant, making sure maintenance and operations are completed. In emergencies, he says, "if the station superintendent isn't here, I become the station director in an emergency. My overall job is to ensure safe shutdown of the reactor and activate the emergency plan."

I ask a question that might seem strange, but Flowers does not hesitate: do nuclear reactors have personalities?

"Yes," he says. "Each unit has a personality. The one here [pointing to the Unit Two control board], is like an old woman. It chugs along and doesn't give us any problems. Unit One always seems to be waiting there to smack you in the mouth.

It will surprise you. It will be running and boom—it trips on you and you don't know why. It will do it for the oddest thing."

Flowers recalls when rain dripping through a small hole in the roof of the turbine building caused Unit One to trip. They have temperaments.

"Unit One gives you more headaches than Unit Two does," he adds. "Even during an outage, Unit One is worse. Unit Two is the hand-me-down kid. It gets everything second."

Unlike Keene, Flowers was not troubled by the NRC investigation into drugs and alcohol.

"It doesn't bother me in the least," he says, speaking for the plant in general. "It's a non-problem here. We've had people come to work drunk and we send them home. It's like other industries. It's no different from other industries, but I would say we have less of it."

Speaking of the control room, Flowers says "the operators can't get away with it. They have to deal with me or others. They can't get away with drinking or drugs. But contractors can get away with it for a while, unless we catch them drinking and we send them off the job." That goes for anyone smoking marijuana too.

"I've had a case with an RO come to work after a few drinks. We don't send them home, we take them home. You know better once you don't get paid that night. I haven't seen anything here in a couple of years. You used to see it once in a great while."

The reason for a shift toward less drug and alcohol use, Flowers believes, is age. The age makeup of reactor operators went from the low 20s to the mid-to-upper 20s.

"They have a lot more training now," he adds. "Years ago, it used to be eight to 10 weeks of training. Now it's darn near a year to complete training to come in to operating." Anyone who works in the control room, including equipment operators, must be licensed by the NRC.

That's the human side of a reactor control room. A large aluminum board with 193 positions arranged in the checkerboard pattern of a reactor fuel core is evidence of the equally important mechanical part. It's a dance between people and technology.

Mark Foley, a technical staff engineer, hangs a brass tag the size of a half dollar on the board as technicians down in the reactor containment building slide a uranium fuel bundle into the fuel core. Each bundle is identified by a letter and a number.

The brute force of nuclear energy needs a delicate balance to survive. Fuel bundles must be spaced exactly right, or there will be no fission. If they are too far apart, no fission. If they are too close, no fission. They must be covered with water, which helps the nuclear interaction. No water, no fission. Water also keeps the fuel from overheating. Without water, residual heat in the fuel pellets from the reactor's high operating temperature causes fuel to melt.

Guarding against an "inadvertent criticality," Foley also is keeping track of neutrons, the driving force of a nuclear chain reaction, inside the reactor, using neutron detectors. Foley is mathematically calculating the reactor's rising neutron count.

Four neutron sources, devices that spray neutrons, intentionally excite the uranium fuel core into splitting atoms and producing more neutrons. The fuel bundles are in water treated with a solution that absorbs neutrons so they don't go out of control.

It's a delicate atomic balancing act.

"The NRC likes to see a controlled startup," quips Foley, meaning starting the nuclear reactor. "Theoretically, we could wait for a neutron from the sky to hit, but it would not be a controlled startup. You could pull everything out [meaning the control rods] and wait for a neutron to come from somewhere. You would suddenly go critical. Theoretically, it is possible."

But that would cause a sudden power surge, one that would potentially damage the reactor system.

"Sudden" is rarely on a nuclear power plant's agenda, as we will see.

Chapter 34: Waking a Sleeping Nuclear Reactor

CRITICALITY: THAT SCIENTIFICALLY MAGICAL INSTANT WHEN A NUCLEAR REAC-
tor reaches a self-sustaining atomic chain reaction in uranium fuel with the help of
one of the four fundamental forces in nature—fission.

In fission, atoms, the basic building blocks of all matter, break apart in a wild
demolition derby that releases tremendous amounts of heat and energy.

This is the moment everyone in the Zion station is planning for, working for.
It marks a transformation in people and machinery, a transformation that happens
while I watch.

In the arcane language of nuclear energy, going critical does not mean anything
is wrong. Criticality is a major event in the life of a nuclear reactor. It's the moment
a pile of quiescent machinery springs into life as a power-generating plant. But it
is a force that must be controlled.

A nuclear chain reaction is controlled two ways. One is with control rods, the
"brakes" of a nuclear reactor. All 61 rods are resting in slots inside the uranium fuel
bundles now in place in the reactor core. The top of the reactor is bolted down and
the containment building is locked tight. The rods are made of a cadmium alloy, a
metal that absorbs neutrons like a sponge, slowing or stopping the chain reaction
as needed.

For two days, the control room staff tested the "worth" of the control rods
to learn if their neutron-killing abilities are as good as predicted to snuff the
chain reaction.

A control rod can rise 228 steps, each step five-eighths of an inch, and makes
audible clicks at each step. The staff tests them, at one point lifting them out of

the core and hitting the "trip" button so gravity slams them back into the core in seconds, just as they should in an emergency. Otherwise, it's slow and easy.

"We pull rods to see if they work," explains Jerry Ballard, a staff engineer. "We do exactly the same thing a second time to be sure things are consistent in the core."

The reactor fuel core is bathed in water heavily dosed with boron, a chemical that also soaks up neutrons. It's like a tranquilizer that prevents stray neutrons from arousing the reactor from its sleep.

While working the control rods, the control room staff begins draining 35,600 gallons of the boron-dosed water out of the reactor and replacing it with fresh water at the rate of 78 gallons a minute. It will take 12 hours to reach the desired level.

It's called "dilution to criticality," akin to draining off the chemical tranquilizer so the reactor fuel core can awake and become more excitable. The slow-motion race toward criticality will take nine hours.

"We're looking for a nice, steady increase in reactivity as we pull rods," says Ballard. The rod-pull procedure lasts three hours. There are eight banks of rods and seven are completely withdrawn at this point. Rods in one bank, D, which have "the most control power," are halfway out.

"You want to creep up on criticality, you don't want it to go zooming in," says Ballard. "You know it's critical when producing neutrons on its own. Once we determine we have reached a critical state, we adjust rods or the dilution level."

"Critical is a lot of neutrons—billions," says Adam Bless, an engineer. A neutron counter registers a sample of 33.9 counts per second. "At 100,000 counts per second, that's when we would be approaching criticality."

The hours pass slowly. At a shift change, the incoming staff talks with outgoing personnel, exchanging information and ending with the shift foreman saying, "Okay, let's go to work." At dinnertime, cartons of carryout food come in: chow mein, egg rolls, and pizza.

"This is the least interesting time here," says Tom Petrak, a technical staff engineer, while waiting for criticality to arrive. "Once everything is settled out, we'll go critical by pulling rods—control bank D. We use boron to make gross adjustments and rods to make fine adjustments."

Suddenly, a man shouts: "Mitchell! Keep her down, boy."

A madness seems to seize the Zion control room as the Unit One reactor is being coaxed from a five-month slumber.

A man I recognize as a union steward named George yells, "Boom!" He creates little sonic explosions by smashing paper cups. A man walks by, shielding his ears from the racket with white plastic cups on the sides of his head.

I ask Keith Dryer, assistant planning coordinator, if this kind of thing goes on all the time? "I'm not sure but what they are doing is for your benefit," he answers.

We turn our attention to a neutron counter, which will prove to be a very valuable piece of equipment as we watch the reactor come alive. The counter is making a slow and steady beeping sound, while a needle indicates a range.

"It adds up all the neutrons it sees, divides it by time, and gives you a number and averages it," explains Dryer. "You've got thousands and thousands of neutrons going around in there, but it gives you a sampling of what is happening. It gives you the basic trends. You keep track of the increasing population in the core."

Time drags on. I notice the control room instrument panel is identical to the one I saw in the Westinghouse simulator center, but this one is real. Red and green lights glow on the instrument board.

"When they are sitting in here with their feet up on the desks, you know things are going great," comments Dryer. "When they're doing things, you know something is wrong."

A red light flashes on the instrument board with an audible radiation monitor warning, "radiation high."

"Oh, my god!" shouts Archie Lucky, a nuclear station operator. "I'll be right back," he says and runs out of the control room. When he returns, I ask why he yelled. "It was a joke," he answers. "You're supposed to act more concerned with a red indicator."

More antics. Messages taped to the side of the instrument panel indicate a quirky sense of humor among staff members. One of them reads, "8/11/79—Hot line to God installed."

It's 5:25 p.m., and the neutron counter is beeping at a more agitated rate.

"There are about four times more fissions going on now than when they started this morning at 10:42," says Dryer. The count is 125.9 and the needle is fluctuating.

It's easy to lose a sense of time because there are no windows in the control room, although there are two clocks, one with the standard 12-hour face and the other with a 24-hour face.

Blue lights suddenly flash. Those lights "warn you that you are going past a point where you must take action, or you will trip it," says Dryer.

Tom Flowers, the shift engineer, is in the control room now, wearing a bright orange sweatshirt and a hunting cap.

"This operation basically is very boring," he says. "Everything is done so slow. There's a lot of tension, too. This is the culmination of six months' work. You see people joking and kidding around when they are nervous about how things are going." Flowers seems to be explaining the horseplay.

Bringing nuclear reactors to criticality seems old hat to Flowers as he looks back on a 20-year career. "I've done eight on this unit, seven on the other," he says. He helped start the Dresden nuclear plant and started four reactors in the navy.

For him, criticality is meaningful "when we start generating power. It means the culmination of long hours of work for everybody. It's a lot easier to tell the corporate people we are making power."

Sitting at the control desk, Flowers is soft-spoken but direct, a gentle man, but one who commands respect and issues orders with authority. There is no doubt he is in charge; his requests are carried out immediately. He ordered a gauge on the control panel changed to make it clear when it reached a normal position. "I don't like when I can't tell it is normal from a long distance," he explains.

Power levels inside the reactor core must be gradual, explains Flowers. If too rapid, the reactor will trip off. "Slow for safety," says Flowers. "We're not in a big hurry."

It's evening now, and neutron activity in the reactor grew from a sprinkle to a shower to a storm to a raging blizzard. As the boron-treated water drains off, four neutron sources made of antimony and beryllium bombard the uranium fuel with neutrons as though prodding it awake. These neutron sources are four canisters, three feet long and three inches in diameter, that fit into the fuel core. Antimony emits gamma rays, which react with beryllium to produce neutrons.

A neutron acts like a cue ball cracking into a rack of balls and scattering them. In a nuclear reactor, neutrons crash into uranium atoms and split them apart, releasing more neutrons that split other atoms. The number of neutrons produced this way grows by leaps and bounds to the billions needed to reach a sustained chain reaction.

An instrument on the control panel, called a source range monitor, measures the growing neutron storm and traces a steady red line on a paper chart; the line is moving steadily to the right. The freshwater gauge registers 35,580 gallons.

The control room staff is tense, riveted to gauges that act as windows to what is happening in the reactor below. "It's not official, but it's close," mutters Flowers, who seems entranced, listening to something others can't hear. More fresh water pours into the reactor.

The line of red ink on the neutron chart veers sharply to the right now. "Who's going to announce it?" asks Flowers, like a schoolmaster, as younger men cluster around gauges. Archie Lucky reaches for a telephone to the station load dispatcher and announces urgently, "Unit One reactor critical! Unit One reactor critical!" Control board gauges and dials that were lifeless only a few minutes ago are moving and fluttering.

"We're critical," breathes Flowers. "Boy! I'm glad," he adds, belying his earlier comment that going critical is a boring process. Everyone around him seems relieved, excited, and happy. "There's fission there," David Farr, a staff engineer, says almost in awe.

It was a dramatic transformation in people and a machine. The nuclear reactor was excited into activity, and the people who caused it to happen also were excited. It was a scientific high-wire act, remarkable for its complexity and its mastery.

Down below, where men and women worked only weeks ago, the containment building is closed like a tomb. The fuel core we saw bathed in blue radiation is aflame now in deadly nuclear fire that no one could see and survive, back at work after a five-month sleep.

Dryer says tests and adjustments will take several days. "You want enough neutron control to maintain the reaction so it doesn't die or get larger. It's like an airplane leveling off to keep a constant altitude." The fine balances come from adjusting the control rods.

Looking back on the outage that took about two months longer than expected because of setbacks and unforeseen repairs, station superintendent Graesser says, "It's not the nuclear stuff that gets you. It's the mechanical garbage that's been around for 100 years," such as the reactor cooling water pumps that needed repairs. "If it hadn't been for that, we'd be home clear. Once you get it heated up and operating smoothly, you've got it made. Then you just let it run. But getting there is the hard part."

Graesser points out that Unit One set a performance record once by running nonstop for 259 days. "That's the way those units operate and make money."

Four months later, gauges on the Unit One control board show that the reactor is working at 99.2 percent of full power. The pressurized water bathing the fuel core is 559.2 degrees Fahrenheit, a normal operating temperature. Lights on the control board are mostly red, and the control room staff has settled down to watching and waiting, a boring routine indicating the reactor is humming along nicely.

But Unit One proves to be as temperamental as Flowers predicted and is plagued with bad luck. After it goes critical on February 6, 1984, it trips or is taken out of service 12 times over the next year. It is out of service a total of 80 days, the longest stretch being 41 days, when the NRC challenges the accuracy of Edison tests on air leaks from the Unit One reactor building.

Before long, like Unit One, I will have some bad luck of my own.

Chapter 35: End of Welcome

TWENTY-EIGHT MONTHS AFTER THE DAY I STEPPED FOOT INTO MY N-GET radiation protection training class, my welcome at the Zion nuclear station ends.

I'm told I will not be permitted to be in the control room to watch Zion's reactor Unit Two go critical. This calls a halt to my open-arms agreement with Commonwealth Edison to watch unrestricted every phase of the life of a nuclear power plant from the inside while escorted.

By this time, January 23, 1986, I've passed my second N-GET training class and watched the reactor head lift for Unit Two, which went as smoothly as the head lift for Unit One.

I appeal the banishment to Terry Rieck, Zion's radiation/chemistry supervisor, who tells me Edison is graded by the Institute of Nuclear Power Operations (INPO) and the NRC on reactor startup, and they are trying to limit the number of unnecessary people in the control room.

The U.S. nuclear power industry created INPO, headquartered in Atlanta, Georgia, in 1979 in response to the TMI accident to promote safety and reliability, setting industry-wide performance objectives and guidelines.

"It is a critical time," Rieck says of startup, and a time when reactor trips are likely. "We had four last year." Tom Flowers "made it sound pretty routine" when I watched reactor Unit One go critical. But in fact, says Rieck, "it gets pretty intense." I can vouch for that.

Edison executives decided that I was an unnecessary person in the control room, and they do not want any distractions.

I suppose after 28 months, I could be accused of overstaying my welcome.

But I'm disappointed to be considered unnecessary and a distraction. The nuclear power folks say they want the public to understand them. I was making a small contribution toward trying to demystify nuclear energy.

Thinking back to the control room antics, ostensibly for my benefit, I can see why Edison executives would not want that repeated. Thinking further, I might have seen early evidence of an existential lack of discipline and professionalism that doomed the Zion station. I will not know that until much later.

At this point, there is only the sting of disappointment at ending my relationship with my nuclear family, despite their antics. It makes me think of the quantum theory premise that observers influence what they are observing by observing it. Maybe so, I'm not going to quarrel with quantum theory.

Before leaving the Zion station, I check records on radiation exposures during my time there. Reports from the NRC say my whole-body exposures for 1983 totaled 40 millirems, but the NRC does not count exposures less than 10 millirems, calling those "minimal." After adding the "minimal" exposures, I got a total of 60 millirems. By nuclear standards, the difference is small.

Using the official NRC figure, I go to Rieck one more time and ask him what a 40 millirem exposure means to my health.

After a moment of hesitation, he shoots back: "Some probability of achieving cancer in your lifetime. I'm not sure of the statistics. You could never detect it and say that was the cause. Your risk increased by something like one in five million. That is the equivalent of a chest X-ray, so coming here was the risk of one chest X-ray."

I've mentioned before that comparing whole-body exposures in a power plant to a chest X-ray might be specious, but that's how some nuclear folks describe it.

Comparing my exposures to his, Rieck says, "that's the level I might pick up in a year, about 50 millirems. My technicians pick up 1 to 2 REMs or 1,000 to 2,000 millirems a year."

Does anyone worry about that?

"I doubt if anyone at the plant worries about it," answers Rieck. "It's of concern, to keep your exposure down and lower your risk. They worry about car accidents and keeping weight down to avoid heart attacks, but not radiation."

Rieck pretty well describes the attitudes of other supervisors toward radiation I interviewed. Some mention "latency," the 30- or 40-year time period before the health effects of radiation appear, versus the risk of sudden death in other occupations.

As I'm leaving, Rieck suggests I double-check the health consequences of my 40 millirem exposure with Rock Aker, another radiation health specialist. "If you picked up 40 millirems, your chances are four in a million in your lifetime of having a fatal cancer," says Aker, based on national radiation health studies.

While some staffers might come to treat radiation casually, radiation health specialists do not. Aker's quick answer to my question is evidence of that. It's their responsibility in a nuclear power plant, and they take that job seriously. I recall that any time I was inside the reactor containment, figures dressed in orange hoods hovered nearby with radiation scanners in their hands. The orange hoods identified them. They were the guardians, ready to answer any questions or sound an alarm if needed.

Because I wanted to keep coming back to the Zion station after the passage of a year, I took my second radiation health class to qualify on November 7, 1984. Bill Harris was the instructor for my second N-GET training in radiation safety, and he presented pretty much the same content that Cliff Nehmer had a year earlier, including warnings on cancer.

"We know that radiation can cause cancer," he said, estimating that a 1,000 millirem exposure might cause one instance of cancer in a population of 10,000. "Any radiation does cause biological damage," said Harris, "but not necessarily fatal."

On contamination, Harris said: "Don't take off running if you have contamination on you. You'll spread radioactive contamination wherever you go."

Cut loose from the Zion station, I reflect on the people I met there, especially the laborers at the beginning of their careers in a nuclear power plant, known then as stationmen and helpers, the "grunts" of the maintenance department.

They were still trying to figure it all out on personal and professional levels.

I was especially drawn to Terry Creekmore, 24, who was a classmate in my first N-GET class. I find him sweeping the floors along with three others. He's a stationman now, sweeping floors, cleaning tanks, including some that are "crapped up," and generally keeping the plant clean.

"I'm hoping to get into one of the maintenance shops—mechanical or electrical," says Creekmore. "That's what I want to do."

How did he react to going into a radioactive area for the first time?

"I was kind of leery and scared to go in," he admits. "After I went in and did my job, I saw no effects. I do my job. Somewhere down the road, something might happen to me, but that's true of any job. That's the bad thing about it. There's a bad thing about every job. Here, it's radiation."

Before joining Edison, Creekmore worked in maintenance for Johns Manville in Waukegan, Illinois, manufacturer of asbestos construction products, and infamous for creating a 150-acre asbestos disposal area in Waukegan. It will eventually become a Superfund toxic waste site and cease operating in 1998.

"At Johns Manville," says Creekmore, "it was asbestos. My father has it in his lungs. But you have to put food on the table."

Creekmore calls himself "a softball fanatic." A shortstop, he's played softball all across the Midwest. The year before, his Dill Brothers team placed fifth among 99 teams competing for the championship in Milwaukee. In the Midwest, softball means a 16-inch ball. In winter, he plays basketball or volleyball.

Like many in the nuclear industry, Creekmore is annoyed at how little the public understands about the industry.

"Around the Christmas holiday, my relatives ask me questions," he says. "For some reason, they think radioactive waste goes to BFI [Browning Ferris Industries garbage dump] on Green Bay Road" near Zion. Low-level nuclear wastes go to designated disposal sites.

The other question he gets is whether a nuclear power plant "is a nuclear bomb sitting in their backyard." Creekmore patiently explains correctly that the uranium in a power plant is enriched 2 percent, while a bomb requires uranium enriched to about 98 percent.

"It's the two main questions I get from relatives and friends," he says.

Not everyone in the plant was as friendly and cooperative as Creekmore. One of the workers called me a "fucker." When I asked George the union steward if I could talk to him about workers at the Zion station, he answered: "I'm not interested." A maintenance man said, "not for a reporter," when I asked if he had a minute. I asked why, and he answered: "I know why you're here. You're looking for a sensational story." I asked if he ever met anyone from the *Chicago Tribune*, and he said his father worked for the *Tribune*.

Overall, most people at the Zion station were not quite that hostile.

Bruce Napierkowski, 32, of Waukegan, an instrument maintenance mechanic, is a science fiction fan and finds a connection with working in a nuclear power plant. He likes to mention at parties that he works in a nuclear plant.

"I enjoy working in it and the challenge of it," he says. "I really like science fiction. It's like a fantasy. You have dirty jobs you don't like to do, like decontaminating tools, working in the turbines with dirty oil and the steam tunnel, where it's always dirty. It's all part of the challenge."

Napierkowski likes to show friends and relatives where he works and arranges tours. "You've got security, but you can sign papers and get them in. I've brought my wife here, my parents, brother, and a couple of friends. They wanted to see it and I like showing where I work. I enjoy the job."

Marvin Ruffin, 44, a maintenance worker, believes the public understands "the big picture" of nuclear power, "that we produce electricity. They don't understand the details and the standards we work here to."

Dave Stachon, 33, of Kenosha, sees what he does this way: "It's not like working in a factory where you can improvise. We can't do that. Everything is done by a set of rules and standards, and we don't vary from them."

Ken Bohning, 23, is a stationman from Kenosha, Wisconsin and does "all the jobs nobody else wants to do." He's an impressive guy, standing six-foot-five and weighing 250 pounds. "The whole thing fascinates me," he says of the Zion station.

Outside of work, Bohning and his wife raise Quarter Horses. He has five of them. "My wife is into showing horses," he says. "My wife got into it. I took her horseback riding and we got into horses."

Mark Rottman, 25, of Waukegan is a stationman who hopes one day to be a reactor operator, "as high as I can go." He describes himself as "a typical American. I have a dog, I like being home," and is a sports fanatic.

In addition to the people, I reflect on the places within the plant. The machine shop is memorable for two reasons. It's a big room filled with heavy machinery and a stunning collection of pin-up girls posted on the walls, another clue that nuclear power is a man's world.

Jobs that seldom get much attention are quality control and quality assurance, in part because they can be difficult to explain and drenched in details in an industry that is all about adhering to details.

"As in any quality control department, the purpose is to have checks and balances—to monitor whether our program is working so that parts are not defective," explains Tony Broccolo, quality control supervisor. "Our responsibility is not to make sure every part is perfect. We don't check every part. We're more of a monitor—a check. In the safety-related part of the plant, there is no tolerance for error."

A federal code of practices dictates 18 points of quality control in nuclear plants.

"We are overseers of the paperwork in this station," says Broccolo. They also check purchasing orders for 600 to 700 items a year, including inspections. The impact on safety in the Zion station is uncertain.

"The quality control job is not to make it safer but to monitor important jobs," says Broccolo. "Monitoring jobs makes those working on it more careful. That is the purpose of quality control. To make sure mistakes are cut down, so that mistakes are caught by quality control rather than by a broken piece of equipment." The Zion station has 11 quality control inspectors who are trained over two to three months to be certified as quality control inspectors, and are recertified every three years with refresher training.

A separate set of independent quality assurance auditors makes sure procedures are followed, says Broccolo, and reports are sent to the Nuclear Regulatory Commission. Quality assurance operates out of Edison's downtown Chicago corporate headquarters. Broccolo calls quality control "our second set of eyes."

To an outsider, this sort of attention to detail might appear boring and just another reason for paperwork. Quality control also is one of the ways a nuclear power plant performance is judged and challenged. It boils down to proving, through that paperwork, that a power station has a right to operate.

Chapter 36: Quality Warnings

QUALITY CONTROL AND QUALITY ASSURANCE, AS BENIGN AS THOSE TERMS MIGHT seem, serve as reliability benchmarks in the nuclear power world. They were reasons Commonwealth Edison was in the headlines again, although not for anything the Zion station did this time.

In January 1984, a three-judge panel of the Nuclear Regulatory Commission's Atomic Safety and Licensing board denied Edison a license to operate its virtually completed $3.34 billion Byron nuclear station, saying it had "no confidence" the station's twin reactors were safe because of serious quality assurance flaws.

This was the first time in the quarter-century history of the U.S. commercial nuclear power industry that an operating license was unconditionally denied.

The licensing board, which reviews applications for operating licenses to assure public health and safety, also cited Edison for "a very long record of noncompliances with NRC requirements."

It finally happened. The legendary builder of atomic power plants had been threatened as early as 1977 with license revocation by an NRC frustrated over Edison's repeated mistakes and blunders. The storied utility appeared to be losing its grip on its nuclear enterprise, once again going where no nuclear utility had gone before in setting a bad example.

This time, the focus was the Byron plant near the city of Byron, 17 miles southwest of Rockford, Illinois. And the problem was quality assurance, the arcane process by which nuclear power plants are built according to federal safety standards, ensuring that all materials used in the plant are suitable and that they are installed according to rigid standards.

In unusually strong language, the federal safety board criticized Edison and Hatfield Electric Co. of Chicago, an electrical contractor working on the Byron plant. Edison, said the board, "made a weak showing, bordering on default, in response to the board's order to present evidence respecting Hatfield. The Hatfield aspect of the proceedings alone requires that we deny the applications for the Byron operating license."

The board said Edison failed to assure the quality of construction at the plant, although it found no "widespread hardware or construction problems." But if they existed, "we are not confident that such problems would have been discovered." NRC told Edison to do a large reinspection of safety-related work.

Edison officials were stunned by the ruling, which was an obstacle to its $8.8 billion nuclear construction project, intended to add three more nuclear stations to the Edison electric generating system. At this point, Edison had seven nuclear reactors operating in Illinois and intended to have 13 by 1987.

"This whole thing is a shock to us," said James Toscas, an Edison spokesperson. "We certainly did not anticipate that the board would find they did not have enough information on which to base a positive conclusion." Edison had updated its quality control programs because of NRC concerns.

"We have a long record of building nuclear plants, and we have not had problems with quality at our nuclear plants," Toscas said. "We don't feel we have quality problems at Byron."

Byron reactor Unit One was 90 percent finished and Unit Two was 67 percent finished.

Problems at Hatfield related to welding on brackets that support electrical conduits and equipment.

"There were, indeed, problems with that welding years ago, but we've reinspected that work and made corrections," said Toscas. "The problems were with Hatfield, which we got on top of."

The safety board named three other contractors responsible for quality problems at Byron. Despite its strong criticism, the atomic safety board said Edison was capable of maintaining reliable quality assurance and was addressing the problems "too late, but it's catching up."

Antinuclear groups attacking the Byron project for safety reasons were jubilant. "It is a monumental and historic decision by the NRC," said Melody Moore of Citizens Against Nuclear Power. "It is the first time in history a utility has been denied a license to operate."

"This decision goes a long way toward instilling public confidence in the NRC," said Jane Whicher, an attorney for Business and Professional People for the Public Interest. "It also sends a message to Commonwealth Edison that slipshod construction practices will no longer be tolerated. This is an absolute victory for the public."

Ultimately, the Atomic Safety and Licensing Board overturned its decision in October 1984 and granted the Byron project permission to operate after a reinspection of more than 200,000 items and components in the plant. Byron reactor Unit One began operating in September 1985, and Unit Two in August 1987.

Byron's quality assurance struggle represented one more odd chapter in Commonwealth Edison's increasingly turbulent history. Attacks by antinuclear critics were not new; they were typical of that time.

The Atomic Safety and Licensing Board's offensive against Edison, so bold and publicly outspoken, was not characteristic of a federal bureaucracy. Matters of this sort usually are handled quietly and diplomatically. In Byron's case, federal regulators were turning up the heat.

But how far could they go? Federal nuclear regulators stress the importance of public safety over costs, even beyond whether a utility produces electricity.

Viewed through a financial lens, the Byron station reveals another story that is tied to federal regulators. In a March 1985 report ordered by the Illinois Commerce Commission, Arthur D. Little, the international management consulting firm, said cost overruns of more than $3 billion in the construction of the Byron station were NRC's fault, and to a lesser extent Edison's lack of construction oversight.

When I saw that figure, I wondered how it was possible to have $3 billion in cost overruns in a power plant whose cost was given as $3.34 billion only a year earlier? Arthur D. Little explained it this way: in the early 1970s, Edison estimated the cost of the Byron project at $800 million, with reactor one operational by 1980 and the second reactor operational soon after. Construction began in 1975 and was expected to last about five years, not 10 years or more.

"Increases in cost and schedule were caused by continually increasing NRC safety and construction requirements and the resulting underestimation of costs and schedule by Commonwealth Edison," said the Little report.

A *Chicago Tribune* article said an Edison spokesperson agreed NRC safety and construction requirements contributed heavily to the delays and escalating costs at Byron. An NRC spokesperson said there was no question that the agency's regulations were partly responsible for delays, but added that Edison was informed of regulation changes in plenty of time to make adjustments.

So, bottom line, Arthur D. Little was counting any cost beyond the original estimated budgeted amount of $800 million as cost overrun, which fits the definition of a cost overrun. This can get big in the uncertainties of the nuclear world, and reminded me of Ed Fuerst's complaints about the role of NRC in the cost of nuclear power.

I thought of Fuerst again while listening to Dr. Joseph Hendrie, former NRC chairman, who gave a speech in Chicago in 1985, saying the agency was "excessively harsh" in regulating the nation's nuclear power facilities.

Hendrie was NRC chairman at the time of the 1979 Three Mile Island accident, when about 150,000 people fled in fear that the plant might explode. Hendrie became president of the American Nuclear Society after President Carter dumped him. ANS is dedicated to the advancement of nuclear energy.

"We really have a regime that is excessively harsh and probably is counter-productive," Hendrie said of the agency he once led. "People on both sides of the fence, industry and the agency, who are sensible and watch this, recognize that on occasion some of the regulators get a little carried away."

The central thrust of NRC regulations, he said, "is to have a full, free, and frank exchange of information. If you want that and then go around and call them liars and fine them $60,000 for a mistake—that is not going to encourage a free, frank information flow."

Hendrie praised Commonwealth Edison for its performance in nuclear construction and operation. "Except for an occasional glitch," he said, "they do pretty well. I wish they had as strong a nuclear utility in New York, my area."

But NRC was in no mood to let up on Commonwealth Edison. In a letter dated January 27, 1997, the agency told Edison that the safety performance of its nuclear stations "has long concerned the commission and NRC staff." The agency requested "information explaining why NRC should have confidence in ComEd's ability to operate six nuclear stations while sustaining performance improvement at each site."

This was another threat to Edison's operating licenses, similar to the one made by NRC regional administrator James Keppler in 1977, 20 years earlier. NRC was concerned about what it called "cyclical declines in license performance," meaning performance on key issues at each power station would improve, then decline, then improve, then decline.

Instead of focusing on each of the power stations, NRC decided to focus its attention directly on Edison itself, the corporate manager.

"Corporate performance can affect performance at individual sites," said the agency, by shifting resources through economic-related stresses like deregulation and company downsizing. NRC also formed a performance oversight panel to monitor Edison and its six nuclear stations and meet periodically with Edison officials to discuss progress.

"In effect, the NRC was saying it is time to fix Commonwealth Edison's problems across the board," explained NRC spokesman Jan Strasma. "Edison has had a pattern of developing programs to fix problems and often adding programs upon programs. Sometimes, before one program is completed, they will develop another program."

NRC was tired of this nuclear shell game.

Chapter 37: Midlife Crisis

MY STORY ABOUT LIFE IN A NUCLEAR POWER PLANT, CALLED "NUCLEAR PEOPLE," appeared in the *Chicago Tribune* Sunday Magazine on June 16, 1985—21 months into my extraordinary opportunity to explore the Zion Nuclear Power Station from the inside.

The magazine's cover photo showed a section of a reactor fuel compartment aglow with the deadly blue Cherenkov's radiation. Inset was an aerial photo of the twin-reactor plant, white-capped waves rushing to the nearby Lake Michigan shore.

"Midlife Crisis," said the cover. "Atomic power 40 years into the nuclear age." The story covered all or parts of 18 pages, including the cover. An inside headline read: "Testing the promise of atomic power inside Zion station."

The story described my first 21 months at the station. I continued visiting the facility for another seven months.

Reactions to nuclear power stories are unpredictable, except for those strongly for or strongly against nuclear power. In this case, they proved to be favorable in unexpected ways.

One of them was a letter from Jack S. Bitel, Commonwealth Edison's operations quality assurance manager.

"Your story captured the feelings of the women and men who work in a nuclear power station," he wrote. I had written about how nuclear people resemble Vietnam veterans who came home feeling disillusioned and embittered. They tend to be hostile and defensive toward people outside the plant. They never accepted me, and it was not unusual to be cursed simply because I'm a reporter. "Your analogy of nuclear people to Vietnam veterans is all too true," Bitel noted. "I have been in the nuclear field since 1956 starting with the Dresden Nuclear Power Station. Back in those days, we were the white knights out to make the atom serve mankind.

"As you point out in your article, today the public, the media, and to a large degree the NRC no longer accept the use of nuclear power. I do feel, however, that when the history of this period is written, it will show that the so-called problems of nuclear power plants were not of an engineering or technical nature, but were only political."

Bitel's letter is interesting because he acknowledges the defensive attitude of nuclear people. His attitude toward politics and government is in line with the beliefs of many in the nuclear industry, including the early belief that they would transform the world with the new technology that was atomic power.

A pleasant surprise was a May 30, 1986, letter from the National Society of Professional Engineers saying "The Nuclear People" won the Anniversary Award in its 20th annual journalism awards competition. Founded in 1934, the National Society of Professional Engineers, with 26,000 members, is considered the voice and advocate of licensed professional engineers in the United States.

Afterward, the Atomic Industrial Forum (AIF) invited me to speak about "The Nuclear People" at an information workshop in September 1986, in a downtown Chicago hotel. Founded in 1953 as a trade association, the AIF would merge and reorganize several times to become in 2011 the Nuclear Energy Institute, a leading organization representing the nuclear industry. It is headquartered in Washington, DC.

I decided to give a slide presentation, showing photos of some of what I had seen inside the Zion nuclear plant. The assignment was an experiment in trust and journalism ethics, I began. Nuclear utilities generally do not trust reporters and often fear them.

Commonwealth Edison was asked to cooperate in the interest of public understanding and accuracy. In short, Edison was asked to take a public relations risk, which it accepted in the pioneering spirit shown in creating commercial nuclear electric power generation.

My focus was on nuclear people, not just on technology. The 1979 Three Mile Island accident taught us that nuclear power is only as reliable as the people running it.

I gave some history of the Zion plant beginning in 1973, and summarized the Nuclear Regulatory Commission's 1982 sex, drugs, and alcohol probe at Zion. I mentioned that plant superintendent Ken Graesser complained that reporters have a warped view of nuclear power because their attention is drawn to the indus-try during abnormal events such as accidents or scandal.

From a journalist's point of view, I made no attempt to go undercover. This was a story that had to be experienced and told clearly because of the highly techni-cal—and highly emotional—nature of nuclear energy.

My Zion assignment ended on January 23, 1986, I told the AIF audience, on the day the Unit Two reactor was scheduled to go critical—when I was informed that I was an unnecessary risk in the control room. So, the Zion assignment from the inside was over.

Every nuclear conference I've ever attended included charts and graphs, so I decided to bring one. A line chart showed one line standing for journalism man-hours. Another line stood for utility anxiety levels.

It shows that no matter how much time a reporter spends in a nuclear power plant—hopefully to build trust and rapport—utility anxiety levels tend to rise in proportion to the time the reporter spends there. I had hoped anxiety levels would drop, and the two lines eventually intersect—with journalism man-hours rising and utility anxiety levels falling—signaling some kind of comfortable working relationship.

I called the chart Bukro's Law on Utility and Reporter Dynamics. In conclu-sion, I asked the AIF audience to help bring those lines together, signifying mutual trust and understanding.

My session at the conference ended with a lively question-and-answer period.

Shortly after the conference, I got a letter dated September 25, 1986, from Leslie S. Ramsey, AIF's education services manager.

"We received rave reviews (4.7 out of 5, highest of any speaker) on that session," she wrote. "You did an outstanding job of helping our participants understand more about the thinking and feelings of reporters as well as the reporter's perspec-tive on them."

Some members of the AIF audience offered written comments on my presen-tation. One of them said: "Good challenge to utilities."

Chapter 38: A Rogue Station's Final Straw

Zion station's doomsday came only a month after the NRC practically begged Commonwealth Edison to fix its flagging safety performance or risk losing the agency's confidence that the utility could safely operate six power stations with 12 nuclear reactors.

The federal regulator had been threatening Edison for years to yank those operating licenses, with mixed results.

A dumb mistake on February 21, 1997, changed all that, putting the Zion station in a death spiral from which it would never recover. Edison's nuclear enterprise suffered too.

The mistake involved rapid, unplanned manipulation of control rods—driving them into the reactor fuel core and pulling them out again—during an eight-minute period of confusion while Zion reactor Unit One was operating. The reactor operator just winged it, trying to recover from his blunder, and nobody stopped him.

That violated nuclear power's prime directive: plan ahead, go slow, move step-by-step in an orderly way, and communicate your intentions with others before acting. The operator should have remembered something else from reactor simulation training: solve the first problem before the second one hits or your world comes crashing down. In this case, the world came crashing down, metaphorically.

Zion's reactor Unit Two was refueling when Unit One stumbled into a quagmire of problems, starting when the containment spray pump took an abnormally long time to start during a test, so it was declared inoperable. It had to be fixed within 48 hours or the reactor must be switched off, although the Zion staff had developed no plan for shutting it down, which is what you're usually required to do in a nuclear plant. Plan ahead.

After the 48-hour deadline passed, the shift engineer conducted a shutdown briefing. About an hour later, the control room staff began cutting power on reactor Unit One at the rate of 0.25 percent a minute. When the reactor reached a 7 percent power level, the shift engineer told the unit supervisor to keep the reactor critical because he expected the containment spray pump to be fixed within minutes.

The unit supervisor and the reactor operator reviewed the steps needed to bring the reactor to 0.025 percent of power. The intent was to put the reactor in idle mode, just above the point of keeping the atomic chain reaction inside the uranium fuel core sizzling. The unit supervisor read aloud the steps needed to reach that mark.

The reactor operator brought the power level to 0.025 by gradually inserting control rods deeper into the fuel core for 3 minutes and 48 seconds, but the power level unexpectedly kept dropping and reached 0.01 percent, virtually killing the nuclear chain reaction. Reactor alarms, including red lights, signaled that the control rods had bottomed out, as far as they could go.

Probably panic-stricken, the reactor operator began withdrawing the control rods for a minute and 45 seconds, hoping to reignite the reactor and reach that 0.025 percent power level they wanted.

A nuclear engineer asked the reactor operator why control rods had been driven in so far. The reactor operator responded, "This doesn't look right, but I am just following procedures." Seven minutes later, the nuclear engineer told the reactor operator he "did not like what the operator was doing." The reactor operator said he was uncomfortable with what he was doing also.

But neither of them communicated their concerns to the unit supervisor, who eventually ordered the reactor turned off.

In the nuclear power world, these events were so egregious, the NRC issued an information notice describing the Zion mistake to all nuclear utilities titled "Unrecognized reactivity addition during plant shutdown." Unrecognized, meaning the agency had never seen anything like this before and never wanted to see it again.

NRC found fault with everyone involved. When the operator realized the reactor was substantially subcritical—instead of stopping, evaluating, and communicating the changes in reactivity—he started withdrawing control rods to make the reactor critical again.

The reactor operator had asked the unit supervisor for advice at one point, and the supervisor responded by rereading the steps for reaching the 0.025 power level, which did not answer the question the reactor operator asked.

The shutdown briefing before the mistake, said NRC, was "informal, poorly planned, and ineffective. Operations supervisors did not provide any direction to the operating crew during the briefing regarding the decision point for proceeding."

Regulators said the control room staff was so fixated on repairing the containment spray pump, they lost sight of the most important thing—operating the nuclear reactor safely. They were more interested in pushing the reactor to make electricity and money for Edison.

Only three days before the blunder, an internal Edison report said, "senior management did not have a good understanding of the significance and depth of issues at Zion," and that Edison "has not continually pursued a safety culture ahead of production budgets." That report was prophetic and showed that at least someone in the Edison organization recognized the importance of having a corporate "culture" that put its nuclear responsibilities ahead of everything else.

Safety consequences of the mistake were low, but the human performance consequences were devastating for the Zion station and for Edison. NRC discovered that 39 people were sauntering in and out of the control room during the February 21 mishap, with another 15 in the immediate vicinity, which was distracting and contributed to a lack of "control room decorum."

This was an astonishing number of people, considering that control rooms generally span about 50 feet by 50 feet or 50 feet by 100 feet. Both Zion reactors were operated from that control room. Nobody explained why all those people were crowding the control room.

How bad could that situation get? What if the reactor operator decided to hit the gas and run it full speed?

"If the operator was completely clueless, the reactor protection system comes into play if there is a rapid power excursion," explained Marc Dapas, regional deputy director of the NRC's division of Reactor Program Management in Lisle, Illinois. Reactors have "trip points" that stop them if power levels are rising too fast, like 10 percent a minute.

"You have operators who follow procedures," said Dapas. "If that doesn't work, you have protective systems that come into play." The Zion mistake never reached that point, but "We took umbrage at the failure of the operator to understand his actions."

The incident brought to light a slew of problems at the plant, from poor training to management's inability to control the control room.

The NRC fined Commonwealth Edison $330,000 for the event, calling it "a breakdown in management oversight and control of operational activities." It caused no injuries, but the fine gives some idea of its outsized severity. And it was seen as one more example of Edison and the Zion staff being unable to get their nuclear act together.

The mistake was the first in a chain of events in 1997 that called into question Edison's vaunted reputation as a premier builder and operator of nuclear electricity generating stations.

It was just the beginning of what would be a tumultuous year for Edison and the Zion station, which would never operate again as a nuclear power generating station despite several ill-fated efforts.

Reactor Unit Two had stopped for refueling on September 19, 1996. Unit One stopped operating on February 21, 1997, the day of its notable mistake. The scientific magic of atomic chain reactions that sparked the Zion station for 24 years was gone, leaving the station a towering hulk of expensive, but economically worthless machinery.

In April, two months after the fatal mistake, James J. O'Connor, chairman and chief executive officer of Edison and its parent Unicom Corp., announced a decision to forego buying $415 million worth of new steam generators for the Zion station as planned.

"It would be a wrong decision [to buy the generators] . . . based on our projections of what the competitive costs of power will be at the turn of the century," said O'Connor. The plant would be closed no later than 2005, Edison officials said, rather than wait until 2013 when the station's operating license expired.

The Illinois state legislature was getting ready to deregulate the utility industry, allowing other utilities to compete with Edison, which long enjoyed a monopoly over its northern Illinois territory, which included the Chicago metropolitan area.

The *Chicago Tribune* heralded the story with a headline saying, "Edison to pull plug on Zion plant." The story said: "In the end it was economics, not safety, that got Commonwealth Edison Co.'s Zion nuclear plant."

Well, not exactly, considering the Zion station's troublesome history, which soon would become even more troublesome, if that can be imagined. But surely, the world was turning for those who expected to keep operating as monopolies and believed nuclear energy would be the world's cheapest source of electric power. Economic forces were proving just as unfriendly to Edison as they were to Zion.

Deregulation was likely to drive down the price of electricity, leaving Edison at a disadvantage because of the high cost of operating its six nuclear stations with 12 reactors. And the Zion station, once called the flagship of Edison's nuclear fleet, was the first of the utility's increasingly controversial nuclear plants to be scrapped.

Another glaring fault against Zion: from 1991 to 1995, the plant operated at only about 60 percent of capacity, well below the national average. It supplied about 10 percent of the utility's generating capacity.

Still, 2005 was eight years away, and the NRC would expect Edison to account for the troubling February 21 incident and show that the utility was taking corrective action. And it did.

By April, Edison tested the skills of all 178 members of the control room operating staff. Six reactor operators found to be incompetent or unwilling to change their work habits were transferred from the Zion station. None of those six were involved in the control rod incident. Edison ordered retraining for 15 Zion control room personnel. Two workers removed from the plant retired.

The assessment and transfers were "aggressive" and "dramatic," Thomas Maiman, Edison's chief nuclear officer, told the full NRC commission on April 25. And "it does violate some of the labor agreements that we have."

This touched off another wave of turmoil. In May, workers who had been reassigned or fired sent a letter to the NRC saying they were being punished for raising safety concerns. This touched a raw federal nerve and brought the Zion conflict to a heightened level of federal concern, described as a "chilling effect" against voicing complaints.

The month of July saw rapid developments that added more upheaval to the Zion station story, and more events never seen at a nuclear power plant. It was the month from hell for ComEd.

As expected, Edison officials had to explain that February 21 event and what it intended to do about it. On July 3, they appeared before a battery of 10 regional NRC officials in their Lisle, Illinois, headquarters.

The mistake "violated a basic precept of reactor operations," acknowledged John Mueller, the Zion station's top manager, but it did not endanger public health and safety. That is not to say Edison believes it was not a significant safety event, he added.

Zion executives never got a satisfactory explanation for why the reactor operator responsible for that big mistake acted as he did, said Mueller, but new processes, retrained reactor operators, new attitudes, and tough new management would turn the plant around.

Unimpressed by these familiar reassurances, A. Bill Beach, the NRC's top regional administrator, answered: "You haven't fixed the plant until operations are fixed." After a four-hour session, Beach grew impatient. "We can't continue to meet on these operational problems," he told Edison officials, and they "have to start operating Zion like a nuclear power plant."

Though the Zion station was cold and idle, NRC and Edison clearly expected it to produce electricity again before its final mournful swan song.

Expecting the plant to be ready to operate again by mid-August, Edison officials staged a four-week demonstration, a dry run by retrained workers to prove the plant was ready to produce power after a five-month shutdown. After 10 days, officials stopped the demonstration, saying it was "unacceptable" and "not meeting management expectations."

"Because we had increased the standards of what was satisfactory, we found people were asking a lot more questions about particular procedures and that would bog [things] down," Mueller explained. Maybe they'd try again in October.

The next time NRC and Edison officials met in mid-July, NRC inspectors reported examples of low morale and poor performance among Zion workers. "This job sucks," a Zion worker told an NRC inspector.

When Beach asked Edison officials for some concrete examples of problems at the Zion station, they said a worker brought a gym bag into the control room.

"I expressed a little concern that I had to listen to 15 minutes of programs and processes with little results, and the gym bag is the result," Beach said of the meeting. "I would have expected there would be a lot more introspection into what kind of problems were being seen." Beach said the utility and the agency had different opinions about the plant's readiness to operate.

Meanwhile, Zion workers were a lot more specific about their concerns and keeping up a drumbeat of dissatisfaction.

"In the first six months of the year, we received 15 allegations from workers who were expressing concerns about possible discrimination for raising safety issues," said NRC spokesman Jan Strasma. "That compares to none from Zion for the previous two years." A member of the electrical workers union alleged that people who brought up issues were reassigned.

Based on these findings, NRC sent a letter to Edison officials in late July saying the leap in these allegations filed with the NRC could "demonstrate that a 'chilling effect' has developed or could be developing" at Zion. Workers told NRC that supervisors yelled and harassed employees who tried to give their opinions, and that new managers did not clearly communicate standards and expectations to employees, leading to confusion. They said workers feared reprisals for raising safety concerns.

Federal law prohibits discrimination against workers for raising safety issues. The six Zion reactor operators who were banished from the station filed complaints with the U.S. Department of Labor, charging discrimination.

While NRC's worries about silencing safety concerns applied to Edison corporate-wide, said Strasma, "Zion is a problem in and of itself."

Thomas Maiman, Edison's chief nuclear officer, acknowledged "a long-standing contentious relationship has existed in the [Zion] operating department for years" between union members and management. The Chicago region has a history of strong, combative, and uncompromising labor unions.

"People are reacting to the fact that we are raising the standards, not only to improve the safety culture but also to improve its production capacity," said Maiman. "Any time you raise the standards, you are making a culture change." He insisted that the company offered several ways for workers to report safety concerns, including anonymously.

About mid-July, 65 current and former members of Zion's operating department—nearly 80 percent of its union members in that department—signed and sent an unprecedented letter to NRC regional administer Beach. Stepping up their attacks on the hot button safety issue, they accused Zion management of intimidating and harassing control room operators and others to silence their concerns about safety. The letter asked for a personal meeting with Beach so operating department members could "speak openly" without Edison management present.

With tempers at a boiling point, the Zion station delivered another thunderbolt described as a first in the U.S. nuclear power industry's history. On July 23–24, 10 control room operators from two consecutive 12-hour shifts staged a protest—some called it a rebellion—by taking off their Edison uniform shirts or covering them with union T-shirts, ignoring repeated orders from supervisors to put their company shirts back on.

The demonstrators told supervisors they were protesting "unfair treatment" and "inappropriate" jobs assigned to the six former reactor operators transferred from the Zion station.

Edison's Maiman countered that a minority of Zion staffers were resisting the utility's efforts to improve safety. Later, Edison said it reached an "agreement in principle on staffing issues" with the union representing the workers, the International Brotherhood of Electrical Workers. The union filed 27 grievances on behalf of all the transferred employees.

Edison was in "very heated" negotiations, said Maiman, over the fate of the six transferred Zion reactor technicians, the 27 grievances and the 10 control room protestors. Eventually, the control room demonstrators were reprimanded but not punished.

In a letter to all 970 plant employees, Zion station superintendent Mueller denounced the control room protests, saying "this type of behavior is not consistent with the expectation of returning Zion to operation."

Confronted with something he'd never seen before, NRC's Beach said he needed to consult with his Washington, DC, superiors about the complaint letter and the control room revolt, calling them "unprecedented." Beach said he was disturbed that the control room operators ignored direct orders from their supervisors to wear their company shirts, but wondered how the station could function without their cooperation.

"They're your first and last line of defense in case of trouble," said Beach, who was planning a closed-door meeting with the disgruntled operating department workers. Several weeks later, he met privately with about 50 union members in a Gurnee, Illinois, restaurant, so they could air their grievances.

After that meeting, Beach told Edison officials: "I have concluded as a result of our inspections and a recent meeting that I had with some of your operators that the potential for a chilled work environment exists at Zion. By chilled work environment, I mean that workers may be perceiving that adverse actions could be taken against them if they raise safety concerns or, at the very least, that their concerns are paid little or no attention by management."

Before Zion is allowed to operate again, said NRC officials, Edison must show it addressed the "chill" problem. Edison officials acknowledged a perception among some workers that such a problem existed. After investigating the charges, NRC in June 2000, fined Edison $110,000 for discriminating against one employee who raised safety issues in 1997.

The "chilled environment" investigation was one more hurdle facing the Zion station in 1997.

At the time, Zion was one of three Edison plants on the NRC watch list of facilities requiring special attention for improvements. The others were the LaSalle and Dresden plants. These three stations had six operating nuclear reactors, two each. Dresden's historic first commercial reactor was in mothballs.

In an assessment of this situation, NRC's Strasma said: "Certainly, the corporate management has been an issue and having three plants out of six on the watch list indicates this is not just an isolated plant-by-plant problem, but a corporate problem. It also is a Zion problem. The problems at Zion and LaSalle bear some similarities, particularly in the area of operator training and knowledge. Yet they are not identical. Dresden is in a different category. While it is on the watch list, we acknowledge they have made a great deal of improvement while on the watch list. Yes, these are corporate problems, but they are also Zion problems.

"It needs to be said that when we sit and talk about problems at a nuclear plant, at no time while Zion was operating prior to the shutdown of the unit do we consider it unsafe. Their performance was not what we would like to see, but they were still in the area of safe operation."

The shake-up rattling Edison workers also was taking a toll on Edison's nuclear managers. After only a year and a half as Edison's chief nuclear officer, Maiman stepped down at the end of August, saying "Zion has been a fooler," keeping its problems hidden and befuddling people trying to get the plant running again. That appears to be an admission that managers never understood problems they were supposed to fix at Zion.

Also falling victim was the Zion station's top boss, John Mueller, a so-called turnaround expert brought in to save the plant. Lasting a little over a year, Mueller was replaced in September by John C. "Jack" Brons, another turnaround expert imported from the New York Power Authority.

During that time, target dates for restarting the Zion station slipped from December to January 1998.

But here it comes again, another historic first. Although a Zion official once called the NRC "the enemy," the agency's demeanor toward Edison could be described as gentle compared with what came next—an evaluation of Zion and Edison by nuclear utility peers that was scathing, even savage.

It's not often that Edison takes a beating from fellow reactor operators who long admired the Chicago utility for its leadership in nuclear energy. They are inclined to speak quietly among themselves. But that admiration grew thin as the nuclear power industry watched in horror at Zion's control room rebellion, a long-standing union-management feud, and a revolving door style of rapidly replacing its nuclear managers.

The Institute of Nuclear Power Operations (INPO) is an industry-financed watchdog over nuclear excellence, safety, and reliability. Headquartered in Atlanta, Georgia, it was created in 1979 as part of the U.S. nuclear power industry's response to the Three Mile Island accident, the worst in U.S. history. INPO's membership consists of 44 U.S. nuclear utilities.

Watchful for performance weaknesses in its ranks, INPO will request a meeting with a utility's executives if its power plants scored poorly according to INPO's own grading system, and the Zion station scored poorly. INPO was not impressed with its corporate parent, Commonwealth Edison, either.

A panel of INPO spokesmen gave a blistering briefing on September 10, 1997, to the Commonwealth Edison board of directors in Chicago, comparing ComEd with other nuclear utilities and expressing dismay at the way Zion station workers were acting.

Compared with nine other nuclear utilities with four or more nuclear reactors, Commonwealth Edison is "the only large nuclear utility that has never had a culture that is conducive to a well-run nuclear program, and its nuclear program has never run well."

That February 21 control rod mistake was "one of the most troublesome events in the industry for the past decade."

INPO called bullshit on the Zion control room operators' contention that they were protesting suppression of safety concerns in 1997. INPO pointed out the Zion operators used the same tactic or ploy in 1995, wearing their union T-shirts over their uniform shirts and opposing professional dress codes ordered by Edison management when a union contract expired.

The 1997 control room action was "just the latest in an ongoing problem of trying to get the Zion operators to wear appropriate dress," said INPO. Complaints about safety appeared to be an effective but devious smokescreen leading NRC to freak out over a "chilled atmosphere." NRC fell for it, but likely had no choice given the live wire nature of nuclear safety.

"Zion is a special case," said one of the INPO speakers, because it has "a history of non-professional behavior by the operators . . . that goes back many years. Their behavior is totally unacceptable in a nuclear plant . . . and no other station has a pattern of non-professional operator behavior that even remotely approaches that of Zion."

These remarks reminded me of my own experience in the Zion station control room while waiting for a reactor to go critical: the madness of popping paper cups, and Archie Lucky shouting "Oh, my god!" before running from the room. Later, I was told it was all a joke and the antics were just efforts to entertain the reporter in the room. It did seem out of place at a moment of extreme seriousness. Without fully realizing it at the time, the high jinks I was seeing displayed an unprofessional attitude that marred the station's suitability for the nuclear business.

INPO's remarks appear in 33 pages of heavily redacted briefing material of that September 10 meeting. Names and "personally sensitive sentences," including a brief conclusion, were blocked out, apparently too damning for public eyes. ComEd provided the NRC with a copy on November 21, and NRC made it publicly available on November 25.

Thirteen INPO managers participated in compiling the report, with the help of three senior nuclear executives from utilities with large nuclear programs. The September 10 briefing was INPO's sixth meeting with Edison.

Taking turns, INPO spokesmen started with findings at the Zion station that clearly mark it as a nuclear rogue.

"The first and most significant finding involves serious weaknesses in safety culture that persist at the station, most importantly in control room operations," said INPO. Several events where reactors were not properly controlled were caused by poor judgment, placing too much emphasis on producing electricity and a lack of teamwork.

Zion shift managers were focusing on administrative duties instead of on their primary role, directing crew members and monitoring plant conditions.

The INPO nuclear experts were especially upset by the February 21 control rod fiasco because INPO spotlighted the importance of closely supervising control rod operations at a November 1995 conference for utility chief executive officers. As required, Edison responded to the conference findings by saying it was taking steps to ensure control room staffs took notice of INPO recommendations. INPO said the February 21 event likely never would have happened if Edison had actually taken the actions it described to INPO after the conference.

"Operator performance and professionalism have been problems at Zion for years," said INPO, and "tension between the operators and Commonwealth Edison managers and supervisors have been reported to us or observed by us."

Aside from the 1995 and 1997 control room protests disguised as safety concerns, INPO pointed to two more examples of unprofessional conduct: a reactor operator in April 1997 "stormed out of the control room because he objected to being assigned as the secondary nuclear station operator, instead of as the primary nuclear station operator that day."

And in December 1995, a reactor operator who was a former longtime local union leader "used abusive and profane language in arguing with the unit supervisor about his order involving changing reactor power." He was removed from control room duties. Other reactor operators "protested this by writing a large number of problem identification forms," some trivial, that overloaded the problem reporting system.

This "feud between the operators and Commonwealth Edison management has been going on at Zion for a long time," said INPO. It erupted periodically in strange ways.

Maybe you had to be inside the Zion station, as I was, to fully understand and appreciate the next problem laid out by INPO. In the 1980s, on a few occasions, "operators locked new shift technical advisors into tall personnel lockers on back shifts," showing a "general lack of respect" toward station engineers and supervisors with engineering backgrounds who "did not come up through the operator ranks."

Here is a generational clash, seen as hazing, that even INPO might not have fully understood. Remember the pride the Zion station staff felt in a tradition of "growing their own" from the bottom up, starting at the lowest rung as high-school-educated janitorial stationmen or equipment attendants working their way up the ladder to become reactor operators.

It looks like resentment by the old guard who came up "the hard way" as the college-educated new guard moved in, and they resorted to sophomoric hazing.

INPO found management weaknesses "at all levels of the Zion organization," along with workforce confusion, conflict and stress made worse by a failure among workers to understand their individual roles in maintaining long-term nuclear plant safety and reliability.

Though INPO did not mention it, the looming early retirement of the Zion station likely played some part in the reluctance of some workers to take ownership for the future of a station that had little future left.

Their managers were not much help because they "spend much of their time in meetings," so much so that workers said they did not see their managers in the plant and, in some cases, "were even unable to identify their manager."

Turning to a Commonwealth Edison corporate evaluation, INPO said "overall performance of most of the [Edison] nuclear stations does not meet industry standards of performance as measured by the INPO plant evaluation assessment . . . and other measures."

In the six years prior to its 1997 report, the nuclear watchdog group found Commonwealth Edison's average assessment declined while the rest of the nuclear industry's average improved.

INPO found that Edison employees did not trust senior executives to have a long-term commitment to excellence in nuclear operations. Instead, they believed "the long-term corporate criterion for the success of a station was cost performance" and only limited commitment to excellence in station operations.

As a result, INPO found a need for Edison executives "to reshape the company culture or 'system' by reinforcing and reinvigorating the chain of command to promote the highest levels of safety and reliability in the operation of each nuclear station." One of the problems: "Dissenting opinions, diverse thinking, and constructive challenging of one another are not highly valued."

Problem-solving often involves creating new programs or plans, but the experts found Edison fell short because it did not follow up on those new proposals because of changing priorities or inadequate resources.

This is where INPO was uncommonly harsh, and insightful, on why ComEd was failing.

"Over the last several years, the company's method of responding to weak nuclear performance or pressure from outside organizations has been to change managers," said INPO. It found a 77 percent annual failure rate among 104 managers serving in the top 30 nuclear positions in four and a half years.

"The careers of many of these individuals have been tarnished as a result," said INPO, meaning that working for Edison could be hazardous to a person's career. It also meant lower ranking managers were reluctant to climb higher into top supervisory positions since the failure rate at the top was so high. The rapid turnover of managers caused unclear direction and a sense by many managers of uncertain support from the company.

Delving into Edison's corporate mindset, INPO found that managers and senior corporate executives simply refused to face reality by "highlighting any possible positive aspects of an issue, and diminishing any negative aspect of the same issue."

This is how corporate heads scold rank underlings. ComEd had become an embarrassment to the entire commercial nuclear power industry, and its leaders had run out of patience with ComEd's bungling.

Then INPO trod into Edison corporate territory seldom openly described. To a degree, this was corporate insider baseball and how Edison managed its nuclear program from its lofty executive suites.

Edison had a Nuclear Operations Committee packed with strong, capable, and distinguished leaders, including three retired admirals. Over the years, the powerful committee spotted some of Edison's disturbing nuclear problems.

"The company has taken some pride in the stature of its Nuclear Committee," INPO observed, and "it has given the company a bit of a false sense of security." The distinguished committee "may have even given INPO a bit of a false sense of comfort about the company's nuclear program because of the legendary status of its members."

But that was an illusion, and the powerful committee became a disruptive force. None of the admirals had experience in the nuclear utility industry, and the committee had "quickly become intrusive into line management and directive in nature." It is clear, said INPO, the Nuclear Committee "is being given a managerial role and that it is undermining the line organization" as Edison reorganized its nuclear management program. Under a new organization recently announced, the committee was expected to play a dominant role "in an effort to shore up a weak line organization."

In a traditional line organization, authority flows from top to bottom.

"It is a fundamentally flawed organizational arrangement," said INPO, "because the committee's long-term damage to the credibility of the line organization will outweigh any short-term positive effects." By line organization, INPO meant Edison's existing managerial organization and those supervising nuclear plants day-to-day. The experts saw a danger that the committee would second-guess those managers or overrule them.

A month later, in its traditional revolving door management style, ComEd hired Oliver D. Kingsley Jr., who had been senior vice president of the Tennessee Valley Authority, to be chief nuclear officer and president of a newly formed Nuclear Generation Group.

Outgoing Edison CEO James O'Connor, making a final appearance before the NRC in Rockville, Maryland, said that even the company's most troubled plants were showing signs of improvement.

"Compared to our track record of a year or two ago, I believe we are beginning to see evidence that we are making progress in addressing the fundamental cyclic performance concerns," O'Connor told the commission. Kingsley was on his second day with the company when he, too, appeared before the commission.

An NRC official said he was impatient to see Edison problems solved because the agency had an additional 13 or 14 inspectors busy keeping an eye on Edison and that was straining the NRC budget.

By the end of 1997, Edison had a record four sites on the NRC watch list of troubled nuclear plants, and both the Zion and LaSalle plants were not operating because of problems. Restarting the Zion station, said a *Chicago Tribune* headline, was "nowhere in sight."

Edison was hoping for better luck in 1998.

Chapter 39: A Naval Commander Weighs In at Zion

A NUCLEAR POWER PLANT SHOULD OPERATE LIKE A BALLET, MUSES ZION'S NEW top boss, instead of like a hockey game.

The clashing and bashing going on at the Zion nuclear station resembles a hockey game.

"We need to turn some things from like a hockey game into a ballet," says John C. "Jack" Brons, a former nuclear submarine commander called in to get the rogue station sailing again. "Both of them are kind of like team sports. One is based on conflict and crashing and the other one is based on orchestration."

Deemed a nuclear delinquent by some of the nation's top nuclear experts, the Zion power station is slated for early retirement. But Brons, the station's newly named site vice president, is charged with bringing it back to respectability.

Regarded as a turnaround expert skilled at invigorating lagging companies, Brons tells me of his plans for the Zion station in November 1997.

What a difference 14 years makes. More than a decade earlier, the plant superintendent (Edison keeps changing titles), Ken Graesser, described the Zion staff as a tight-knit group that socialized in their off-hours and shared an interest in boats and fishing.

That's not what Brons inherited. He is trying to restore peace in a station that is more like a battleground with feuding factions.

"My sense of being here is that people were in violent agreement with each other, but never stopped long enough to realize it," explains Brons, a white-haired, blunt-spoken man with the soft Rs of a New York accent.

"Instead of recognizing they were in agreement, that the concerns of the bargaining unit [union] people were virtually identical to the concerns of management, they fought each other," says Brons. Once the combatants were pulled apart and their issues examined, "all of a sudden it began dawning on people that we were part of the same plant and had the same ideas and the same concerns."

To overcome the hostility, the former military man embarked on a peace offensive to build trust in the Zion station through politeness and good manners.

"Trust cannot be demanded," he points out, "it's got to be earned. And so we're on a campaign now because my approach to building trust is first to begin appreciating each other. You go to the dry cleaner and say 'please' and 'thank you.' So, we're frankly beginning that step toward trust by learning to say 'please' and 'thank you' out here. They're things you do in any family, and I'm convinced we can move from appreciation to respect, and from respect to trust over time. So these are the kinds of things we are doing with our people."

Though his approach might appear Pollyannaish, Brons is no stranger to corporate strife and political in-fighting. In June 1993, he was ousted as president and chief operating officer of the New York Power Authority (NYPA), the largest state public power utility in the nation, in what might be described as a political coup.

He seems an unlikely savior of the Zion station, since two NYPA nuclear plants, Indian Point and FitzPatrick, went on the NRC's list of troubled plants, one of them for "weak managerial processes." It could be argued that NYPA's nuclear program was in disarray and Brons was partly responsible.

"The facts are true, and I agree it could be argued that way," Brons says. "I think we made substantial forward progress and achieved a level of recognition in the country that was pretty solid." As for the circumstances under which he left, Brons says, "It was very clear it was a political action that caused me to be asked to step aside. Nevertheless, I'm proud of what I did. . . . I'm not a politician; I am not always clear on what it is [to be a politician]." He denies the NYPA nuclear program was in disarray.

Brons joined Edison in May 1994, after Edison officials carefully checked his background and the controversial way he left NYPA after serving 13 years there, rising to the top. Brons joined NYPA after a 21-year career in the U.S. Navy, where he served as deputy commander of a submarine squadron. He was a graduate of the U.S. Naval Academy, Rensselaer Polytechnic Institute, and Harvard Business

School's advanced management program.

Confident of Brons's leadership abilities, Edison hired him to be directly and exclusively involved in its nuclear program. He landed at the Zion station on September 3, 1997, as the top boss, a super qualified man with multiple degrees at a place that once celebrated the skills of high school graduates.

Almost immediately, Brons had to deal with the plant's renegade behavior. NRC requested a meeting to discuss four cases of employees coming to work with alcohol on their breath, and none of them had been tested for fitness for duty, as required. NRC issued a notice of violation and proposed a $110,000 fine. Brons agreed they "had clear and straightforward rules on this," and that would be communicated to the staff.

In the new plant manager's first plant-wide action, Brons announced a four-week demonstration period intended to show the Zion station would be ready to restart one of its reactors on or about December 17, 1997. "It might sound silly," the new manager quips, "but we are going to schedule an outage," though the plant is not operating and technically in an outage.

It's a bold act for a plant that has been nothing but trouble, like trying to shake off gum stuck to your shoe. But Brons sees possibilities.

"Although a demonstration period is not required for a station approaching restart, we want to take this extra step to further show first to ourselves, and to the [Nuclear Regulatory Commission], how well we work together as a team here," says Brons. It would start with testing the control room staff on reactor simulators, then move into the rest of the plant to focus on cooperation between departments.

Highly optimistic, Brons clearly is putting his heart and soul in this effort, probably a trait that accounted for his success in the navy and at the nation's largest state public power utility. But the Zion station is a heartbreaker. Even all of Brons's enthusiasm and expertise can not overcome the station's habit of leaving a trail of broken dreams. The December reactor startup will never happen, but not because Brons doesn't try.

The man has insight, a way of seeing the human side of problems. It's worth harvesting some of those crisp thoughts. Brons has a reputation as a turnaround expert. What does he turn around?

"People," he answers. "We're dealing with people who, like people in all other walks of life, keep searching for the level of comfort. When is it that I can say that tomorrow will be like today? Like it was maybe 150 years ago? Or in the Middle

Ages or some time past that none of us recognizes? Certainly the pace of change is accelerating, the knowledge of standard-raising. When we did things in this country 50 years ago, one didn't care whether they did them better in Sweden or Japan. But we have to care today about whether we do things better than anyone else in the world because they have instant communications."

What is his take on that infamous February 21 event, when the reactor operator hastily lowered and raised the control rods helter-skelter? Brons is sympathetic.

That event demonstrated, says Brons, "that we couldn't work very well together to get what I honestly believe were common and mutual objectives. . . . We asked the people on shift that day to do an extraordinary amount of things in a very short period of time. It's my belief that we demonstrated that we in maintenance, we in engineering, and we in senior management didn't understand our role in supporting the operators in the control room. And as a result, we pushed them into a position where we set them up to fail."

Essentially, it was a failure of teamwork.

Brons is pushing hard to restore the cooperation that makes a nuclear power plant more than "a bunch of cold iron and something that works to produce power." It's important, he believes, to get the Zion plant running to "provide voltage support for our transmission system," which was why the plant was strategically located in Zion.

"It's also important because, clearly, our business is safely to produce power for our customers and when we're shut down, we are not doing that. We're a drag on our customers. . . . We deliver power, we help keep the lights on."

Though the beleaguered Zion station in its early years was proudly dubbed the "flagship" of Edison's nuclear fleet, Brons resists that description.

"I don't think Commonwealth Edison today seeks to have a flagship," he says. "We seek to have a flagfleet, if there is such a word. I'm part of the ComEd fleet, and we want them all to be good."

Brons points out that the Zion station, which he now commands, played a part in his early training in nuclear power.

"I have history here," he says. "When I came out of the navy, I went to work for a utility in New York State. As part of starting for that job, I was sent here to Zion in 1980 to certify as a reactor operator and a senior reactor operator." It was a nuclear homecoming of sorts.

Brons, like Graesser before him, sees nothing mysterious about nuclear energy.

While reactor physics might qualify as super technology, a nuclear plant contains a lot of machinery that dates to the turn of the last century, he says, like pumps, electrical engines, and steam turbines.

But the widely publicized reactor operator revolt at the power station was making nearby residents uneasy, even fearful, proving to be another "people problem" confronting the new station manager known for his turnaround skills.

If there was any honeymoon period to settle into his new job, it was short. Two months after his arrival, Brons was grilled by members of the Lake County Board, consisting of 21 elected officials, on safety and the future of the lakefront station. The Zion station sits in Lake County, situated in the far northeast corner of Illinois.

Citing recent news reports, Commissioner Mary Beattie of Lake Forest said, "We are frightened; our people are frightened." Brons told the board he understood the apprehension and said it was the result of "a lack of harmony at the Zion station." But that was addressed, he said, by a string of improvements since he took over.

Among them was the completion of the month-long demonstration project which showed that the station's staff and management are competent and trained to do their jobs. The plant would operate again when Edison and the NRC were satisfied with its performance, he said, but no date was set.

After the meeting, Brons says, "I don't think there is a really good reason for them to be afraid, but nevertheless they are because we have not performed well, and people don't know how to translate that to their own particular safety in the community. . . . We have a lot of people that are concerned and skeptical, and I can't blame them." Unlike some in the nuclear energy business, Brons does not mock people for their fears about atomic power.

The Three Mile Island nuclear accident also inspired fear and skepticism, that time on a national scale, forcing major changes in the nuclear energy industry.

"Three Mile Island did a number of things for the industry," observes Brons, "and some of them were very positive. One of them was the man-machine interface."

This can be described as the advent of user-friendly reactor control boards, unlike those in the early 1980s that were not laid out in a logical way, making them confusing. "TMI fixed some of that stuff," he says. The goal was to make control boards more readable and understandable, like the dashboard of a car.

"You can glance at your dashboard of an automobile and know pretty readily without reading each instrument whether you have the right oil pressure and what speed you're going," he says. "We've moved things around" on the reactor control boards.

Training is another aspect of improving human performance. "We teach people that when they put their hand on a switch," says Brons, "to stop, think, act, and review, because we still have some switches where pump A is here and pump B is right next to it."

Computer technology was added, although Brons was apprehensive about that, fearing that reactor operators would become so dependent on computers they would "forget how to read the dashboard."

"I'm happy to tell you that didn't happen," he says. "I don't know if I'd call it a special breed of person, but the people we have operating these plants *like* these kinds of systems and machines, and they are more attuned to the systems on the control panel." Computers are best at displaying trend patterns, he says, "that aid their memory in seeing where things are going," helping "those very intelligent human beings that run the plant to see and understand better what's going on there." This reminded me of the time Tom Flowers insisted on making a gauge more visible so he could read it from a distance.

On whether Brons sees the NRC as helpful, he says he does not see it as "the enemy," although "on some given days I would say it doesn't seem real helpful. But overall, I really do. When the plant's overall performance is not good, we all need somebody from the outside to hold us accountable for the things we are not doing well. And the NRC does that."

The commander treads lightly about the "chill" factor that unsettled the NRC over reports that workers were punished for reporting safety issues.

"A chilled environment does not require that the facts be true," Brons says. But the heated controversy over whether workers were thrown offsite for raising safety issues could cause workers to have doubts about raising safety concerns, "so it is a genuine issue for me, whether or not it's true. And we've been working very hard on that issue."

After a series of intensive interviews and conversations with his staff, "we have almost unanimous agreement onsite that it is okay to raise a safety issue and I need not worry about that."

Most important in the trust and safety debate is whether anyone is listening, and Brons believes he scored a victory soon after his arrival at the station. "I'd been here two weeks and I was in the electrical maintenance shop," recalls Brons. "I just went in there at break time, and the next thing I know there was 20 people standing around me just shootin' the breeze. One of them looks at me in the eye and he says, 'Well, Jack, can you open our door?'"

A door that conveniently led from the outside and into the electrical shop and the health physics department had been welded shut several years earlier for security reasons.

About two weeks later, at a ribbon-cutting ceremony, the door was reopened and the man who raised the issue had the honor of cutting the ribbon.

"I brought it up," says Brons, "because I can't imagine a more golden opportunity handed to anybody who wants to demonstrate we want to hear your concerns, and we really want to open doors for you." It was a management victory for Brons.

The future of nuclear energy worries everyone in the nuclear power business, and these issues are beyond the abilities of one nuclear power plant manager to solve. Brons was confident of nuclear fuel's advantages over other fuels, but that became less certain. He is certain, though, that Commonwealth Edison showed the industry how to build nuclear plants economically for years.

"Edison did a very good job of building its plants relative to the rest of the industry, at relatively low cost per installed kilowatt," he says, "and that's a credit to the company in my view. I wasn't here, but those of us outside the company kind of envied Edison's performance. While other companies build nuclear plants for $5 billion to $8 billion, Edison built them for less.

"I think there is a lot of evidence out there that a fleet of nuclear power plants can be operated economically, and that's a part of why we are thinking of ourselves as a fleet as well.

"I personally believe that nuclear power is an asset," he says, as the world begins to reckon with the environmental impact of all forms of power generation. But he favors one form of power generation above all.

"Hydro is best," he says, referring to hydroelectric power produced with the help of fast-flowing water in rivers and dams. "I came from a company that had lots and lots of hydro. I love it. You know, I went out to a hydro plant on a rainy day and the plant manager said to me, 'Look Jack, we're refueling.' And that's a wonderful concept. But we don't have a lot of that around here because we don't have much of the height differences. So nuclear is a good answer to our environment."

That's part of the problem with understanding nuclear power, it was sold as an energy miracle. Its history is loaded with myths and promises. In its early days, advocates said it would be an endless source of power.

"I agree that somehow that notion was created," says Brons. "The endless energy relates more to the fuel source, that we have enough uranium. And of course, in the early days we talked about recycling [burned-out nuclear fuel]. Frankly, I don't think it is such a terribly bad idea. But it is not something we do here.

"But more to the point, I think there was a concept that these plants would be built, and they would run somehow magically 40 years to the day without anybody lifting a finger, and that was an absolutely foolish, silly notion. It's difficult for me to imagine where it came from. But I agree that it existed. These are machines and people, and they require maintenance. They require replacement, just like you've got to periodically put a new oil filter in your car. You've got to put new filters in this power plant."

Far back in the history of nuclear power is an unforgettable statement by Alvin Weinberg, an American nuclear physicist who called nuclear energy a "Faustian bargain," a deal with the devil, requiring eternal vigilance.

"I don't think it's a Faustian bargain," responds Brons. "In the navy, of course I was in the navy nuclear program. But before I got in that, since I've got a fair amount of gray hair, I really started in the navy before nuclear was such a big buzzword, and I was a navigator on a destroyer. When I studied navigation at the naval academy, there was a big sign over the building where we studied navigation saying that eternal vigilance is the price of good navigation. And so, I agree that eternal vigilance is the price of safe nuclear operation as well, but I'm not sure I would agree that that is a Faustian bargain. Eternal vigilance is probably the price of good journalism. It is probably the price of good retailing. All these things are human endeavors."

Here's Brons's vision for the Zion station: "The vision that we're trying to create here is, not this day, not next month, but maybe four or five years away, where all these publics come to Zion to take something away, because we are viewed as having set leadership standards, and when you want to find out how things are done, they come here."

That same day, I get to see how things are done, 12 years after I last set foot inside the Zion station. It's an invitation to see how things have changed under new standards established since the Three Mile Island accident—and to see how it's done in this new day of Zion station operations.

Escorted, I enter the Zion station through the turbine building, where I immediately notice a major change. The big, torpedo-shaped turbines lying end to end are color coded. Unit One turbine is painted beige, Unit Two turbine is painted blue. Valves are color coded and labeled.

We walk to the Zion station control room, where a new way of communicating is evident. Only employees who need to be in the control room are allowed inside, explains Tim O'Connor, a tall man with a rakish mustache, who is operations manager. A sign over the control room door says, "Stop."

"Permission to enter the control room," says O'Connor at the control room door. Inside, Mark Bode, licensed shift supervisor, answers: "You have permission to enter the control room." O'Connor responds, "I understand I have permission to enter the control room."

This is called three-leg communion, intended to avoid confusion. Making a request, answering the request, and confirming that the answer is understood.

Inside the control room, O'Connor explains: "Standards here have clearly risen. The control room is a highly professional environment, one that does not have distractions." He calls it "church-like."

Unlike that long-ago visit to the control room, I hear no joking or shouting. In fact, such things are banned. A 1988 NRC review of the Zion station mentioned that corporate and plant procedures specifically prohibit "practical jokes and other distractions under penalty of disciplinary action, including discharge."

Another noticeable change was a newly created space, separate from the control room, called the Operations Work Control Center (OWCC). This is where training, tests, project management, paperwork, and other administrative duties are conducted, away from the control room.

"We filtered that out," as much as 80 percent, says Bode, to minimize the number of people in the control room and avoid distractions. The use of the OWCC is described as a big change in the way Zion operates, inspired in part by the February 21 control rod errors.

It's "one of our key corrective actions we told the NRC we thought was the right thing to do as a result of the February 21 event," says O'Connor.

The control room itself is staffed by 14 people, including four senior reactor operators. "Their job is to manage the reactor and safety," says O'Connor. They work in 12-hour shifts with rotating crews including supervisors and equipment operators.

The newly motivated Zion station staff put a lot of effort into intensive standards training sessions.

The past standard, "do good and avoid evil," proved to be too general. "We needed to be a bit more clear, so it could not be misunderstood," explains O'Connor. This evolved into the three-legged communication standard.

"Now there is no misunderstanding between crews and shifts," says the operations manager. "It's one of the things we've been criticized for—cyclical performance." The performance overhaul involved four key steps: creating the OWCC, training, establishing solid corrective methods, and incorporating those steps into the way workers perform.

As operations manager, O'Connor is responsible for making those changes happen, and recognizing when they fall short.

"You have to change what is in their hearts and in their heads," he explains. "My job is to define expectations. You're having to teach your people, reinforce and create new habits. You do it in a systematic way—a procedure." It's like adopting a philosophy. "We don't do things except by procedure. If it doesn't work, we fix it."

Breaking bad work habits can be difficult, but O'Connor says "that comes from intervention. You just have to say this is not what I want and confront and mix it up. You have to have a vision of what good looks like. That's how I influence change."

Prior to coming into the control room, Bode hands me a sheet of instructions, seven pointers on "control room access and conduct." The first pointer, right at the top, reads: "All business conducted in the control room SHALL be conducted in a professional and business-like manner." A red carpet in front of the control room instrument panel is highly visible.

"Do not step on the red carpet," Bode informs me. "That is the operator's private territory. You do not step in the zone without his permission." Along with the instruction sheet I get a big, round green badge about six inches in diameter marked "visitor."

My eyes wander over the reactor control panels. The beige one on the right is for reactor Unit One and the blue one on the left is Unit Two. I think of them as old friends. They remind me of the night I was next to Tom Flowers, staring hard at those gauges and dials, looking for signs that Unit One was coming back to nuclear life after its outage slumber.

Glowing red lights show some machinery is operating. "We're not supplying power to the grid," explains Lane Oberembt, the shift manager, but some cooling water pumps are running.

The control panels look like they've had a face-lift. They are color coded for easy recognition, like Brons described. Colored pathways lead between a switch and a dial of the same color, showing that the switch is connected to that dial. Patterns of pathways show the connections, like a dashboard, that reveal a reactor's pulse beat, readable from a distance.

Oberembt hopes a reactor might be ready to operate by mid to late December.

"We will make electricity because we operate the plant safely," says operations manager O'Connor. "That is the only way we will operate this plant." He sees a measurable change in the culture of the station as they test systems "to give us a good feel the plant will do what it was meant to do when we fire it up."

It's an optimistic view, a young leader's view. It involves "changing the culture" of a rogue power plant. It also shows some of the nuclear updates required in the post-TMI era, from the inside.

I'm in the Zion station only a short time, after a 12-year absence. From my viewpoint, one of the biggest changes since that time is an absence of hostility. I see an eagerness to communicate, to tell their story, a hopeful story of getting beyond the Zion station's reputation as a problem-plagued rogue.

It is like a high-strung runner stepping up to the starting line, eagerly waiting for the starter's shot.

Chapter 40: Shoes on Shiloh

KEYED UP AND READY TO SHOW ITS BETTER SIDE, THE LEGENDARY ZION NUCLEAR station and its 25-year run came to a sudden and crashing end on January 15, 1998.

It was one last opportunity for journalists to use the tired cliché that Commonwealth Edison was "pulling the plug" on the troublesome power plant, despite months of efforts to change its vexing ways and become harmonious.

"Our decision to close Zion station is based on economics," insisted James J. O'Connor in announcing the decision to put the plant on a path toward oblivion, known as decommissioning. A path intended to wipe the twin-reactor station off the face of the earth in record time. The decision, he said, was reached "with a lot of certainty and sadness." Sadness for employees and communities hit by the decision.

By this time, O'Connor was chairman of Unicom Corporation, an energy holding company formed in 1994 with one primary subsidiary, Commonwealth Edison, in a corporate image makeover. O'Connor stepped down in 1998 for underperformance.

The top executive did not mention that only the day before, the Nuclear Regulatory Commission rebuked Edison for the sloppy way it was managing safety at its six nuclear power stations.

"That's a joke—the NRC was breathing down their neck," said Howard Learner, executive director of the Chicago-based Environmental Law and Policy Center.

O'Connor carefully avoided any mention of the anti-management control room protest by 10 reactor operators that seemed to indicate a loss of control over some of the Zion station's most vital workers. Or that blistering report by the Institute of Nuclear Power Operations that singled out Edison as having the worst-run nuclear program in the nation.

But economics is the way Edison wanted to explain the decision, and no doubt that played a role, since the economic advantages of nuclear energy were changing compared with other forms of energy, and at a time of utility deregulation. Nothing that complicated is usually done for a single reason.

"A thorough analysis of the projected costs to produce power at the station and the expected price of electricity in a deregulated market led us to one conclusion: Zion station will not be able to produce competitively priced power in a deregulated marketplace over the remaining useful life of the plant," said O'Connor in a prepared statement. Edison already had announced it would close the station 15 years earlier than allowed by its NRC operating permit, rather than buy new steam generators. When it was built, the Zion station was expected to last at least 40 years.

Closure appeared to be a quick, last-minute decision, with the NRC snapping at Edison's heels. O'Connor said the Unicom and Edison board voted the night before to abandon operations at the Zion station.

Earlier that day, Edison and NRC tangled again over the longtime "cyclical performance" problem, where one nuclear plant improves, but another takes a nose-dive. This time, the major offender was the Quad Cities plant near Moline on the Mississippi River. At the meeting, Edison officials admitted that management at Quad Cities had lost its focus on basic safety issues, joining two other Edison plants that were not operating.

Clearly frustrated, the NRC regional administrator, A. Bill Beach, asked, "What else do we have to do? What other tool do we have? Have we used all our tools?"

Diehards hoped Edison would come to the meeting saying the Zion station was ready to operate, but Oliver Kingsley, Edison's new nuclear group leader, dashed that hope. He told the NRC that the Zion station was a long way from starting. "I don't even think it's close," Kingsley said. But the NRC was still worried about the Zion station and other Edison plants.

"We are in a place where there is zero room for any more surprises," said NRC's Beach. "The surprises are over!"

Wow, was he ever wrong. O'Connor unleashed a blockbuster surprise by announcing the next morning that Edison was closing the Zion station permanently, affecting hundreds of its workers and dropping a heavy economic blow to the northern Illinois suburbs.

"Closing a station is clearly one of the most difficult announcements that a chief executive officer has to make," said O'Connor at an 8:05 a.m. teleconference that memorable day, January 15, 1998, for the company's 17,000 employees, including the 801 at the Zion station. The CEO acknowledged the hard work and effort by Zion station workers in the past year to restart the station, and he thanked them.

"It was literally like someone got hit in the solar plexus," a hard punch to the stomach, recalled Ronald Schuster, who was a Zion station radiation safety specialist watching the morning teleconference.

Jack Brons, the Zion station's top executive, was having lunch in the Star Lite restaurant on Sheridan Road in Zion with eight somber station workers after the closure announcement. "Some of these people have been here years and years," he said. "It's part of their lives. They built it. They felt good about it when it did good, and they felt bad about it when it did bad."

In the *Chicago Tribune*, Peter Kendall put it this way: "Lashed by regulators and stripped of its monopoly, Commonwealth Edison Co. at last gave in to mounting pressure to permanently close its Zion Nuclear Power Station, which once held all the promise of atomic energy."

Surprised but adamant, NRC's Beach said: "I don't see that it changes anything. We continue to go about our business. . . . It doesn't lessen the concerns we talked about yesterday. Mr. Kingsley has one less plant that he has to have on his radar screen, so that is going to give him the ability to focus more on five and that should help." NRC deleted the station from its watch list of troubled plants the next month.

Wall Street's coldhearted investment world rejoiced, saying the Chicago utility should have closed Zion months or years ago to cut its losses and move on. A month earlier, Illinois governor Jim Edgar signed legislation guaranteeing that Edison would recoup much of what it invested in its nuclear plants, a nice incentive to get out of the nuclear business. Edison critics saw the closure as an end to "perverse incentives" given to electric utility monopolies that stuck ratepayers with high payments and few alternatives.

In his startling closure announcement, O'Connor said something else surprising. He expected the federal government to remove all the radioactive spent reactor fuel from the Zion site by 2014, only 16 years away. Then, as now, the Department of Energy had not put to rest a thicket of environmental and political concerns to figure out where to put the nation's first graveyard for spent nuclear fuel rods and other highly radioactive waste. He gave no evidence for that statement.

On another level, the closure rocked a community long accustomed to the jobs, income, and tax revenue that the Zion station gave them, while also providing relief among those who always had their suspicions about nuclear energy and would be glad to see it gone. It was like the American mill town that lost its mill, only this was the atomic age version of that story. For decades, nuclear power was good for communities in northern Illinois. It was a mighty economic engine.

The sting of losing the power plant was still fresh as residents tried to make sense of it all. Atomic power has that odd ability to cause unrest, whether a nuclear reactor is coming to your neighborhood, or if one is going to the scrap heap.

Zion mayor Chuck Paxton said the Zion plant was making him nervous until the NRC stepped in to straighten things out. "We don't like to see them shut down and having the problems they've been having so long," he said. "It seemed kind of like a family thing. They were disagreeing about so many things, some of them minute."

Peggy Taylor was shoveling snow off the walk in front of her home on Elizabeth Avenue, about a mile from the Zion plant. She was hoping the plant would operate again, but pointed out that she got mail periodically from state and federal emergency management agencies urging her to plan for a nuclear emergency, since she lived so close to the plant.

"We have a plan of action for whenever it blows up—my husband, myself and the four kids," said Taylor. But she was sympathetic toward Zion workers because "a lot of people around here depend on it for a job, and the economy needs it." Her reaction reflected caution, but generally a cordial relationship with the power plant shared by many people in the community.

Barry Hokanson, the county's planning and development director, said a nuclear power plant going belly up in the community "is not a common event," and he was looking for guidance. He quit a year later.

Pamela Newton was chairwoman of the Lake County Board's Planning, Building, and Zoning Committee. "This is a significant hit for us all," she said, and she was worried about the future of the power plant site. "Right now, it is a nuclear graveyard on unusable land sitting in our county. If it is removed from the tax rolls in two or three years, that will be devastating to the community."

In his closure statement, CEO O'Connor said Edison would pay $19 million in property taxes in 1998, plus another $30 million for the next two years, for a total of $49 million. He mentioned nothing beyond that.

For Zion station employees, speculation turned wrenchingly real as Edison started cutting the workforce. Friends, family, and co-workers were walking away from those twin silo-like containment buildings for the last time.

Among those released was Bill Stuckerath of Antioch, who worked as a maintenance coordinator in the plant for six years. He was chewing pizza and sipping beer with about two dozen former plant employees in a Winthrop Harbor restaurant.

"People are glad it's over and ready to move on," said Stuckerath, who found employment elsewhere. "I'm going to miss the people, my job. I liked my job. It's a big loss to me."

"Today wasn't so bad, but tomorrow will be," said Lee DuBois of Kenosha, Wisconsin, a plant engineer for 10 years. "Not seeing the people you've built relationships and camaraderie with when they're all gone."

Displaced workers were going to get six months of career counseling, placement, and salary. Gathering in a transition center, they were offered an incentive to find another job within 45 days, when severance payment is highest—50 percent of annual salary and 2 percent for each year worked. After the first 45 days, severance would decrease.

Some workers transferred to other Edison plants; others were laid off in staggered fashion to meet the needs of a station that was going out of business.

And that's when it happened, an eerie ritual and tradition unique to the Zion station. One by one, the just-laid-off workers rolled down their car windows and tossed shoes out on Shiloh Boulevard as they drove away for the last time on the main and only road between the plant and downtown Zion.

On one April morning, dozens of shoes, sneakers, high heels, and even cowboy boots littered Shiloh Boulevard for half a mile, a custom described by some as leaving their soles behind and adding to the legend of "shoes on Shiloh."

"That's not unexpected," commented Don Kirchhoffner, Edison's director of external communications. "It's an ancient tradition," he joked.

Well, not exactly ancient. Plant folklore has it that the tradition started more than 20 years earlier when a supervisor told a departing worker that his "shoes would be on Shiloh." The worker took it literally and tossed his shoes on Shiloh as a parting, maybe defiant, gesture. A bittersweet farewell and the beginning of a tradition.

"It is partly sad and somewhat humorous," admitted Kirchoffner on the day the boss was letting workers off early, rather than finish their shifts. Two groups were released in April, first 70 and then another 130, "the largest number who have ever left Zion station at one time," said Kirchoffner.

The departures were peaceful and without bitterness, and "why not?" he said. "It's bittersweet and goodbyes. It's friends. They are saying goodbye and moving on." The cast-off shoes were collected and donated. "We will take those shoes and make sure they don't go to waste."

"I left mine on Shiloh," said Ginny Kennedy, a clerk for nine years in various departments of the power plant. "Toward the end, a lot of people said it's getting old. It's just a tradition."

A Zion resident, Kennedy didn't believe Edison's official version that the plant was terminated for economic reasons. Instead, she believed it was the conflict and turmoil. "It was during all these problems with personnel, the ongoing fights between management and the union, that we got shut down. They told us it was financial, but I don't think we totally believed it was financial.

"When the control room operators took off their shirts and the February 21 incident, when the control room operator put the control rods in and tried to take them out, that's what shut down Unit One for good. That fiasco was the beginning of the end right there."

Kennedy made important points, and I agreed with her, especially about the so-called rebellion by 10 Zion reactor operators. It looked like Edison management was engaged in a never-ending conflict with union members. Closing the plant would end it. The incident about removing company shirts reminded me that when I watched control room operations, describing Tom Flowers wearing a bright orange sweatshirt and hunting cap in my *Chicago Tribune* Sunday Magazine article with photos, nobody in the control room was wearing company shirts. They were clad in T-shirts and all sorts of informal street clothes. Maybe someone at Edison saw those photos and decided the staff needed a more professional look.

As the shock of the closure announcement wore off, the mood of local government officials was turning angry, noticeable at a special meeting of the Lake County Board on January 30, 1998, called to ask a panel of Edison vice presidents and managers why the closure was necessary. The board consisted of 21 elected officials.

Seated in tiers in the council chamber of the county building located in Waukegan, county commissioners heatedly pressed for answers.

"Don't tell me you couldn't upgrade the plant," said Carol Calabresa.

Another board member, Martha Marks, said: "We're talking about a piece of land that is right in the middle of a state park. You are going to leave us with a white elephant. How long before we can use that land?"

Bonnie Thomson Carter said: "I believe your credibility has been tarnished."

James LaBelle said: "I think your company acted too fast to close Zion. We're going to fight and fight hard. We have to have some relief."

Mary Beattie said: "That's a terrible, terrible thing to do to a community."

Tempers were hot.

"It's been a painful decision for us too," answered Lou Delgeorge, an Edison vice president. "We all have personal friends at Zion, and have worked there as well. The decision to close Zion was an economic one. . . . You have a plant that is losing money by operating." Edison can only wait until the federal government claims the radioactive spent fuel, he added. "It's fair to say Commonwealth Edison does not have control of that process."

It was the first in a chain of sometimes emotional public meetings devoted to official explanations, from Edison or the Nuclear Regulatory Commission.

In May, a tantalizing proposal aimed at using the Zion station, rather than destroying it, surfaced. Amoco Power Resources Corp. offered to study whether it could be used to generate electricity by burning natural gas. But that would mean running gas lines into the plant, with the potential for explosions or fires, near radioactive reactor fuel stored nearby. The plan flamed out.

Like an army given new marching orders, Zion workers turned their attention to the new buzzword—decommissioning, defined as the safe removal of a facility from service and reduction of residual radioactivity to a level that permits release of the property for unrestricted use and termination of the plant's operating license. It must be completed within 60 years.

The first step in Zion's decommissioning began the day Edison announced the plant ceased operating, explained Michael J. Wallace, Edison's senior vice president of the nuclear generating group. The second step, the next day, was when Edison sent a letter to the NRC certifying cessation of operations.

Over the next three to six months, Edison winnowed the plant staff down to a decommissioning workforce of 180. All the uranium fuel from reactor Unit One already was in the spent fuel pool. The utility had 45 days to move fuel from Unit Two to the spent fuel pool. Then Edison would send another letter to the NRC saying the plant was permanently defueled.

"The work then goes on from that point through the next two years to drain the systems that have radioactive water in them, perhaps chemical cleaning to remove radioactive deposits, and get all of the liquid radioactive waste out of our systems processed and off our site," said Wallace. It could go to a radioactive waste dump in Barnwell County, South Carolina.

Edison gained some experience with preliminary decommissioning at the Dresden nuclear plant with reactor Unit One, which operated from 1960 to 1978. Once the reactor fuel was removed, the reactor was placed in SAFESTOR for long-term maintenance and monitoring until final dismantling.

On May 28, an NRC panel of experts gave a briefing for city officials in Zion City Hall. "Although it is a unique and new situation in this community, decommissioning has occurred elsewhere," said Anthony Markley, project manager in the NRC's Office of Nuclear Reactor Regulation. "This is not a new experience," he said, and "once fuel is out of the core, there is no turning back."

Twenty-one reactors were in some phase of decommissioning in the United States at the time and three were finished. The Zion station, he said, had the distinction of being the first two-reactor site to be demolished. Basic rules for decommissioning, said Markley, are "Don't hurt anyone; don't drop things; don't spill things," but Commonwealth Edison was responsible for destroying the plant.

Zion Park District Director Al Hill asked Markley what would happen if Edison did not have enough money to tear down the plant, and the NRC official said Edison could raise money from ratepayers.

Joseph Trexler, an Edison spokesperson, said the utility expected demolition to begin between 2010 and 2014. NRC expected Edison to meet its 2014 target date for dismantlement.

By mid-1998, the Zion station was like an old racehorse that was hitched to a plow, its racing days over. At a cost of about $10 million, the station's two idled electrical generators were converted into devices known in the trade as synchronous condensers to produce "vars" necessary for maintaining voltage on an electrical power system. No nuclear power is involved. Zion was called a voltage-stabilizing facility now.

By June, Commonwealth Edison was struggling to meet demands for electrical power, forced to the brink of "rolling blackouts" by severe weather, unexpected loss of power from the Quad Cities station, a high-priced bidding war with other utilities for power from the national electric grid and bad judgment. The utility misjudged how much reserve power it needed when summertime power demands soared. Fearing its system was overloaded, Edison shut off power to hundreds of factories and other businesses and asked residents to shut off their air conditioners.

Now Commonwealth Edison had a new chief executive officer, John Rowe, and it was his turn to appear before the Nuclear Regulatory Commission. This time, in late June, Rowe was asked to explain why reactors at two power plants were shutting off unexpectedly repeatedly, known as "scrams," at the Zion station before it closed and at Quad Cities.

After a scolding from federal regulators, Rowe said "we cannot write guarantees" against power shortages in northern Illinois of the kind seen in the last few days. ComEd's performance as a power provider was proving no better than its work operating nuclear plants.

Days after NRC officials said they were losing patience with Rowe and his utility, on July 6, 1998, he said he had lost patience with nuclear power. Rowe announced a "new strategic direction" for ComEd involving the sale of all six of the utility's coal-burning power plants and the first open admission that building the nation's largest fleet of nuclear power plants was a mistake.

Rowe made it clear the utility had lost faith in nuclear energy, the brand of power that brought Edison fame and, for a while, fortune.

"With the bright light of hindsight, it was clearly a mistake," said Rowe, becoming the first ComEd CEO to say the nuclear plants, once the pride of the company, had become an embarrassing and costly burden. If he could, he'd sell the nukes, too.

"I keep exploring options," he said. "I do not know how to sell the nuclear plants in an effective way at this time." Edison didn't want to make electricity, either, according to the announcement of a change in direction.

Trying to position itself in a deregulated market economy, Edison wanted to make money by selling power generated by others and by transmitting power over lines Edison controlled. This was how the utility saw itself operating in 10 years, scrambling, like every other utility, to find a place in the future. The company yearned to become an energy retailer, leaving production to others.

This sudden switch in directions is reminiscent of the Institute of Nuclear Power Operation's observation that Edison created new plans or programs when confronted with problems, only this time it was on a much grander scale. It was reshuffling the entire organization in a game of corporate casino.

With the Zion station on its way out, Edison had five operable plants worth about $4 billion. Still the largest nuclear utility in the nation, ComEd had no stomach for being a national competitor in what Rowe called "the new-generation marketplace."

It also would not have the Nuclear Regulatory Commission breathing down its neck if it were rid of those stations, some of which were still substandard performers. Only a few months earlier, three of them were completely shut down. Edison's golden Midas touch for atomic power was gone.

And its skills as an energy provider were not improving. The next year, mid-1999, a series of major power outages struck the Chicago area, including a widespread blackout during a 100-degree summer heat wave. Scores of Chicago's downtown buildings went dark. This time, an infuriated Mayor Richard M. Daley was doing the scolding, a force much more potent in Chicago than a mere federal agency.

Daley, his face flushed with anger, said at a news conference that he was "sick and tired" of Commonwealth Edison, and had called Rowe, telling him he should work 24 hours a day and seven days a week to rebuild the company's aging and failing infrastructure. "You ought to get outside contractors and outside engineers and get it done immediately," he told Rowe.

The chastised utility CEO agreed his company's performance was "totally unacceptable" and "a personal disgrace to me." His response to the mayor, he said, was a contrite "No excuse, sir." It's the kind of response learned in the military. No excuses.

Rowe, hoping his corporate roll of the dice would pan out, revealed his most ardent wish publicly in a strange and unexpected way, during one of those routine NRC public hearings on Zion decommissioning.

The public hearing on April 26, 2000, at the Zion-Benton Township High School was sparsely attended as the public wearied of the routine explanations. The phalanx of NRC and Edison officials outnumbered those in the audience.

Gray-haired and wearing a gray suit, Gene Stanley, Edison's vice president of nuclear operations, gave the routine explanation for why Edison was closing the Zion station: It "would be unable to produce competitive power in a deregulated market."

The estimated cost of demolishing the plant, he said, was about $904 million, and the utility would "have sufficient funds to complete the decommissioning of the plant."

Minutes later, he said something that was not routine, answering a question from Zion resident Therese Vogelsberg. She'd heard that Edison was planning to change the Zion plant's name, and she asked why Commonwealth Edison had applied for a name change at Zion station when it was being shut down.

Stanley responded: "We are planning to merge with another corporation, PECO. With the merger, expected to become complete by August 1, we'll change the name of Commonwealth Edison to Exelon. You will see the name of Commonwealth Edison removed. You will see new branding and the name Exelon go up."

Actually, Exelon became the parent company of Commonwealth Edison, and CEO Rowe got his fondest wish, to be rid of those storied nuclear stations.

In September 1999, Unicom Corp., the parent company of ComEd, and PECO Energy Co. announced an agreement "providing for a merger of equals." It created Exelon Corp., a new utility holding company based in Chicago with a total value of approximately $31.8 billion. PECO is based in Philadelphia.

The new utility holding company became the nation's largest electric utility based on its approximately 5 million customers and total revenues of $12.4 billion. It became the nation's fourth largest power generator.

The operating licenses for 20 Edison and PECO commercial nuclear power plants were transferred to Exelon Generation Company, which became the nation's largest nuclear power plant operator, with 27 reactors at 15 locations, including the two Zion reactors. The NRC approved those transfers in August 2000.

Unicom already was the nation's biggest nuclear utility and PECO was number two. Together, they controlled about 20 percent of the nation's nuclear generating capacity, allowing them to pool the nation's two largest nuclear fleets to make them profitable.

Stripped of its nuclear power stations, ComEd became a power distribution company, ending its legendary history in commercial nuclear power.

At the time of the merger, CEO Rowe said: "My immediate priority is, and must be, bringing Commonwealth Edison's distribution service to levels that meet those achieved by other metropolitan utilities." His customers were furious about waves of power blackouts, and Rowe no doubt remembered the tongue-lashing from Chicago's Mayor Daley. That spurred him to fix Edison's shaky transmission and distribution system, and the merger with PECO gave him cash to do that.

That simple question from Therese Vogelsberg gave Zion residents their first acknowledgment that a major shift in the Zion station's ownership was in the works. It was an unorthodox, even a shabby, way to inform a community that had long been loyal and dependent on the power utility of a major pending change.

After dropping that bombshell at the Zion public hearing, Edison's Stanley went on: "The law requires the Department of Energy to take custody of spent

fuel. . . . We expect the first shipment of spent fuel in 2022 and the last in 2025."
Again, Edison officials were dangling the tantalizing assurance the federal government would move in and carry away those despised and feared spent radioactive
fuel bundles soon. But the Department of Energy still had not settled on where to
put all that radioactive junk.

After Stanley finished his remarks, Ginny Kennedy, a longtime clerk at the
Zion station, came to the microphone to complain that her former Zion supervisors were deaf to her concerns about fire hazards and improper record keeping,
such as leaving records in parts of the plant that were closed.

"When these things happen, it makes me feel uneasy and unsure about things
at Zion station," said Kennedy, to which Stanley responded: "We appreciate you
have identified things to be corrected."

Kennedy went on: "The same management people you had there before are
there now. Why did I have to go over their heads [to file complaints]?" Her remarks
showed that the long simmering friction between station management and workers lingered even now.

"We appreciate you bringing that up," Stanley told Kennedy. "We think the
management team there is adequate to shut down for the long term."

Bruce Jorgensen, the NRC's decommissioning chief for the Midwest region,
described the condition of the Zion station this way: "They disengaged systems
and drained them. Rooms are barred and bolted shut. There's no equipment there.
Major portions of the plant are empty, cold, and dark. In the summer, it's empty,
warm, and dark."

Though that sounds like some kind of ghost station, I was invited to see the
plant from the inside one last time, in August 2000.

Hammered by criticism and strife, the Zion station changed many times in the last
several years. It's truly difficult to predict what I'll see this time. But I'm eager to
find out. It's a homecoming of sorts. I've seen it in good times and bad.

Signs prohibiting alcohol, controlled substances, firearms, and explosives still
greet visitors as they approach the front gate, and warnings against trespassing. But
the security guards who once shooed away uninvited strangers are gone. The guard
house is empty as I drive past.

"Welcome to Zion Station," says the sign hanging from the chain-link fence
topped by barbed wire, still surrounding the lakefront station. A newer sign

identifies Commonwealth Edison as "a Unicom Company." The parking lot, once filled with hundreds of cars and trucks, now has weeds growing from cracks in the pavement. Only 47 vehicles are parked there now.

Inside the front entry, gone are the conveyor belts and metal detectors that once scanned the belongings of workers shuffling their way to work under the watchful eyes of armed guards. They're gone too.

I'm greeted by Ray Landrum, Zion's operations and engineering manager, and David Knox, ComEd communications manager. A security guard appears, dressed in a gray jumpsuit and black boots. He points to Landrum and says, "Don't leave his eyes," and hands me a phone number to call in case I get lost in the maze of empty rooms and buildings.

The Zion plant now employs about 50 workers, plus security guards, a huge drop from the 800 when the station produced power. The station is not considered an operating plant, but NRC says it is "under the decommissioning umbrella."

In the last two years, the plant's staff painstakingly shut off pumps, motors, and valves; drained pipes; and disconnected millions of electronic and electrical connections.

"All the systems that supported the reactor system have been shut down," says Landrum. "If it doesn't support the synchronous condensers, it has been shut down." An ex-nuclear navy man, Landrum has been working in the Zion station since 1973, the year it started operating.

"Fortunate for me," says Landrum, "I was able to stay. I bring a lot of history, a lot of memories." He says hearing some machines running "provides some satisfaction that we're producing something," even if it's just megavars.

Dave Bump, the Zion station's decommissioning plant manager, explains the plant is not yet in decommissioning phase. "We're going to be in SAFESTOR between now and sometime in 10 years from now, to fully decommission and get rid of nuclear fuel and greenfield the plant. We'll have to get DOE to take the fuel."

"We are still important to the distribution system of Commonwealth Edison," says Bump. "The synchronous condensers are a great enhancement to the reliability of the system." They are expected to operate until about 2003.

The armed security guard walks us to the doorway of the auxiliary building, where each of us places a hand on a biometric plate that takes a picture to identify the person. This is new; the old identification tags are not enough. We go to the radiation safety office to register for entry into the fuel pool building, which is radioactive. We get helmets, earplugs, industrial grade safety glasses, and something else that is new: electronic dosimeters replaced the old pencil-like dosimeters we would hold up to the light like a telescope to see a scale inside.

The new dosimeters are called digis, explains Landrum. He's wearing one. It looks like a black plastic pager with a small window showing digital readings in millirems. Digis are exchanged monthly, the new ones color coded in different colors.

On our way to the fuel pool, I ask Landrum if we're likely to see Cherenkov's radiation? "No," he says, "that has to be recently burned fuel. When you were here last time, when you turned off the lights, you saw the blue radiation. You won't see that now. The amount of radioactivity is low."

Inside the fuel building are familiar caution and magenta and yellow radiation danger signs. Landrum and I step up to the fuel pool, surrounded by a hip-high yellow metal rail fence set about six feet back from the edge of the spent fuel pool.

"There are the racks," he says, pointing to the box-like underwater compartments holding the radioactive fuel bundles, each about 12 feet long. "You don't want to lean too far over this," he warns against the radioactive "shine" from the fuel assemblies or falling into the pool. "Don't lean over too much," he repeats. The racks are dark with glistening edges.

This is the main feature of the Zion plant now. Edison spent $3.1 million to isolate it from the rest of the plant. It's called a "nuclear island" because it has its own cooling water and ventilation systems. The pool, 40 feet deep by 33 feet wide and 60 feet long, holds 2,226 uranium fuel assembles that are still emitting heat and radioactivity that will be dangerous for thousands of years. The 1,041 tons of spent uranium fuel are all that the plant burned since it began operating in 1973. They are immersed in 580,000 gallons of water, which covers the fuel bundles by 12 feet.

The pool temperature is 94 degrees Fahrenheit, fairly cool since the blazing hot fuel was placed in the pool in 1998 and began cooling off. The surface of the pool is still, with no ripples. Two pumps feeding 15 gallons of water a minute into the pool whine in the background.

"This is our top priority, safe storage of fuel," says Landrum as we look into the pool. "We still have an operation because that gives us technical specifications [to meet]. It's a lot less than an operating reactor, but this is what we must maintain. This is our business. The NRC is really interested in this. NRC was onsite two weeks ago. This is where the danger to the public is, although it is very safe. This is what we must safeguard."

I mention that some people in Zion hope the station will operate again someday. The reaction is decisive.

"No way," says Knox. "We have no license to operate the facility. We'd have to go through the whole licensing process. We've taken the systems out of commission, and it would take a lot to put them back. They're done."

Landrum agrees but looks at it in a more practical way: "We've drilled holes in bottoms of pipes to drain them and taken out and cut pipes." They've literally cut the plant to pieces. It's in unworkable shambles.

Our next stop, the control room, verifies everything Landrum said. Control room operators often emphasize the importance of being able to understand the condition of a nuclear reactor with a glance at the control board from a distance.

One glance at that control board, and even I could tell that the Zion station is dead. Gauges, dials, lights, and switches are dark or lifeless. I look at the source range monitor that charted a roaring neutron storm the night reactor Unit One went critical, remembering how important it was. It's dead. All of them motionless, killed. Useless. Except for the small part for the synchronous condenser, a temporary patchwork.

"If you see no light, we shut off the equipment," explains Landrum. Enunciators on alarm panels are disabled. "As you drain and de-energize the system, you no longer need that alarm system. . . . None of this stuff is needed, none of it is functional."

Nobody in the control room bothers with the rigid control room rules and three-legged communication I saw that day when Jack Brons was trying to restart the Zion station. Everything is informal now, with only a skeleton crew to attend a control room that once monitored the pulse of the power plant.

"This was a hub of activity," recalls Landrum. "The phones were ringing, and alarms were going off. Now 90 percent of the control board is not working. There was a lot of change. There is a lot of emotion among the people who left and the people who stayed. You deal with it."

I notice Landrum is wearing earplugs. Not me. When did he insert his? It reminds me that during my time at the Zion station, they would hand you earplugs, but often didn't tell you when and where to use them. It was one of the things about the plant I never quite figured out, entering high-noise areas with my earplugs in a pocket.

Upon leaving the spent fuel pool area, there is no step-off pad routine or a need to change clothing. The changing room in the auxiliary building still has bins of anti-contamination clothing. A sign says, "If you are contaminated, do not attempt to decon yourself." Seek the help of a radiation protection technician.

Landrum and I walk into radiation monitors and put our hands into scanning slots and wait for a green light allowing us to proceed. My notepad and pen pass inspection too. But Knox sets off alarms repeatedly, causing radiation protection specialist Paul Gotz to come to his aid. Gotz asks Knox if his suit is polyester, and he says it's a wool blend. Gotz vacuums Knox's trousers and passes a frisker over him repeatedly until Knox stops touching off the radiation alarm.

Landrum explains that uranium is a natural material that gives off radon gas, which can become embedded in clothing. Joking, Knox said he was the only one among us to become "crapped up and set a bad example."

Before leaving, we check our digis for radiation exposure. Mine reads 0.000 millirems. "We sent you into where there was nothing," says Gotz, "so that figures." We walk through another radiation monitor on the way out without any trouble.

Chapter 41: Radioactive Waste City

WHEN THE ZION STATION STOPPED OPERATING WITH THE LOSS OF 800 JOBS, THE City of Zion went into an economic tailspin that changed it into a struggling, blighted community.

The biggest hit was the loss of $19.5 million a year in property taxes paid by Exelon Generation Co., of which the city got $1.2 million, which accounted for half the city's budget. The rest went to other Lake County taxing bodies like schools, parks, police, fire, and social services. It was 54 percent of the township's tax base.

As the plant was torn apart, its assessed valuation dropped from $682 million to $34 million, giving the City of Zion $130,000 a year and turning it into a cash-strapped town.

Another $41.9 million in payroll taxes to Zion and surrounding communities also left town, along with the former power plant workers who quickly sold their homes to absentee landlords. Of the 6,700 housing units in Zion, 66 percent became ill-kept rental units.

Ninety-six acres of the power plant land were converted into a storage site for all two million pounds of nuclear fuel ever used there. City officials called it a radioactive waste dump preventing development of their most valuable asset— Lake Michigan shoreline property. Some called it Zion's "gold coast."

Imagine yourself the mayor of the City of Zion during this cascade of horrors in a city intended to be paradise on earth.

Two mayors, both colorful in their own ways, took the brunt of that civic onslaught and fought to find paths out of the multiple dilemmas brought on by the plant closure. Their mission: save Zion.

One of them is Billy McKinney, a former Chicago Bulls basketball player, a Michael Jordan teammate in the 1985–86 season, with 40 years in the NBA as player and front-office manager. He's Zion's first African American mayor, elected in April 2019.

The other is McKinney's predecessor, Al Hill—stocky, white-haired, and affable. Director of the Zion Park District for 28 years, he was elected mayor in April 2015. He exudes a patient calm, likely born of his years as a parks and recreation manager dealing with rambunctious children and adults at play. He has an easy smile, but frowns when talking about something that upsets him, like nuclear waste.

Zion has an ugly history with nuclear waste, and Hill has gone as far as South Korea to tell that story at a forum for the safety and prosperity of nuclear power plant cities in 2018.

"Our community is staggering," Hill told the forum. "The closure and decommissioning of the plant has had a negative impact on local taxes, local employment and our ability to maintain sustainable economic development. We were crushed by the loss of nearly half of our property tax bases in 1998."

And it all looked so promising at first, Hill said at the South Korea forum and at several press conferences to explain his town's dilemma. "In 1968 nuclear power was a new technology that was to provide low-cost electric power. This was good for Zion, good for Lake County, good for Illinois, and good for the entire country. The City of Zion cooperated with Commonwealth Edison on this exciting new adventure."

The cost would be giving about 300 acres of the city's lakefront property to Edison for a nuclear generating station with the understanding, said Hill, that Lake Michigan shoreline property surrendered for the power plant would be returned in pristine condition for development once the power plant's operating license expired.

"That was the deal," says Hill. "Unwritten, but that was the deal. There was never an understanding that once the plant closed, the Zion community would play host to a radioactive dump that contains 2.2 million pounds, and I'll say that again, 2.2 million pounds of nuclear spent fuel rods. That was not part of the deal." He fears those spent fuel rods might remain in Zion for 100 years, also not part of the bargain. The radioactive waste storage facility is inside Zion city limits.

Federal bureaucrats call those spent fuel rods "stranded nuclear waste." Stranded, as in no place to go except where it is. The Nuclear Waste Policy Act of 1982 gave the U.S. Energy Department responsibility to build a permanent disposal site for radioactive wastes, but the department has failed so far.

Hill points out that all nuclear power plants eventually will be closed and decommissioned, insisting: "The age of nuclear power is over, and the age of decommissioning is upon us."

Zion city officials are hanging their hopes on the Stranded Nuclear Waste Accountability Act, first introduced into Congress in 2016, and repeatedly introduced as Congress fails to act on them each year.

The bill directs the secretary of energy to pay $15 per kilogram of spent nuclear fuel to communities where nuclear power plants ceased generating electricity and store spent fuel onsite. Zion is one of 13 communities that fit that description and would get $15 million a year based on the $15 per kilogram formula.

"Someone needs to either get the fuel rods out of here or compensate this community for becoming a de facto interim fuel storage facility," says Hill. "We should be compensated. We also believe that the federal government should do the compensating."

The reason for that is simple. Zion needs the money. The tax burden to pay for city services did not change; it jumped from the utility to homeowners. The property tax rate skyrocketed 143 percent to make up for the loss of utility taxes. Tax on a $100,000 house in Zion doubled to $6,000. "We can't draw businesses because the taxes are so high," explains the former mayor. "And the taxes are so high because we can't draw businesses." This is another kind of nuclear meltdown.

"We have had a perfect storm," says Hill. The plant closed and people moved out, some of them abandoning their homes or selling them to companies that put them up for rent. "When 2008 hit, we got clobbered again." The housing crisis that year cut the estimated value of Zion property in half. It was a one-two punch.

Relations with Exelon Corporation soured as utility deregulation took hold, changing tax practices.

"Plants were assessed originally on construction cost, minus depreciation, plus new construction," explains Hill. "When deregulation hit, it went to fair market value. A willing buyer and a willing seller. Nobody wanted to buy a nuclear plant." Exelon contended the value of the Zion station property "went to zero. We said wait a minute."

That "wait a minute" moment led to a lawsuit filed against Exelon by the City of Zion and five other taxing bodies when they could not agree on taxing property to store that 2.2 million pounds of spent fuel, known as the Independent Spent Fuel Storage Installation (ISFSI), built at a cost of $40.7 million. This is what Zion officials call a radioactive waste dump.

The ISFSI was a high-priority task of the decommissioning project and also is known as a waste isolation island. The island consists of a 7.75-acre cement pad on the southwest corner of the 96 acres enclosed by a security fence. A concrete security building also stands on the island, which was finished in August 2013. Then the really serious business there started.

Decommissioning crews moved the spent fuel assemblies out of the reinforced concrete spent fuel pool and inserted them, while underwater, into half-inch-thick stainless steel cylinders 71 inches in diameter and 15 feet long. Each cylinder holds up to 37 spent fuel bundles and is welded shut. The stainless steel cylinders then slide into concrete casks, known as MAGNASTOR casks, which stand 19 feet tall and are 11 feet in diameter. Each cask weighs 161 tons, as much as 75 average cars.

By January 2015, all 2,226 of the Zion station's spent fuel bundles, plus other highly radioactive parts of the nuclear reactors, went into 64 casks, which are lined upright in rows on the 7.75-acre cement pad. That nest of casks, called a nuclear bowling alley by some, contains 99.9 percent of the total radioactivity on the Zion station property, according to the NRC. It is an inferno of radioactivity, but that does not mean radioactivity in other parts of the station is harmless.

Guarded 24 hours a day, the casks are designed to withstand earthquakes, floods, tornadoes, fire, extreme temperatures, heavy snow, tipping over and "man-made disasters." They are air-cooled through vents on the sides of the casks and inspected every eight hours.

The ISFI is several hundred yards west of Lake Michigan, the source of drinking water for 9.5 million people in four states, and next to a Commonwealth Edison electrical switchyard. It also is 1.6 miles as the crow flies from the center of the City of Zion, which is among the reasons the nuclear waste island is so contentious. It's walking distance from the business district.

Exelon argued that the ISFSI should be taxed as personal property, not at a higher rate as real property. Hill accuses Exelon of being a bad corporate neighbor.

"They fought us all along the way," he says. "They tried to change the definition of real property versus personal property. They tried to define steam generators and turbines as personal property, like pictures on the wall. We had to take them to the Illinois Supreme Court. We won. It cost a lot of money. All they are about is their bottom line."

Some officials wanted to tax the hated spent fuel stored in Zion, but could not.

Larry Wicketts, Zion township assessor, explained that Illinois state statutes define personal property as anything that can be moved. "We're not allowed to value the fuel or the canisters [for tax purposes] because they are [Exelon's] personal property. If they wanted to move them, they could. Just take a helicopter and move them. That's how they got there. It is considered personal property because they can be moved, sort of like the roller coasters at Great America. They can be unbolted and moved away. It has hurt our community. It hurt all the taxing bodies, the decrease in funds."

Wicketts, a rugged looking man with tattoos and quick with figures, also sees some humor and irony in the notion that Exelon would want to move its highly radioactive fuel. "How many places in the country would say, 'Boy, we want those fuel rods?'"

The end result was a carefully crafted 2015 settlement agreement requiring Exelon to make annual property tax payments totaling $7.65 million to all the suing taxing bodies from 2014 to 2019 for what became known as "the station parcel." City officials wanted to levy "impact fees" for the stranded atomic wastes, but reached a settlement with Exelon instead.

Describing himself as angry, but later softening that to feeling frustrated and irritated, Mayor Hill has advice for communities considering hosting a nuclear power plant. "Make sure there is a clear benefit to having a nuclear power plant in your community. What happens to the taxes? What happens to the spent fuel rods? They are in place to make money. They are not in place to make benefits for the local community. The only ones concerned about the local communities are the local communities. Get everything in writing that can't be changed later on."

Meanwhile, the City of Zion was changing. "It's a tough town now," admits Hill, as its demographics changed. Chicago, to the south, was tearing down crime-ridden low-income housing projects—Cabrini Green and the Robert Taylor Homes—about the time Zion lost its nuclear power plant jobs, causing a northward migration of displaced Chicago residents to Zion and neighboring Waukegan, North Chicago, and Round Lake.

"Poor people are more expensive to provide services for," says Hill. "Part of that is associated with the power plant. So many people moved in after the plant closed. . . . Sixty percent of our property is rental units. A healthy community is 23 to 24 percent rental." The mayor describes the city's population as culturally diverse with

30 percent African American, 30 percent Caucasian, 30 percent Hispanic, and 10 percent Asian.

Although the city has 3.8 percent of Lake county's population, it gets 35 percent of low-income housing assistance vouchers from the county.

"I think people move to a community for three reasons: safety, work, and housing," says Hill. But the Wisconsin license plates of many people working by day in Zion show they do not live in Zion. "There is a perception that Zion is a rough town," says the former mayor. "If you drive around and look at our housing stock, much of it was developed in the 1950s and 1960s. Our rental units are not conducive to professional people. Those are two of our challenges: safety and housing." He wants those people driving home to Wisconsin after work each day to live in Zion and spend their money and taxes there.

The largest taxpayer in town now is the Midwestern regional medical center for Cancer Treatment Centers of America, but its tax payments are a fraction of what the electric generating plant paid. "They are not anywhere close," says Hill, "maybe 20 percent of the nuclear plant." Some find irony in a cancer treatment hospital moving into a town where some residents feared the radioactive power plant might cause cancer.

Always eager for new sources of revenue, Zion still can't shake loose from some of the dictates imposed by its original founder, John Alexander Dowie, who outlawed alcohol. His scolding ghost kept the town alcohol-free for much of the last 100 years, true to its founding principles.

"I don't want any bars in town," Hill declares. "We have enough trouble." But Zion is not quite the dry community it once was. Zion issued its first controversial liquor license in 1999 to the park district's 18-hole Shepherd's Crook golf course. Since Hill was the park district director at the time, "people looked at me thinking I'm Satan incarnate" for allowing alcohol.

Half of the golf course was in the dry city and half of it was in wet Lake County, creating a crazy-quilt pattern of wet and dry golf holes. "We served beer and wine," says Hill. Golfers could get beer in the clubhouse. They could not drink on holes one, two, three, four, and half of five, but could on the other half of five and holes six, seven, eight and the ninth tee, though not on the rest of the ninth hole. Drinks were not allowed on holes 10, 11, and half of 12, but they were allowed on the other half of hole 12 and on 13, 14, 15, 16, and 17, though not on the 18th hole.

Since then, these restrictions have eased and Zion restaurants serve alcohol with food, while some supermarkets like Jewel and Piggly Wiggly sell liquor. But Zion has no liquor stores or bars.

A Midwesterner, Hill was born and raised in the oil refinery and steel mill region of northern Indiana and worked in the steel mills as a young man. It's a region that breeds tough men and women. He worked hard to recruit his successor, Billy McKinney, who had vowed to avoid two things in life: going into politics and living in Zion, his hometown. Events and Hill proved to be persuasive.

The transition from one mayor to the other hardly skipped a beat as McKinney picked up on an agenda set by Hill, leading with the radioactive fuel rods, property taxes, economic development, and absentee landlords—plus some unorthodox ideas of his own.

"The top issue, number one, is getting compensation for those nuclear fuel rods," says McKinney, who is bald and athletically trim, his face covered with a short stubble of beard. "The second item of importance is economic development. That is bringing in more businesses to our community. Last year, 2019, we had 19 new businesses come to town." He's wearing a gray zippered sweater over a black-and-white-striped shirt with black slacks, a dapper but informal look.

McKinney sometimes breaks into the cadence of a motivational speaker, practiced during his time in professional basketball. It is a time displayed brilliantly in his Zion City Hall office, where his framed red and white Chicago Bulls jersey (number 7) and shorts hang behind his desk, with the word "mayor" carved into a wooden desktop placard. Adorning the walls are photos of his playing days and the togs he wore while playing with the Kansas City Kings, Utah Jazz, Denver Nuggets, San Diego Clippers, and Chicago Bulls.

Here is a competitor, although a gentle and soft-spoken one, and noticeably short for a professional basketball player, six feet tall. Those photos on his walls show some of McKinney's teammates towering over him, up to seven feet tall. It never occurred to McKinney that he was short for basketball, or as he put it, "vertically challenged."

"If you can't be tall, you can be fast," he says. His quickness on the court earned a nickname, "The Crazed Hummingbird."

Although being mayor in Zion is a part-time job, in "Crazed Hummingbird" style, he works feverishly full-time, paying back the community where he grew up.

"People will tell you I could live anywhere and go anywhere, but I realized after 40 years in the NBA, I missed being home. I missed being in a place I could be me, where I was accepted in the community, not because I was part of a pro team or a professional franchise, but because I was the person I was my entire life. I have a lot of equity built up here. I've always been proud being from Zion, and proud of being able to represent Zion on a bigger stage. For me, it's been, whether I realize it or not, the path I blaze is a way of telling anyone in this community that you don't have limits on what you achieve in life."

Born William Mervin McKinney III, the new mayor admits his homecoming was not entirely planned. He was working for the Milwaukee Bucks and living in Milwaukee, a 45-minute drive from Zion, when his mother died in 2012, leaving the mint green and taupe split-level family home to McKinney and two sisters. It is the home where his southern-born mother raised a family of six children single-handedly.

"I was originally going to fix the house and sell it," says McKinney. "And I realized during the process of renovating the house that I could not sell the legacy that my mother left for us. A mother of six kids, sixth grade education, instead of putting us in an apartment building when we moved here in 1962, she found a way to put us in a house." His parents divorced when McKinney was seven.

Statistically, the young McKinney "probably should have been robbing stores or been in jail. Instead, my mom was my role model." McKinney was an honor roll student at the Zion-Benton Township High School and attended Northwestern University on an athletic scholarship, becoming the first African American player to start on Northwestern's baseball team and the most valuable player on the basketball team for three consecutive seasons.

Mayor Hill tried repeatedly to get McKinney interested in government and politics. "It took him three times," said McKinney, who resisted because of the polarized nature of politics these days. At dinner, Hill made the winning argument by saying: "Billy, I've got to tell you. Politics is really talking to people and persuading them to look at things from your perspective, and that's something you've been doing your entire natural life."

"When he said it that way," says McKinney, "I came away from dinner saying I'd give it a try." Hill appointed McKinney commissioner of building, property, and zoning for a two-year term, and that put him on the path to becoming mayor.

Something else happened shortly after McKinney moved into the family home that motivated him, adding a layer of passion to the job.

"When I moved here," says McKinney, "I was looking to buying the house next to me. People living there was an educator with a son with a disability. They left when the landlord would not fix up the property. When I went into the house, I would not let my dog live there. That upset me."

Upon becoming mayor, McKinney helped launch a controversial rental house certification program that insured life and safety issues are monitored and repaired when necessary to protect renters.

"The city was sued over this by about 80 landlords," says the mayor. "The city prevailed in that suit. But approximately seven of those landlords live in Zion. So they were all absentee landlords. That was one of the issues we were having. We have absentee landlords not properly maintaining their property."

Zion city officials now conduct regular landlord forums. "The ultimate goal is we want those landlords to be successful in town. We want their properties to increase in value. We want them to be able to put the best tenants in their apartments or in their homes, who will take pride in those homes and reduce some of the small percentage of bad element we have in town."

That helped to solve problems with malfunctioning heating systems in apartment buildings or rental homes. City officials got three or four complaints a week about lack of heat in the winter. Today, says the mayor, it's more like a few calls a year.

Time brought some improvements. Zion had 900 vacant homes in the city in 2010, after the 2008 real estate crash. By the end of 2019, it had 300 vacant homes, down in part because the city applied for grants to tear down 22 abandoned buildings. A 22.2 percent hike in water rates raised $900,000 a year for water and sewer replacement.

Though residents still complain their property taxes are too high, McKinney patiently explains that only 15 percent of their property taxes goes to the city. Like a cheerleader for the city, the mayor points out "what they get for their 15 percent—the 120 employees working on behalf of citizens. Fire, police, building, and accounting departments."

Litter is another issue close to McKinney's heart. "When I first moved back here, I was shocked about the city I was seeing that was so different in terms of cleanliness and tidiness from the city I grew up in."

That led to Clean Up Day, a program asking residents and volunteers to restore civic pride and natural beauty in the city's diverse neighborhoods. Residents are urged to mow their yards and to remove garbage cans from the streets once they are emptied, which is required by ordinance.

"When you go to apartment buildings, they have large containers to use," says McKinney. "Obviously, a lot of residents would not be good basketball players. They miss their containers." While walking his dog, the mayor picks up debris, hoping neighbors notice and follow his example.

"Somebody calls the city and says trash is blowing across my back yard," he says. "Well, pick it up! Why does somebody have to come from the city to pick it up? In the time you took to make that call, you could use that time to improve the neighborhood without city help."

Like Hill, the mayor before him, McKinney does not appear to have a good working relationship with Exelon, Commonwealth Edison, and the company hired to demolish the power station. Persuasion requires a certain amount of communication, but McKinney has found little opportunity to communicate.

How would he describe their attitude toward the city?

"It's hard to say what an attitude is when you don't have ability to communicate with people," he says. "We have to sit down. . . . Some would say dismissive. We have to continue to knock on the door. We're not going to go away."

McKinney attended decommissioning meetings conducted by the demolition company and the NRC. "My perception of attending a couple of those meetings is they were checking the boxes. They weren't answers to questions citizens had, or citizen groups had. There was not enough conversation for me to say if those meetings were productive other than them telling us: here's what we are doing. . . . There are no answers. We're here to deliver the message."

Economic development does not appear as frustrating to the new mayor, who is prodded by residents asking, "Why don't we have a Starbucks or Five Guys?" Looking for an answer, McKinney got an eye-opener while attending a conference in Las Vegas on attracting businesses.

"Companies look for demographics, median income, things of that nature, that fit the profile of their company," McKinney learned. "At this point, while we don't have the demographics or median income, we have to prepare the city for when it arrives. We will be there to have a company like Starbucks to look at us."

Meanwhile, Zion city officials are going in directions that might have infuriated founder John Alexander Dowie.

Illinois legalized recreational marijuana beginning January 1, 2020, and rang up more than $1 billion in legal cannabis sales the first year. The Zion City Council quickly approved cannabis sales, but did not designate dispensary sites.

"They could be huge," says McKinney eagerly. "If we were to get a dispensary, it could help with our budget significantly." The move toward marijuana sales came a few years after the city allowed video gaming.

"Some residents were opposed to it," said McKinney, "but from a city standpoint, gaming had added another $250,000 to $300,000 a year to the city budget. I would have to add this part to it as well. The speculation when we allowed gaming was that there was going to be violence surrounding it, crime, and that hasn't been the case." The city has four or five downtown gaming cafes.

Zion also is sniffing at the edges of an idea that would make John Alexander Dowie howl: more open alcohol sales. Studies show Zion residents spend more than $36 million a year to buy alcohol products.

"They're getting it from surrounding communities, Beach Park, Wadsworth, Waukegan, Winthrop Harbor," observes McKinney. Despite their prohibition heritage, those kinds of dollars make city leaders drool. "When you start looking at those numbers, of how much our residents spend on alcohol, and the city is not getting the benefit of that revenue. It's here, but we are not taking advantage of it."

The mayor is not going so far as to encourage liquor sales. "There has to be a discussion among everyone in the city about it, here at city hall, with our commissioners, before we think about making that step. Just like gaming, cannabis, alcohol—people can be opposed to it from a philosophical standpoint. But we have to look at every decision made from a financial standpoint and look at what is in the best interests of the city to move forward."

But is McKinney for it? "I did not say if I'm in favor or not. It's a point of discussion. I'm not a drinker. I don't drink, don't smoke cigarettes, nor ever taken illegal drugs. Now cannabis is legal. I don't smoke pot, either. I have to separate my own individual feelings from it and look at it as mayor and look at examining these options that could be available to the city from the revenue-generating standpoint."

As an athlete, McKinney grew up at a time when drinking and smoking were not part of the athletic culture. "You are always preserving your body and not putting poison into it," he says. "In addition to the alcohol, my father was an alcoholic. That's the reason I was never interested in drinking. I saw what it did to him as a man and I saw how it impacted our family."

Talking about preserving his body, poisons and the evils of alcohol, McKinney sounds exactly like John Alexander Dowie, subscribing to his views.

"Some of them I do, absolutely," admits McKinney. "What you find with religion is everyone has their own interpretation and a commitment to what they believe is right or wrong from a religious, philosophical standpoint. I grew up in the church. My mom was devoutly religious. I consider myself a spiritual man, but by no way, shape, or form do I consider myself perfect. I'm very flawed in many ways, as we all are."

Any discussion of Zion's future invariably involves the Lake Michigan shoreline occupied by the Zion nuclear power plant and its offshoot, the concrete pad containing 2.2 million pounds of radioactive waste.

For some of Zion's leaders, the site is a glistening mirage, a vision of what could be there in the future, especially if all that nuclear waste was gone. It's tantalizing. It's something to think about, to dream about.

"The lake is our greatest asset," declares McKinney. "To be able to enjoy some time at the lake, with either retail down there, or a miniature version of Navy Pier [a Chicago lakefront amusement venue]. Shops. A golf course down there. There are so many possibilities. A place for entertainment for our residents to enjoy, being so close to the lake. They've got great hiking trails down there. Hosea Park is down there. It's an undiscovered park by some people in town. We've got a great park system here, a central part of my growing up."

The little City of Zion is renowned for having more parkland per capita than any other park district—33 acres per thousand residents. That's a legacy of John Alexander Dowie. The present generation, if motivated in that direction, could add lakefront acreage to its generous park system.

McKinney has his dreams for the lakefront. Larry Wicketts, Zion township assessor, has his.

"Until it is a clean, pristine property, it will not compare with other properties on the lakefront," which command $500,000 to $600,000 for a full acre overlooking Lake Michigan in some places, Wicketts says.

"It would be nice to have some commercial and condos," he adds, dreaming. "We'd have to stay away from industrial. It's a beautiful piece of property. If you were up on the second floor [of a condo], looking at the lake, it's a beautiful property. It would be an economic boost for the community. How about a casino? That's everybody's dream. A high-rise overlooking the water. We need something to get the tax rates down."

Former mayor Hill has his own dream for the lakefront: "A hub for an eco-village that provides high-density residences [apartments and condominiums] and supportive shops and services for the lakefront area. It would be developed with high-performance green buildings and infrastructure." He sees the setting as "a restored lakefront ecology located between two great metropolitan areas accessible by train and auto and offers developers and institutional organizations an opportunity that exists nowhere else in the region."

Imaginations swing from the restorative powers of natural beauty to economic necessity. Economic necessity usually wins.

Adding his dreams for the lakefront is Bob Feffer, acting superintendent of Illinois Beach State Park, which borders both sides of the Zion station property.

"Hopefully, they'll let us run bike and hiking trails along the lake to link the north and south sections of the park." In the past, bicyclists could skirt the landward side of the power plant, "but not after 9/11," because of fears of terrorists. Cyclists going from one section of the state park to the other had to swing wide to 29th Street, then back to the lakefront bike trail.

As the Zion demolition project came closer to being finished, Feffer gained an ally in his hopes for opening the lakefront for more recreation opportunities.

Waiting in the wings for the final stages of this decommissioning saga is Anthony Orawiec, Exelon Corporation's senior decommissioning manager, who admits he is not a disinterested onlooker. "I have a love of the community of Zion," he confesses, because he too, like Mayor McKinney, grew up there. His kin reach clear back to John Alexander Dowie's day.

"My grandmother worked in the [Zion] curtain factory," says Orawiec. "My uncle was a pastor of the Christian Catholic Church," founded by John Alexander Dowie. Orawiec attended school in Zion, from kindergarten through high school. While living in Beach Park, his four children attended the Zion-Benton Township High School.

"I raised them in the community that is Zion, and Zion [power] station is part of that history," says Orawiec. His heart is in the final outcome because he is a member of the Zion community.

"We actually are looking forward to that day in taking the site back. It is a very worthwhile project as we watched EnergySolutions execute their comprehensive plan," he says. As part of the Zion decommissioning project, Exelon Corporation temporarily surrendered the two reactor operating licenses and property ownership to the decommissioning company, a first-of-its-kind maneuver. "We right now are focused on becoming the licensee and storing the fuel safely. When the site comes back, it comes back as a [spent fuel storage installation].

"What we will do with the land is under careful consideration, and we have not reached any decisions," says Orawiec. "We will keep the stakeholders informed. Even though we turned over the [operating] licenses, we have reestablished the importance of the community to us."

He recognizes a bike trail crosses a portion of the Zion station site and connects with Illinois Beach State Park. Under an agreement with the Illinois Department of Natural Resources, biking and hiking trails are allowed over station property. About 100 acres of the station property were improved for wildlife habitat and native plants.

"On the north end of our site, a small park, Hosea Park, is owned by the city," observes Orawiec. "We allow people access to our land to enjoy the park. We are contemplating reuse and engaging with the community."

These efforts appear to dovetail with some of the dreams expressed by city leaders, to open Zion's lakefront for public use. It's a small beginning toward some big hopes and ideas. It's a beginning in a new direction.

Chapter 42: How Green Is My Greenfield?

Dreams of lakefront casinos or condos had to wait until that tower-ing power plant actually vanished, and guesses when demolition would start were vague.

Demolition was slated to begin sometime between 2010 to 2014, and finish by 2028, according to pronouncements by Edison and NRC officials at various public meetings on decommissioning.

The dreams seemed more like shimmering mirages during all the talk about steps toward decommissioning, which really was just a foggy idea lacking the details found in documents known as the License Termination Plan, which is where the project really gets down to serious business. It's the actual blueprint for what is taken away and what is left behind, and something always is left behind.

It all seemed a bit hazy until August 2010, when Exelon Corp. announced a $4.6 billion spending program for 101 projects in Illinois—an economic stimulus program for the state.

The first project in line was demolishing the Zion nuclear power station, begin-ning September 1, 2010, according to the announced timeframe. It was the first firm date for dismantling the plant, surprisingly sudden, and the first of a string of unusual developments involving the Zion station. Edison's, and now Exelon's, nuclear story was always told in superlatives, and this was no exception.

"The $1 billion, 10-year project will be the largest nuclear plant dismantling ever undertaken in the United States, requiring an average of 200 skilled workers each year, most of them local, and a peak workforce of 400," said Exelon's press release. The Zion station also was the first twin-reactor power plant in the United

States marked for destruction.

The Nuclear Regulatory Commission called it "one of the largest demolition and cleanup projects in the history of nuclear power." Edison constructed the Zion station at a cost of about $583 million; demolishing it would cost twice that much.

The method Exelon chose to destroy the Zion plant also was unusual. In a "first-of-a-kind arrangement approved by the Nuclear Regulatory Commission," Exelon transferred both of Zion's reactor operating licenses to ZionSolutions, a subsidiary of EnergySolutions, a Salt Lake City, Utah, nuclear services company with a low-level radioactive waste disposal facility near Salt Lake City. ZionSolutions would dismantle the power station and transport radioactive material and parts 1,500 miles to the Utah waste facility. Exelon also gave ZionSolutions a $900 million trust fund for the project. That money came from Edison customers who paid one-tenth of a cent for every kilowatt-hour of electricity they used (about 15 to 20 cents on an average monthly electric bill) until the end of 2006 to cover the cost of decommissioning the Zion station.

"At the completion of the project, responsibility for the site will transfer back to Exelon, and the 200-acre site will be available for other unrestricted uses," said Exelon. "Throughout the process, Exelon will retain ownership of the plant's used nuclear fuel, which must remain on the property in a secure facility."

Under the arrangement, ZionSolutions will return the two reactor operating permits to Exelon after destroying the Zion power plant and removing radioactive hazards. The company that holds the reactor operating licenses has complete authority over the plant property.

Exelon had the trust funds to demolish the Zion station, and the know-how to do the project itself. Why did it turn to EnergySolutions and ZionSolutions?

"The crux of why we did that and the advantages were pretty clear," answered Anthony Orawiec, Exelon Corporation's senior decommissioning manager. "EnergySolutions' and ZionSolutions' expertise lie in the safe and efficient demolition and decommissioning of nuclear power plants." They also "have a full complement of people who previously dismantled nuclear power plants in the United States."

Exelon, he said, focuses on producing clean, carbon-free power. "Our desire is to focus on our core business."

Exelon prepared the Zion station for decommissioning by certifying it had permanently ended operations and removed all the uranium fuel from the two reactors. Then, between 1998 and 2010, the plant sat in SAFSTOR while the plant cooled down and some of the radioactivity decayed away, opening the way for decommissioning and demolition.

EnergySolutions "provided a path to license termination and freeing the use of the land" sooner than expected under previous demolition plans, said Orawiec, because it was ready to take the job and "we are able to take the site back considerably ahead of schedule."

ZionSolutions finished the demolition project 12 years earlier than Edison had projected under previous plans. Key to the rapid progress was an innovative "rip and ship" method that skips a time-consuming and costly mistake made at other decommissioning projects: separating radioactive materials, which must go to a licensed dump, from nonradioactive materials which can be used as clean fill material or go to an ordinary industrial landfill.

Sorting seemed to make economic sense if nonradioactive material and metals culled from the separation process could be salvaged and sold. But experience at other decommissioning projects showed it was not always that simple. Disagreements between a utility and a decommissioning company over what is radioactive and where it should be dumped can stop the project, while demolition crews stand idle at great expense waiting for a final decision on what to do next.

That did not happen at the Zion station demolition project. ZionSolutions held both reactor operating permits and had complete authority over the demolition project, subject to NRC regulations. Under the "rip and ship" method, anything that might include radioactive materials is considered radioactive waste and shipped in bulk by rail to disposal sites in Utah or Texas.

On September 1, 2010, the NRC issued license amendments showing that the Zion operating licenses were transferred from Exelon to ZionSolution as of that date. Once that transfer was made to ZionSolutions, "this company is solely responsible for the safe decommissioning of the plant," explained Viktoria Mitlyng, an NRC senior public affairs officer. EnergySolutions, the parent company of ZionSolutions, also obtained a $200 million letter of credit for the Zion project, which serves as a performance bond in case of default.

EnergySolutions provided an easement for disposal of low-level radioactive wastes at its Clive, Utah, disposal facility in the desert 80 miles west of Salt Lake City. ZionSolutions estimated that 90 percent of Zion's radioactive waste would go to the Clive facility, which is licensed to accept one class of low-level radioactive waste. More intensely radioactive wastes go to Waste Control Specialists LLC, near Andrews, Texas, which is licensed differently.

Zion is a first of a kind in the way Exelon turned the reactor operating licenses over to a decommissioning contractor, expecting to get the licenses back when the job is finished. "It is a further evolution of the decommissioning industry," said Orawiec.

Another example of the changing nature of the nuclear power industry was Exelon's 2019 outright sale of its Oyster Creek Nuclear Generating Station in New Jersey to Holtec International of Juniper, Florida, an energy industry equipment supplier with a decommissioning division of its own. In January 2021, Holtec expressed interest in building a "new generation" nuclear plant at the Oyster Creek location.

"Unfortunately, many nuclear plants in the country have shut down," said Orawiec, but that has opened the way for a growing business in finding new uses for that idle property, estimated to be about 35 sites across the nation. "There is an interest to be an owner of the site in the industry," said Orawiec.

This changing nuclear industry landscape appeared to take a strange twist when the NRC on May 8, 2013, announced that it approved "indirect transfer" of both of the Zion station's reactor operating licenses, including the license for the spent fuel island, to investment fund entities of affiliates of a venture capital firm, Energy Capital Partners II LLC in Short Hills, New Jersey.

A venture capital firm with nuclear reactor operating licenses? Further investigation showed that Energy Capital Partners had acquired EnergySolutions and ZionSolutions, becoming the parent company of both firms deeply involved in the Zion station decommissioning. It was a multistep process involving some corporate finesse in which EnergySolutions, corporate parent of ZionSolutions, was acquired by Rockwell Holdco, which was formed to acquire EnergySolutions, which is held by entities of Energy Capital Partners.

The NRC approved the indirect transfers after deciding that ZionSolutions staffing to continue decommissioning had not changed and that financial requirements were met.

License transfer requests became routine as deregulation of the electric utility industry led to a growing number of mergers and acquisitions, resulting in requests for the NRC to transfer nuclear power reactor operating licenses. NRC reviewed more than 100 license transfer applications between 1999 and 2017. Most of them came from companies in the energy industry, not venture capitalists.

Elizabeth Archer, a spokeswoman for Exelon Corp., said she was not sure of Energy Capital Partners' motives for becoming the owner of EnergySolutions, "but this is their business model." And, she said, Exelon still expects to get the Zion station's reactor operating permits when decommissioning is finished.

Those were a few more ways that the Zion station proved to be a trailblazer and a history-maker in the nuclear industry. But it would be a mistake to think the demolition project got off to a smooth start. That is not the Zion station's style. Any fanfare was cut short.

Decommissioning already was under way on September 28, 2010, when John Rowe, Exelon's president and CEO, received a "stop work order" from Zion mayor Lane Harrison for failure to pay a 1 percent permit fee on the massive project, which amounted to $10 million.

In stilted, lawyerly, bordering on Elizabethan, language, Harrison's letter to Rowe read:

"It is with great dismay we are compelled to emit the instant letter and STOP WORK ORDER for failure to obtain the necessary permit(s) pertinent to the decommissioning/demolition of the Zion Nuclear Power Station (the 'Project') and construction of suitable nuclear waste storage facilities by Energy Solutions, Inc./Zion Solutions LLC ('ES/ZS'). In short, ES/ZS has arbitrarily determined that our published standard permit fees of one percent (1%) of the estimated project cost are 'exorbitant' and non-applicable. As the owner of record of the property you are hereby noticed of the permit violations."

EnergySolutions and ZionSolutions considered $10 million "unsupported and improper" and subject to negotiations in the presence of lawyers, especially since "this is the city's first experience with decommissioning" and permit fees for such a project. Offended by the implication that the city was ignorant or naïve, city officials pointed out that ES/ZS had never done a demolition project of Zion's magnitude and the City of Zion "deems such an argument in support of evading the standard permit fee treatment is at best disingenuous and hypocritical." The permit fee was not negotiable.

In the Zion station's long tradition of turmoil and belligerence, even its final chapter got off to a rocky start. Exelon and Zion city officials said they could not recall what broke the impasse.

Al Hill, by this time ex-mayor of Zion, said to the best of his recollection, Exelon "wrote a check to the city for $5 million. The city settled with them." The settlement prohibited the city from collecting "impact fees" for the stranded spent fuel rods at the spent fuel island. That freed ZionSolutions to wreck the Zion nuclear station.

An office building and warehouses were demolished in 2015, followed by the turbine buildings in 2016. Two others, including one for handling fuel, came down in 2017.

I was wrong in imagining that an old-fashioned wrecking ball would raze those mighty towering reactor containment buildings. The tool of choice, which I watched from a distance, were industrial-size jackhammers, their blunt noses mounted on long, retractable mechanical arms extending from manned tractors.

They looked like mechanical scorpions, their "stingers" striking at the base of the containment buildings with a hammering thud, thud, thud to a chorus of mechanical clanks and clinks. The concrete walls, though several feet thick, shattered and crumbled under the scorpion-like attacks. Little by little, the structures shrank under this attack from ground level.

The License Termination Plan describes the attack in more technical terms: "The primary method that will be used to completely remove the concrete is through large-scale demolition using hydraulic-operated crushing shears and jackhammers fitted to large tracked excavators. Concrete structures will be fractured and crushed by these tools. As the concrete is reduced to rubble, the embedded [steel] rebar will be exposed and segregated from the concrete rubble."

Remote-controlled equipment is used to demolish highly radioactive areas too dangerous for human contact. Hydraulic crushing shears finished the job of turning big pieces of concrete into small pieces. The most radioactive outer layers of concrete walls were shaved off a quarter of an inch at a time with pneumatic drills and other tools.

Terry Printz, the Zion station's operations supervisor from 2008 to 2010 and one of the last remaining 40 or 50 Edison employees still there, saw demolition begin.

"I saw the early stages of it, where they were starting to get the big stuff out, the radioactive parts of the plant. Where the big turbines are and the generators. They were heavy things to take apart and shift offsite. I was there when they were starting to put holes in the containment. Then I was transferred to the corporate office."

Printz also saw something that disturbed him. "I went into the augmented dispute resolution process designed by federal law for people to raise concerns without retribution or losing your job," said Printz. "There was a settlement. I'm not allowed to talk about what the problems were. Shortly afterward, I was asked to leave or quit."

Printz reached out to Edison's human resources department and stayed with the company working on power upgrade projects from 2010 to 2012, when he retired. "I have no evidence of things done improperly," said Printz. "I raised some issues, it went to the [dispute] process and that's all I can talk about."

Scholarly looking, with a high forehead and a quizzical expression, Printz looks like the engineer that he is. He's the kind of brainy guy you might have encountered years ago on campus who carried a slide rule, when those were used, that marked students as budding engineers. Printz started his 36-year career with Edison and Exelon as an entry-level engineer at the Zion station, and eventually became a reactor operator.

Printz described his time with the utility as a series of ups and downs. The 2008 stock market collapse affected the trust fund intended to pay for dismantling the Zion station, for example.

"They decided there was not enough money in the fund to start dismantling the plant," recalled Printz. "At that point, they ceased operations and got rid of people. They had to stop decommissioning. Then right around 2010, the market came back and they believed they had enough assets to get started. I left in 2010. I was not there long when decommissioning was getting off the ground in earnest."

When the NRC complained about Zion performance issues in the mid-1990s, "I was part of a bunch of managers," said Printz. "We don't know what's wrong but we're going to move people around and change the culture. I was involved in some of that, and the union stewards. We changed the culture quite a bit in the 1990s. I remember one individual I respect. He said they would never close the plant, and we want more benefits. I said I'm not the guy who can give them to you and maybe they could close the plant. Then they made the decision to pull the plug."

Printz also was at the Zion station at the time of the so-called control room rebellion that disturbed much of the commercial nuclear power industry.

"We call it a rebellion," mused Printz. "It was an unfortunate event. From my view, a lot of people wanted the plant closed. It was near Chicago. A lot has changed since. A lot of the guys there as operators, they got transferred to Byron and Braidwood, and the feedback is they are some of the best people in the world. I don't know what happened."

Even after his retirement, Printz stayed connected with the Zion station demolition. With his home just four miles from the Zion station, "I'd ride past on my bike and the containments were coming down and saw it from a distance, exciting stuff." He said it almost wistfully, as though sorry that he had not been part of this historic project, although, "I feel I contributed to the job in a small way in the decommissioning. I feel I did a good job."

Printz stayed connected with the project by having breakfast with colleagues. "I will say I talked to guys who worked there throughout the decommissioning process. In general, it went pretty well. As far as I know, nobody died. You could laugh, but any major construction projects [can have fatalities]."

Then Printz said something I had never heard before: six or seven construction workers died while building the Zion station. "There were people walking high up in containment who fell through grating pretty high off the ground. You don't survive that," he said. The containment buildings were 200 feet high. Printz started working at the Zion station only a few years after construction was completed and could have heard about such accidents.

I contacted Edison's media relations department and spoke to Jorge Cabrera, senior communications manager, attempting to confirm the fatalities. Cabrera said he was unable to locate any information on that.

Printz was not the only one who took a personal interest in watching the demolition project that started on October 1, 2010, 12 years after operations ceased at the Zion station. I wanted to see it too, and periodically drove to Zion to see how it was going, and saw something like a real live time-lapse show as the walls came tumbling down.

January 9, 2016—Six years after decommissioning started, I drive down Shiloh Boulevard toward the plant, wondering if the familiar twin concrete containment domes topped with turquoise-colored trim will be gone? No! There they are. But they look dirty. The southern silo has black smudges on the walls. Black streaks run down the sides of the north structure. Can't help wondering what a stickler for ship-shape details like Ed Fuerst would say about their appearance.

November 9, 2017—The two silos look dirtier, shabby, unkept and abandoned. A sorry end to what once was called the best of the atomic power-producing world, the flagship of Commonwealth Edison's nuclear fleet. A triangular-shaped structure juts from the base of the north side of the north building, a white fabric stretched tight over metal frames. It's a portal through those massive concrete walls that once encased a nuclear reactor. A locomotive passes, pushing a line of rail cars while a backhoe nearby transfers dirt from a pile in the parking lot to an open truck trailer.

September 10, 2018—This time, the domed structures look radically different. The north building is leaning landward, away from Lake Michigan, at about a 20-degree angle, looking as though it might topple over. The buildings are shrinking as they are chopped away at the bottom. "People don't go near it," says Rudy Espino, 26, of Round Lake, Illinois, a truck driver who has been hauling crushed stone to the site for almost two months. "They have it closed around it." Espino makes four trips a day from the Milwaukee area.

Soon after, both of the containment buildings are upright, but shorter. It must have taken considerable skill to get that leaning building upright again, while chopping away at its base. Now the buildings are shrinking evenly.

October 25, 2018—The two containment buildings are gone. All that remains of the landmarks that stood for 45 years are piles of rubble.

March 13, 2020—The site of the former nuclear power plant is outlined by a chain-link fence, surrounding mostly empty space. A line of trees marks where the silos were. Lake Michigan is glistening in the sunlight beyond, waves shuffling to shore. It's quiet now. There's no activity, no demolition sounds of that striking mechanical scorpion. Inland is a broad swath of tan-colored cattail marsh, beaten down by the recent winter snows.

It went from two 20-story towers to rubble in 10 years.

Before all that happened, the Zion station's License Termination Plan dictated the course of decommissioning, based on radiological studies of the site and leading to turning the plant site into a green field suitable for unrestricted future use, including housing, recreation, or farming.

The goal of decommissioning, often stated by NRC, is removing a power plant from service "and reducing residual radioactivity to a level that permits release of the property [for public and private use] and termination of the operating license."

"The NRC is currently reviewing Zion's final radiological survey results to ensure the site can meet the NRC's release criteria," said NRC's Mitlyng in an email in July 2021. "When the NRC is satisfied that the site can be released for public use, the license for operating the spent fuel installation will be transferred to Exelon. It will be the only remaining area of operation and NRC oversight." By January 2023, the radiological surveys were not complete.

The final step in this process is a Final Status Survey (FSS) report showing that no part of the power station property emits more than 25 millirems of radiation a year after the cleanup operation is finished. That's the magic number deciding whether the property is safe enough for unrestricted human habitation or activities.

The License Termination Plan, written by ZionSolutions, all eight chapters and thousands of pages, is surprisingly readable and informative. The termination plan is a storehouse of information about the Zion station and offers a vivid description of the radioactive conditions left by a nuclear power plant.

What remains is the atomic age's equivalent of the ash piles and toxic wastes left behind by an earlier industrial age. Both are sobering reminders of the environmental price paid for "progress."

The termination plan describes the geological history of the Zion area, together with a history of the plant, the buildings on it, hazardous spills, types of radioactive material contaminating the buildings and soil, radiation levels, decommissioning costs, and a description of the natural resources and wildlife in the area. Plus many revisions and lists of tables and charts containing mathematical equations that most people would not understand. It's high-tech wizardry.

This kind of exhaustive detailed information on a plant about to be torn down makes sense. It's wise to know the locations of highly dangerous radiation before sending demolition workers there.

One of the arresting details in the termination plan was finding the Zion station property covers 331 acres, contradicting Edison's longtime description of 250 or 260 acres. The double chain-link security fences surrounding the power plant itself enclosed 87 acres instead of 96 acres as some described. It looked like someone took time to take measurements, assuming they are right.

Chapter two shows that a nuclear power plant leaves a big radioactive footprint. It lists every known spill of radioactive and hazardous material, 305 of them, from 1974, after plant operations started, to 1999, based on records and interviews with more than 300 plant personnel up to 2014.

Sixty-four of the spills involved radioactive liquids or spent resins, most of them in the reactor containment buildings, the auxiliary building or in open areas. More than 100 gallons of radioactive resins, used to filter radioactive contaminants and minimize corrosion in water systems, might have leaked into the soil.

The documents identify 132 incidents of "the loss of control of radioactive material," meaning they ended up in trash or dumpsters. Seven incidents involved "personnel leaving [the] site with radioactive material on their person or in their vehicles." Considering all the radiation monitoring devices at plant exits, this is astonishing.

Reactor Unit One's steam generator was leaking up to 500 gallons of radioactive water a day, possibly spreading contamination to other parts of the plant.

Both reactor containment buildings housing the Zion nuclear reactors were intensely radioactive hotspots, not only because of leaks and spills, but because ionizing neutron radiation from the reactors caused nearby concrete and metals to become radioactive.

All 87 acres inside the security fences were found to be contaminated by radioactivity, including 11 major buildings on the station grounds, such as the auxiliary building, the fuel-handling building, the turbine building, the service building, the water crib house, and storage tanks. Using technical jargon for contamination, the license termination report called these areas "impacted." The auxiliary building was contaminated because it was "routinely flooded with contaminated water during operations." Power plant workers would have said these areas were "crapped up," using their preferred lingo.

Radioactive contamination also was found outside the fenced plant grounds, including the employee parking lot, an open area on the plant's south boundary facing the state park, the beach on the Lake Michigan side of the station and a stretch of Shiloh Boulevard that was used for training.

Waste discharges flushed radioactivity into Lake Michigan, but the report said they did not exceed federal limits.

Two of the thickest and most detailed chapters of the License Terminal Plan are chapter five, the final radiation survey plan, and chapter six, compliance with radiological criteria for license termination.

Chapter five identifies radionuclides found at the Zion plant and their radioactive half-lives. A table lists 26 radionuclides of concern. Five of them are "dose significant" for demolition workers and account for 99 percent of the exposure dose from radioactive concrete, soils, and elsewhere. They are cobalt-60, cesium-134, cesium-137, nickel-63, and strontium-90.

"The vast majority of dose is from Cs-137 [cesium-137] at 97 percent," which has a radioactive half-life of about 30 years, says chapter six. Cesium-137 is one of the more common fission products in the nuclear chain reaction of uranium-235 used as reactor fuel. "The next highest dose contributor was Co-60 [cobalt-60] at 1.7 percent," which has a half-life of 5.3 years.

Chapter seven covers decommissioning costs, although some parts are blacked out as proprietary information. Edison said the decommissioning project would cost $1 billion. The termination plans break that down into specific activities. The cost of removing the radioactive contamination alone, called radiological decommissioning, including packaging, transportation, and disposal, is $309 million. The estimated volume of radioactive wastes produced by decommissioning is 1,500 cubic meters or about 529.7 tons. The disposal cost of transporting that waste 1,500 miles to Utah totaled $3.7 million. The cost of a traffic fatality during transportation would be $3 million if such a thing happened. A rail shipment from Zion to Clive takes 23 hours.

The time worked to remove that radioactive contamination is calculated at 1,391 man-hours. The labor cost for that work is $329,209. A laborer costs $66.78 an hour, a supervisor costs $90 an hour, and a radiation protection technician costs $55.59 an hour. Equipment costs were $98,975. The cost of restoring the Zion station site after decommissioning is $39.4 million.

Those costs apply only to disposing of radioactive materials and structures. They do not apply to removing nonradioactive material or structures.

Radiation exposures to workers and to the public, when trains loaded with radioactive waste pass them on the streets, also are seen as costs. The dose rate to workers used in the report was 3 millirems per man-hour, assuming they worked in areas only after the removal of major sources of radioactivity. The worker dose-cost given for the 1,391 man-hours worked to remove radioactive contamination was $8,346. The monetary dose-cost to the public from three rail shipments to Clive, Utah, was $69. A REM of radiation exposure to a person is cost-calculated at $2,000.

Following ALARA (as low as reasonably achievable), the guiding principle of radiation safety, ZionSolutions said occupational exposures to decommissioning workers were three times less than originally predicted. It cited the use of remotely operated diamond wire saws and visual and audio monitoring for this achievement.

ZionSolutions takes credit for disposing of 22,390 curies of low-level radioactive waste from Zion. Another 469,000 curies of waste—some of the most powerfully radioactive parts of the nuclear reactors—were packaged into four dry storage casks that remain on the guarded Zion waste isolation island.

These figures come from documents filed by ZionSolutions with the Nuclear Regulatory Commission and are publicly available on the commission's Agencywide Documents Access and Management System (ADAMS) record-keeping system. ADAMS was a major source of information on the Zion decommissioning project for this book, researched between 2016 and 2021. Revisions might have been filed after that period and are not accounted for here.

A cost analysis for a project the size of Zion prepared in 1999 by TLG Services, Inc., of Bridgewater, Connecticut, for Edison, said labor typically accounts for about 39.65 percent of the total project cost, radioactive waste burial for about 17.17 percent, and 43.18 percent for everything else.

Chapter eight of the termination plan is an environment report covering natural resources and wildlife in the immediate decommissioning area. It was a delight to read because of instances of tender care shown toward wildlife encountered by decommissioning crews.

The only threatened or endangered species seen at the power plant site was the Blanding's turtle, listed as a threatened species by the state of Illinois. "During the decommissioning process, Blanding's turtles have been observed, rescued, and protected," said the termination plan. "Blanding's turtles awareness signs have been posted and inspections are performed to ensure that the Blanding's turtles are protected" in accordance with Illinois Department of Natural Resources and Illinois Environmental Protection Department recommendations.

ZionSolutions worked with local groups to rescue a den of snakes, which included several western fox snakes, found during work on a rail crossing north of the station site. They were relocated to a hibernaculum for rescued snakes. The western fox snake is not threatened or endangered, but it is considered an important part of the ecosystem.

Piping plover shorebirds and the Massasauga rattlesnake were known to inhabit the area in the past, but they were not seen. It appears the ZionSolutions wrecking crew had a heart, and treated the environmentally sensitive area and wildlife with a gentle touch.

The environment report described the region's climate; hydrology; nearby land use; water use; water and air quality; aquatic ecology; terrestrial ecology; occupational safety; socioeconomic impacts; environmental justice; cultural, historic, and archeological resources; aesthetics; noise; irretrievable resources; traffic and transportation; and regulatory governance of decommissioning.

Groundwater sampling found tritium, radioactive water, at radioactivity levels less than federal drinking water standards allow. The City of Zion purchases the city's drinking water from the Lake County Public Water District, so it was considered unlikely any private drinking water wells exist in the city. Demolishing the Zion station was not expected to change groundwater characteristics.

Overall, decommissioning impact on all environmental issues was listed as "small" because of the relative short duration of the project. Since 90 percent of the station's radioactive waste was hauled away by rail, barges were not needed. As a result, the Lake Michigan shoreline was largely undisturbed by decommissioning.

Aesthetics would improve, said the report, by providing a more open view of Lake Michigan when the power plant was gone.

Remaining after decommissioning are an Edison electrical switchyard, a security gate, parking areas, the Chicago and North Western rail line, haul paths, the Independent Spent Fuel Storage Installation, fences, and a sanitary sewage system lift station for the spent fuel monitoring building. These are in an area zoned for industrial use. An open marshy area lies west of the switchyard in an undeveloped area containing overhead transmission lines and corridors maintained by Edison.

The Nuclear Regulatory Commission, meanwhile, continued inspecting the power station and producing periodic inspection reports mentioning violations and accidents.

The most recent inspection report, January 21, 2021, found no violations.

A report dated March 18, 2020, found two violations of NRC regulations. One involved improperly placing demolition concrete debris in an area that was declared clean by a final radiation survey. The other cited ZionSolutions for failure to submit complete and accurate information in a final status survey report.

A July 18, 2017, report revealed "several near-miss industrial safety accidents." On December 29, 2016, a steel plate weighing about 5,000 pounds fell about 15 feet onto a manned excavator in Unit One containment, but caused no injuries. On March 10, 2017, a wire rope fell from the top of the containment dome and hit a worker standing below, sending him to the hospital for treatment. On June 5, 2017, a vehicle struck the supporting cable for an electrical line, which snapped in two. The severed live wires landed on a manned excavator and an engineering building, but no one was hurt.

But those appeared to be exceptions. Others said, "The inspectors determined that the licensee and supplemental workforce conducted the decommissioning activities in accordance with the regulations and license requirements." Another said, "Workers followed work plans, surveillance procedures, and industrial safety protocols and were aware of job controls specified in work instructions." Another said workers adhered to radiation safety controls spelled out in radiation work permits and general radiation safety policies.

The License Termination Plan lays the groundwork for the work ahead. Armed with information on where radioactive dangers lie, the project shifts into a stage of decontamination and radioactive waste removal. The goal, in ZionSolutions' words, is to render structures "cold, dark, and dry," and suitable for demolition workers to enter to do their jobs.

When a structure is ready for demolition, a documented survey and a formal turnover is made by the radiation protection group for the demolition group, validating that the radiological conditions in the structure are suitable to begin work.

After demolition or decontamination, a final radiation survey decides whether various sites in the power station can be declared safe enough to be released for public use. This final survey also might detect areas that need more work before they can be declared safe, meeting that 25 millirem number or less.

Most of the Zion nuclear station will vanish from sight, but large parts of it remain underground. This might come as a surprise to people dreaming of lakefront casinos and condominiums on the old plant site. Unless a way can be found to coexist in an area presently zoned for industrial use, this is the reality spelled out by ZionSolutions:

"The decommissioning approach for the [Zion station restoration plan] requires the demolition and removal of all impacted buildings, structures, systems, and components to a depth of at least three feet below grade." All systems and components down to and below the three-foot level will be decontaminated, disassembled, demolished, segregated by waste classification, and disposed of as clean demolition debris, clean salvage, or radioactive waste.

Major portions of the power plant remain underground, below the three-foot clearance level. But further demolition activities on those underground structures focus on concrete walls, especially those thick, steel-lined concrete walls of the two reactor containment buildings.

In both containment basements, all concrete will be removed from the inside of the steel liners. Those interior walls, facing the reactor, are radioactive. The steel liner and concrete outside the steel liners remain in place. In the auxiliary and turbine building basements, all the interior walls and floors will be removed, leaving only the reinforced concrete floors and outer walls of those structures. They will be hollowed out, leaving open voids.

The lower 12 feet of the fuel pool inside the fuel-handling building remains, including the concrete fuel transfer canals after the steel liner is removed. Portions of four other structures remain underground, including the wastewater treatment facility, the main steam tunnels, circulating water inlet pipe, and circulating water discharge tunnels.

Open basements and spaces left by this destruction will be filled with concrete rubble, soil, sand, or other material, including soils from the Zion municipal landfill. Demolition debris will be segregated for recycling, reuse, or disposal. Radioactivity-free concrete will be considered for use as fill. Buried piping one inch to 42 inches in diameter will remain in the ground when demolition is complete. They are expected to contain minimal radioactivity, but will be capped or filled with grout.

The demolition contractor, Brandenburg Industrial Services Co., will decide where nonradiological debris such as steel, pipes, and turbines end up. Most of it will be recycled and sold. The price of these materials was included in its proposal to tear down the turbine building.

Look carefully at the License Termination Plan and you often see the words "clean wastes" and "radioactive wastes." These are among the most important words in the plan because they decide the trajectory of the demolition project, its costs, what debris remains in Zion, and what is transported out. Radioactivity is the deciding factor, and it's a tricky concept.

The Nuclear Regulatory Commission recognizes three categories of radioactive wastes: high-level waste that usually means commercial spent reactor fuel, transuranic waste containing specific concentrations of specific radioactive elements, and four subcategories of low-level radioactive waste defined by increasing levels of radioactive curies per cubic foot of waste. All these categories of nuclear waste have some number of long-lived radionuclides.

Low-level waste, says NRC, is "defined by what it is not." Meaning it is not spent fuel, transuranic waste, or byproducts of uranium processing. Low-level radioactive wastes include anything from slightly radioactive trash like mops, gloves, and booties to radioactive metals from inside nuclear reactors. A cost-benefit analysis would be performed in some places to decide if radioactive underground concrete should be decontaminated and abandoned in place, or removed and hauled away.

An estimated 3.2 million pounds of class A waste, the least hazardous waste, will be hauled to Utah.

After all that is done, the entire site will then be covered with three feet of clean soil, replacing the top three feet of surface and everything on it that was cleared away.

One of the major problems with radioactive waste is that much of it will be dangerously radioactive for hundreds or thousands of years, requiring isolation from the human environment. The Zion nuclear plant is a link in that chain of events imposed by the atomic age, without a final resting place.

Looking to the future, when a "resident farmer" might occupy the plant site, the restoration plan comments that it would be unwise to attempt to sink a drinking water well into this field of underground rubble and basements filled with debris, which could collect rainwater. The plan offers radiation exposure scenarios to that future farmer who used or drank water from an onsite well. It cites hazards from inhaling airborne radioactivity and from consuming plants grown with irrigation water or meat and milk from livestock eating fodder from fields irrigated with onsite well water, which could be radioactive.

The water table in the area is 12 feet from the surface, so that as much as 37 feet of groundwater will cover some of the deep underground structures left behind.

As the decommissioning project approaches the finish line, it's a good time to ponder winners and losers in the game of commercial nuclear power. Commonwealth Edison got rid of its fleet of unmanageable and economically feeble nuclear power plants. Exelon ditched the troublesome Zion power station. The City of Zion lost its main source of property tax revenue. Potentially, the city might reclaim the power station's lakefront property and find a revenue-producing use for it.

It appears EnergySolutions took a gamble that did not pay off as intended. *Chicago Tribune* reporter Julie Wernau in 2012 reported that EnergySolutions underbid the Zion decommissioning project by about $100 million, gambling that publicity would help snare similar work around the world. The company was struggling financially, replacing its chief executive and chief financial officer twice in two years.

"We undertook Zion for strategic, not financial reasons," David Lockwood, EnergySolutions chief executive and president, told the *Chicago Tribune*, doing his best to make a financial Black Swan sound like a good idea.

The company revealed that it grossly underestimated dismantling costs. It lowered profit projections from Zion by 10 to 15 percent in March of that year, and by 5 to 10 percent in June, partly because it overestimated how much money the decommissioning trust fund would earn.

In 2015, Wernau reported that EnergySolutions was running out of money, although the company in 2011 told investors it expected to earn about $200 million from the Zion project at profit margins of 15 to 20 percent.

Lockwood was ousted from EnergySolutions in 2018, and in 2019 he became a member of the Department of Energy's Secretary of Energy Advisory Board. EnergySolutions was swallowed by venture capital firm Energy Capital Partners.

As a former superintendent once pointed out, the Zion station can be a fooler.

Chapter 43: A Zion Walkabout

PEOPLE ARE POWERFULLY DRAWN TO ZION'S LAKEFRONT, EVEN WHILE A PART OF it was an off-limits demolition zone with occasional heavy-duty trucks rumbling past.

The demolition itself was a spectacle. Views of the inland ocean that is Lake Michigan, with its invigorating breezes, were irresistible. Wait on Shiloh Boulevard, or in nearby Illinois Beach State Park, and you'd encounter people passing by. Most of them had memories or opinions about what they saw.

John Austin and his wife, Tricia, were walking their two dogs on Shiloh Boulevard, toward the lake and slightly north of the power plant.

"When I was a kid," said John, "me and my buddies would hang out at the beach. People don't swim much [here] in Lake Michigan anymore." The area was a playground then, but that changed, Tricia pointed out. "I'm concerned with the question of radioactivity, decommissioning, and what kind of impact it will have. Cement tombs will go into the ground. How long will that last?" She glanced at her husband and commented, "He's shaking his head." The couple from Winthrop Harbor clearly had differing opinions. John insisted federal regulations "make it pretty tight."

"We make decisions on data now," responded Tricia. "We don't know about 10 years from now." Cancer Treatment Centers of America became the biggest industry in Zion. "Having the power plant and the cancer treatment center is ironic," she said, because of public fears that nuclear power plants might cause cancer. The plant also was a cause for jokes in Zion.

While Tricia attended Zion-Benton Township High School, "our baseball team would play other towns. Our guys were so much smaller. The joke was it was the nuclear plant" that stunted their growth. The opposing team was "so much bigger."

Another time, when one of the power station's containment buildings was tilting, Rich Stasatis and his wife, Rhonda, of nearby Beach Park drove up to the same location on Shiloh Boulevard in their open-top mini sports car. They wanted to see the leaning silo everyone was talking about.

"In hindsight, we shouldn't have developed it," said Rich, who retired as a technician at the cancer treatment center and lived in Zion. "It cost us more money than it made us. We still have all the wastes. Does that make sense?"

Looking at what remained of the power plant, Rich added: "It would be nice to get our lakefront back. It's such a neat hunk of land." Lightening the conversation, Rich joked, "I do glow in the dark and don't need a night light," and his four granddaughters "all have two heads."

As we were talking, Armand Sheffield, 80 years old, also of Beach Park, drove up to join the conversation. "I helped build it," he said out of the open window on the driver's side of his car. For four years, he was an electrician in the Zion station. "It looked good until they tipped it. I took a picture of that and sent it to my friends who worked here too. They'll say it's an 'oops.'"

You can learn a lot about the Zion community by walking along Sheridan Road, the main drag, and talking to folks you meet along the way.

The 2008 recession, combined with the generating station closure, left Zion's downtown business district along Sheridan Road riddled with empty storefronts, their glaring vacant glass windows looking shocked and forlorn.

Walking down Sheridan Road, Matthew Rios stopped and looked across the street to a dome now used as a bandstand. It was all that was left of the magnificent 350-room Elijah Hospice, later known as the Zion Hotel, torn down in 1979.

"I will never forget it," said Rios, who was born and raised in Waukegan with a brother. "One summer afternoon, we were going to the beach. They were dismantling it. My mom wept. It was something to see."

Rios, wearing a baseball cap and a shadow of beard, hoped to start a family in Zion, but he was concerned that the power station demolition might pollute Lake Michigan drinking water. It was a concern he'd learned as a child, when his father worked for the Johns Manville Corp. plant in Waukegan, maker of asbestos products. Its 150-acre asbestos disposal site on the shores of Lake Michigan became a federal Superfund toxic waste site.

"He'd come home and say, 'We dumped stuff in the lake and I'm not happy about it,'" said Rios. "He eventually left his job. He was concerned about his family." The father told a supervisor about his concerns, but, according to Rios, the supervisor said, "Don't worry about it and do your job. There are 25 guys who want your job."

Today, signs posted at Illinois Beach State Park warn visitors that asbestos-containing materials have been found on the beach. They should be reported to park headquarters but left in place. Such materials look like white rocks and concrete rubble.

The economic downturn in downtown Zion is not finished taking casualties. A sign on the Zion Antique Mall's front door read: "We're retiring after 32 years in Zion, but we still have many nice antiques and collectables to sell." Cash only.

The store was a riot of figurines, knickknacks, baubles, tableware, old magazines, art, and furniture.

"This place used to be booming," said Don Bourdeau, 84, referring to Zion and surrounding communities. Bourdeau is co-owner with his wife, Sharon, of the building containing the antique store facing Sheridan Road in the heart of downtown Zion. It's hard to say if the business district is recovering, said Sharon. "We still have empty buildings, but probably not as many as there used to be. The Dollar Store and Walmart do well." As for others, "some days they don't get a single customer. . . . A lot of our customers are from the [cancer] hospital. They buy little things. We don't move a lot of merchandise."

The Bourdeaus are looking for somebody to buy the building and its antiques shop, but, says Sharon, "It's hard to sell a place with all the vacant buildings" nearby.

The City of Zion got $1 million from Exelon for downtown sidewalk improvements, said Don, but "they almost put us out of business" while the project blocked passage to the store. In another act of good intentions, federal and state emergency agencies designated evacuation routes in case of a serious accident at the nuclear plant.

"It cracked us up," said Sharon. "They don't know what traffic is like here. You might as well get out of your car and walk." Sheridan Road is the only major thoroughfare through town. As for the power plant, "it's one of those things you wish had never happened."

As often happens in Zion, if you talk to the residents long enough, you discover they have direct connections to the John Alexander Dowie days and recall stories about the quirky nature of those days.

"My mother was the first girl born in this town in 1901," said Don. "She told me a lot of stories." At the age of 17, Don's mother got a job at Zion Manufacturing, which had a strict dress code. Employees had to wear garments in blue, gray, white, brown, or navy. "She wore a gray dress with a red scarf around her neck and they fired her. She begged for her job back." The girl's mother, Don's grandmother, successfully pleaded with city officials to return the teenager to work. Sharon believed Zion Manufacturing might have been one of the country's first shopping malls.

"When you did your laundry and hung your unmentionables on the line, you had to hang them so the undies did not show," said Don. "You had to put them between sheets or pillowcases so you could not see the underwear."

In another case of Zion quirks, Don's father was driving with a friend in the 1930s when their car had a flat tire. "They lit up [cigarettes] while changing the tire," said Don. "A Zion policeman stopped and took them to the police station and fined them." Non-residents were fined for smoking and told never to return to Zion. The town's squeaky-clean reputation continued far into the future. A *Chicago Tribune* reporter whose police beat included Zion recalls, "We all used to laugh about nothing criminal ever happening there."

North on Sheridan Road was evidence of new kinds of industry moving in. The Hive Gaming Café was lit by garish, bright yellow lights, with a black-and-yellow sign over the door. A few blocks to the north was the town's biggest industry, the Cancer Treatment Centers of America medical center on Elisha Avenue, about a block off Sheridan Road.

With gaming and modern medicine moving into town, the church established in Dowie's "City of God" seemed like an important place to visit. But I was about to find out just how much that had changed as well.

Not many years ago, the location of Christ Community Church could be found simply by listening for the crystal-clear voices of children singing "Praise the Lord!" on a Sunday morning. Like many things in Zion, that is a thing of the past.

The church dating to 1896 originally was known as the Christian Catholic Church or the Christian Catholic Apostolic Church. The name changed in 1996 to reflect its nondenominational status. It's described as an evangelical nondenominational church following Protestant doctrine.

Its location changed too, in a park-like setting distant from the central city. This remote location probably would not have pleased founder Dowie. He wanted the church to be the center of everything in his religious utopia.

The front window of today's church facing the city is shaped like a towering, peaked bishop's hat inset with hundreds of shaped glass panels in blue, red, and white with a large cross in the middle—giving the inside audience a dazzling display of multicolored light.

Kenneth Langley, senior pastor of the church since 1997, is nothing like Dowie in appearance. Langley is tall and slim, and his short-cut gray hair is parted down the middle. He's likely to wear open-necked, short-sleeved knit shirts with casual slacks while preaching. Services begin with an eight-piece band of guitars, drums, and singers. Dowie was short, bald, stout, and clad in robes, with choirs of thousands of singers behind him while preaching.

Upon finishing a sermon, Langley will say in an earnest, down-home sort of way: "Let's ask for God's help. Pray with me."

It's unfair to ask Langley if Dowie would recognize the church and the city he started a century ago, if he suddenly appeared and walked the streets. But I can't resist.

Gamely, Langley answers in his direct Midwestern style: "Probably not. Dowie was a visionary and so we were blessed by that."

Given his anti-science bent, would Dowie disapprove of a nuclear power plant in his town, especially one that brought the town so much grief?

Mulling that one for a moment, Langley responds slowly at first, then catches his stride: "I don't know how to read how Dowie's mind worked. He was ahead of his time in some ways and also willing to think outside the box. I assume when you say anti-science, you have in mind he had antipathy toward medicine. You are right there. He would have embraced a different kind of technology. He started so many different ministries here. The whole Zion project is so different, maybe he would have embraced something like nuclear energy."

The church also is a casualty of the economic slump and the power plant's closing; church attendance dropped from roughly 550 a decade ago to 450. The preschool program ended when attendance dropped to 87.

"I would say the financial difficulty that Zion is facing, due in part to the plant's closing, has led to a number of our people to seek greener pastures across the border," says Langley. "Better job opportunities. Lower taxes, better schools. This is their perception. We have faced the challenge of people leaving Zion. That challenge is not unique to us or to churches. We use the phrase 'Illinois Exodus.'"

Langley calls Dowie "one of the original religious entrepreneurs. He was one of them. He was a gifted person in many ways and had a good mind, until he lost it."

At the Christian Catholic Church's height, when the charismatic Scottish Australian preacher held the attention of the world, Dowie had about 20,000 followers in the United States alone. By 2008, that dropped to 3,000 in the U.S. and Canada. Dowie's evangelical ministers reached into Japan, Philippines, Guyana, Palestine, Indonesia, and the Navajo Nation. But Dowie hit the jackpot in South Africa, where the Christian Catholic Church packs a punch far beyond its weight religiously.

"I know it sounds like an exaggeration," says Langley, "but there are secular scholars in South Africa who estimate 15 to 20 million people call themselves children of Zion. It's phenomenal. They are talking about Zion, Illinois, the Christian Catholic Church." They are known as Amazeoni, people of Zion.

"That, to me, is a fascinating story," says the pastor. "I have visited Africa a couple of times and they treated me almost like the pope, which for a guy like me, an ordinary guy, it's mind-boggling. The king of Swaziland is a Zionist. We had an opportunity to meet the king and speak to a stadium filled with thousands and thousands of Zionists on Easter. It's a remarkable story. It doesn't seem real, actually. You talk of Dowie possibly being disappointed if he saw Zion today. People there think of Zion as being a paradise on earth. A pilgrimage here would be a rude awakening."

It could be argued that Dowie got his wish. Millions of people believe the city he created on Illinois farmland is heaven on earth, if only in their imaginations. But Dowie was a spellbinder, then and now.

To learn more about him and his beliefs, a visitor to the City of Zion only needs to go further north on Sheridan Road and turn left on Shiloh Boulevard, then one block to Elisha Avenue.

There it stands, Shiloh House, a 25-room Victorian mansion built for John Alexander Dowie that radiates his lavish, spendthrift lifestyle. The four-story red brick and sandstone residence with white painted upper porches and trim looks like a Swiss chalet, in stark contrast to the humbler common brick and farmhouse-style structures nearby. It was built in 1902 for the princely sum of $90,000, then furnished for $50,000 with luxury articles imported from Europe.

The abode became seriously dilapidated after passing through several owners, until Zion jeweler Wesley Ashland bought it in 1967 for $18,500 so it could be preserved and restored by the Zion Historical Society. Now a museum, it houses Dowie's archives, historical society headquarters, and the chamber of commerce welcome center. On May 12, 1977, it was added to the National Register of Historical Places.

Lorna Yates, a slim, stylish woman, is Shiloh House's highly enthusiastic greeter and guide, with her own personal connections to the city's early history.

"My great-grandparents came here in 1905," she says, establishing her credentials as an offspring of the city's pioneers. "My grandmother was five years old when my great-grandparents came here."

Yates is president of the historical society, and admits, "I enjoy the history, the time period, the look of the time period. We're trying to maintain the house and the history of the house. Zion is very unique, one man's vision."

But surprisingly, she is no fan of John Alexander Dowie and is outspoken about it.

"He was living large and bankrupted the town," she says, spending $2,000 a month on his personal expenses. "Power, money, greed. He fell from grace. Money corrupts. Half the town believes he was a big, fat crook. He was spending money wildly."

Yates compares Dowie with modern televangelist preachers who fell from grace, like Jimmy Swaggart and Jim Bakker.

"Many people came to him because of his reputation as a divine healer," says Yates. "Get off smoking, drinking, he'd tell them," which is very modern health advice. "Some were healed, others died," she says matter-of-factly. "People thought there would be a Christian utopia, and that did not happen."

If not a utopia, Dowie's house was quite comfortable and accommodated his five-foot-four stature. As Yates shows me around the house, she stops at a tall cabinet. She reaches down to its base and pulls out steps that rotate out of the base, used by Dowie to reach the top of the cabinet. She rotates the steps back into the base, where they vanish from sight.

The first floor features a grand foyer and staircase with adjoining parlors, a dining room, and kitchen. The second houses bedrooms, including a master bedroom with separate beds for Dowie and his wife, where Dowie died in 1907. The third floor was servants' quarters, but today that's one of the most interesting parts of the dwelling. It houses artifacts dating to Zion's beginnings, including samples of lace produced by the Zion Lace Factory. Copies of *Leaves of Healing*, Dowie's newspaper, feature stories about "God's witnesses to divine healing," including Mrs. Rebecca H. Potts, "miraculously cured of hernia." The room includes copies of Dowie's sermons and writings.

You get an idea of Dowie's passion and drive by reading Dowie's remarks on May 6, 1902, to the Zion City Council while presenting the corporate seal, which he crafted personally, for adoption.

"I don't think that the danger of Zion from outside is worth considering," Dowie began. "THE ONLY REAL DANGER THAT CAN EVER COME TO ZION IS FROM WITHIN." All that is in capital letters, as though he had begun to detect the internal strife that would cut short his everlasting utopian dream.

The corporate seal he devised featured a dove, emblem of the holy spirit. A cross representing redemption, salvation, and healing. A sword, which is the word of God. And a crown representing righteousness and rejoicing. Above them appear the words "God reigns."

Dowie might pass away, he said, but not the seal. "I hope [it] will never pass away from Zion City, until the end shall come and a new heaven and a new earth to be created."

That corporate seal appeared throughout the city for 90 years on the Zion water tower, street signs, city hall, and the shoulder patches of city employees—from 1902 until 1992, when the U.S. Supreme Court upheld a lower court decision that the city seal violated the principle of separation of church and state and the Christian symbolism must be removed.

Opposition to the symbolism began the day in 1986 when Robert Sherman, a local atheist leader, happened to drive through Zion and was shocked by what he saw. "I felt like I was in a theocracy," he told the *Chicago Tribune*. "Like I was in a foreign country." He called the Zion corporate emblem "the most blatant abuse of religious symbols by a governmental unit in the history of mankind. Really." You could say his zeal matched Dowie's.

First Sherman asked the Zion City Council to remove the offensive symbolism. When that didn't work, in 1987 the Illinois chapter of American Atheists filed suit against Zion City, saying the symbolism was unconstitutional. The case went all the way to the U.S. Supreme Court, which settled the matter.

Zion's mayor at the time, Howard Everline, said, "People have told me to tell him [Sherman] to go to hell. I just won't do that." The mayor might have been simply observing the Holy City's longtime ban on cursing and being disagreeable.

Part III: Nuclear Power—Past, Present, and Future

Chapter 44: Pioneering Commercial Nuclear Power

WHAT WERE THEY THINKING?

It's a question often asked in consternation or admiration.

What were they thinking when leaders of Chicago-based Commonwealth Edison Co. decided to take a chance on a new technology, nuclear power, to produce electricity? And who were those leaders?

Commonwealth Edison blazed a much-admired trail over 40 years—from 1960 to 2000—as the nation's premiere atomic power utility.

The beginnings of such great endeavors sometimes are lost in the mists of time or indifference as new bosses take over with their own agendas. But not in this case. It can be traced to a triumvirate of mild-mannered but forceful men who exercised powerful corporate influence, before committees made decisions. The three men participated in the birth of commercial nuclear energy and helped make Edison the largest producer of nuclear-generated energy in the United States.

They were Thomas G. Ayers, Wallace B. Behnke, and Gordon Richard Corey. I made a point of interviewing them in 1998 and 1999 and preserving my notes for an opportunity like this. Ayers died in 2007 and Corey in 2012. All three already had retired when I tracked them down and interviewed them.

Corporate leaders often are judged by how well they react to the challenges of their time. For Edison, the most pressing challenges in the 1950s were coal miner strikes and the formidable, beetle-browed John L. Lewis, president of the United Mine Workers of America. The scowling labor leader was a driving force in organizing millions of industrial workers and was president of the miners' union for 40 years.

The United Mine Workers of America strike of 1946 was one of the 10 biggest labor strikes in U.S. history, when some 400,000 miners walked off the job in 26 states. They demanded safer working conditions, health benefits, and better pay. Eventually, the demands were met in a compromise reached with the help of President Truman, but only after an earlier proposed settlement was rebuffed in hardball negotiations.

Coal provided more than half of the nation's energy from the 1880s to the 1940s, and Edison's growing reliance on coal was troubling.

"We were concerned terribly about the amount of coal we were burning—16 or 17 million tons of coal a year, and doubling every 10 or 12 years," recalled Ayers. "It seemed to us there ought to be a better way."

That's how Behnke saw it too. "I remember when Commonwealth Edison first took a serious interest in nuclear power in the early 1950s. One of the principal driving forces was the need to find some competition for coal. It was an economic matter. Those were the days when the United Mine Workers were striking quite regularly and John L. Lewis was in charge. The large coal users in the United States were searching for a large alternative." Edison was having trouble getting a steady supply of coal because the coal miners "held us at ransom" through strikes.

Edison was looking for a better way, said Ayers: "That's why we went into the first program the Atomic Energy Commission and we had on the peaceful use of atomic power." Commonwealth Edison took a giant leap into the unknown world of nuclear energy.

As Edison's chief operating officer, Ayers probably had the most direct personal influence on the company's direction when the Atomic Energy Act of 1954 allowed private companies to own and operate nuclear facilities. ComEd contracted with General Electric Co. to design and construct the Dresden Unit One reactor near Morris, Illinois, for $423 million. Construction began in 1956 and the station started operating in 1960, the first privately financed commercial nuclear power generating station in the nation. A third of the contract price was shared by a consortium of eight companies.

The U.S. government is credited with building the world's first full scale atomic electric power station in Pennsylvania, which began operating in 1957 and produced power distributed by the Duquesne Light Company.

"We just said we wanted to build a Model T and see how reliable it was," said

Ayers. "We tried to keep it simple." Scientists at the Argonne National Laboratory near Chicago were helpful. The science and engineering national research laboratory had experience with nuclear energy dating to the Manhattan Project of World War II.

"We were interested in getting an input from a fellow who has to operate the plant, not a scientist telling us what to do," explained Ayers.

Among the frontrunners in exploring the uses of atomic energy was the Enrico Fermi prototype liquid metal cooled fast breeder reactor spearheaded by Detroit Edison near Monroe, Michigan.

A breeder reactor produces more fissile fuel than it consumes, and is cooled by a liquid metal, such as molten sodium, as implausible as that might sound. Mercury, commonly used in thermometers, is another liquid metal. But hey, this is advanced nuclear science.

The experimental Detroit reactor had a partial fuel meltdown and, after repairs, was shut down by 1972. It was the only one of its kind.

Speaking characteristically bluntly, Ayers said the Fermi plant developer "was adding everything new that came along, and in the end he had nothing. It was too complicated and nothing worked. He was a damn fool."

The Dresden station was a winner and set the stage for Edison's nuclear future for the next 40 years. As Edison's president in 1964 and CEO and chairman from 1973 to 1980, Ayers helped to set that stage. Ayers described himself simply as "into the operating side of the business."

"I took a great interest in Dresden," said Ayers. "Dresden proved to be everything we hoped. It was very reliable. It was easy to operate. It could throttle loads faster than coal. . . . We found this newfangled thing called nuclear could get up a hell of a lot faster, it followed the load well, and those are factors we liked."

Coal-burning power generating stations were more difficult. As demand for electricity dwindled toward the end of the day, boilers had to be turned down and banked in readiness for powering up again.

"Then getting them up in the morning was a hard job," said the executive. "You had to run like hell. In coal-fired plants, you work the ass out of it to get the last drop of energy out of the coal. You have high-pressure plants, a thousand degrees and high pressure and high temperatures.

"The nuclear plants were low-temperature plants. The thing that everyone worried about was safety. That is something that has to be number one when you build a plant like that. On the other hand, the record of the nuclear plants is better than any industry [in America], except for Three Mile Island. That was an accident caused by an error on the part of the operators." Ayers commented later about the 1986 Chernobyl nuclear disaster in the Soviet Union. The 2011 Fukushima Daiichi nuclear accident had not yet happened.

Edison paid close attention as General Electric Co. built the Dresden station using a boiling water reactor design.

"We learned a lot," said Ayers. "GE built four units on a turnkey basis," meaning move-in ready and at a fixed price, "and they lost their shirts. They thought we didn't know what we were talking about."

"We watched them every day and saw all the mistakes and saw how costly it was to repair them. We tried to learn how to do it, and I think we did."

Companies with long experience building conventional power generating stations thought they could easily switch to building nuclear stations, but they were mistaken. "You've got to be careful," said Ayers. "It's a whole different technology. You have to have a degree of safety that you are not used to in a plant."

But when it comes to safety, Ayers believed anyone in the power generating business "would say they'd rather sleep in a nuclear plant than a coal-fired plant."

When people asked why Edison didn't simply turn to burning natural gas or oil, Ayers answered: "We always took the position there was a finite amount of gas, and it ought to be saved for household uses, and not burned in the utility business. That's how we got into Dresden in the early '50s."

Behnke added that oil-burning power plants also proved vulnerable. The 1973 Arab oil embargo forced Edison, he said, to "do something so we would no longer be pushed around by the Persian Gulf." The utility stepped up its investment in atomic power, "largely on the basis of lowest, long-run cost. That was the principle driving force."

Ayers might have been the among the first at ComEd to lock horns with the Nuclear Regulatory Commission.

"I'll tell you a little story," Ayers confided wryly. "Every year, the nuclear regulatory people would have to have a session with the head man, which was me. They used to give us not the best marks. The tests of [reactor] operators had a lot of physics in it. I'd say you had to pass a sophomore physics class to pass that. We thought it was more important to have simulators."

Ayers and Behnke saw how pilots practiced their skills on the Link Trainer, the world's first commercially built flight simulator, and believed that simulator training was important for reactor operators too.

"So this fellow from NRC, I said I don't understand why you are not pushing simulators more. He said he wants them to know what's going on in the plant. I thought he should know what the flashing light down the line means. He said we'd rather have you know more about the physics. I said who does the best job? He said the operators at Three Mile Island. The next time I saw him [after the 1979 accident] I said, tell me more about those Three Mile Island people. He said, 'Mr. Ayers, you're a mean man.' No, I said, I'm trying to get your attention. The simulator is more important than somebody knowing about the physics going on in the plant. I was pissed off about the arrogance of the thing."

But Ayers believed it was possible to disagree, but still get along. As he put it, "You can make them love you and still get your point across." Others might have disagreed.

"I think the thing that utility operators didn't like was the degree of supervision by the government," he said. "I always thought we should quit bellyaching about it and get on with your job. I just said shut up and do what you have to do the way it has to be done."

The executive was annoyed, though, by NRC officials who said nuclear control room operators should wear ties and uniforms to show pride in their work. "I said I never had to have pride in a uniform to think I was proud of what we are doing. I always thought that was a dumb idea. I was always a working man."

Ayers saw no reason to believe the nuclear power industry was on its way out in the United States, despite its up and downs. "It would be a tragedy," he said, and saw the lulls as temporary. "I think you will have to utilize the various heat sources you have. I don't know who is whispering in congressmen's ears now. There are a lot of dumbbells who get into this business, but you have to look at a broader perspective. Prices have gone up like hell, and I'm a suspicious son of a bitch. The kind of regulation going on now has discouraged companies from going into it."

Economics was a leading cause for Edison choosing the nuclear path, but economics was the reason given for closing the Zion power station. Did Ayers regret leading ComEd into nuclear energy?

"I don't have any misgivings," he answered. "The way of making power that is cleaner, kills fewer people, is well within our ability to manage shouldn't be shut off. It's not the right answer in every situation, but I think it has a great deal to offer to America and to the world. Around the world, not everybody was as careful as we were."

For example, Ayers and his Finnish wife, Mary, were visiting friends in an upper region of Finland when the utility executive was invited to see a Russian-built nuclear power plant.

Upon seeing it, Ayers observed, "You don't have any containment building," and was told the Russians said it was not needed. "I said bullshit on the Russians. They are probably 99 percent correct, but if you have an incident, then the containment has a lot to recommend it to the people at large. They had a dumb accident in Chernobyl. They were doing work at 4 a.m., they made a mistake and had a serious one."

The Chernobyl accident, involving a reactor with no containment building, is considered the worst nuclear disaster in history in terms of cost and casualties.

"The Russians have a very low opinion of life, so you have to be careful," said Ayers. "They are great scientists, but they are free and easy."

The Finns asked Edison to help them design a containment building for their nuclear power station, spent six months at Edison, and constructed the safety structure.

Though nuclear power plant construction costs are high, Ayers saw the low cost of uranium fuel balancing that out. "Definitely, we had to get involved in the economics of it," he said, but "you had to equate high capital cost with very low fuel cost" and the cost of getting a kilowatt-hour out of a power generating plant and to the distribution system.

"I don't think we were wrong with what we knew at the time," said Ayers. "There is an awful lot of loose talk about nuclear being terribly expensive. The answer is what it costs to produce a kilowatt-hour."

Through finance, Behnke became another architect of Edison's nuclear enterprise. As vice chairman, he pulled the corporate purse strings and became a leading nuclear power expert by managing billions of dollars that went into building the utility's network of nuclear power stations—ultimately a $13.8 billion investment.

"I came on board in 1965, started in 1965 as assistant to the president, heavily involved in nuclear licensing affairs with the Atomic Energy Commission," said Behnke. "I was heavily involved in the period of 1965 to 1976."

"The people I knew on the Atomic Energy Commission were serious, competent people who were working hard to be sure that this new technology took advantage of all the most recent experience, particularly in terms of safety," said Behnke. "There was a lot of careful work in the early days. As it always turns out with brand new technology, there were some surprises that you did not anticipate and is part of what drove the cost up—licensing delays and requiring changing plans."

Economics was a driving force in steering Commonwealth Edison toward nuclear energy, and "economics in my judgment will continue to be the principal determinant in the long-range future." Nuclear plants bought and paid for "should operate quite economically on the margin," he believed, while new plants must "compete with oil and gas-fired, combined cycle plants."

"I think there is room for nuclear in the [nation's energy] mix, particularly in extending the life of existing plants," he said. "It will be driven by economics. That means the companies are able to run the plants efficiently and safely at competitive costs. In fact, it is interesting to me that there have been sales of nuclear plants to electric power producers. I suspect they bought these plants with the notion they can run them quite profitably."

Taking note of the Unicom merger with PECO Energy, creating Exelon Corp., Behnke said, "It's encouraging that the new management of the company is taking a look at the future and trying to position itself in some kind of forward-looking strategy."

Both Ayers and Behnke recognized that one of nuclear power's strength was being more environmentally friendly than coal, at least with air pollution.

This is a good place for Gordon Richard Corey to step into this story. Unlike many other nuclear utility leaders, Corey boasted of his friends at the Natural Resources Defense Council and the Sierra Club. They might have disagreed heartily about nuclear energy, but they liked each other personally.

"I consider myself an environmentalist at heart," said Corey, who was Commonwealth Edison's chief financial officer in the 1960s and 1970s, retiring as vice chairman in 1979. "I hate to see us using up our precious oil and gas just to make electric power, and yet that's what we're doing."

An impish man, Corey agreed with Ayers and Behnke that nuclear power was the way to go, but he arrived at that conclusion in his own, individual way.

"I used to lecture on this at MIT [Massachusetts Institute of Technology]," he said. "Two things. One is, we're dealing with a fuel supply that is virtually inexhaustible. You can say we run out of uranium, yes, but ultimately we would be using the breeder reactor. With the breeder, you virtually never run out. That is a statement somebody would argue with me about for an hour or two, and we probably would still disagree."

Looking at the problem from the available number of BTUs in the world, Corey believed that breeder reactors that create reactor fuel from spent reactor fuel could produce an unlimited supply of BTUs. Oil and gas can't. "We have a lot more gas than we thought 15 or 20 years ago, but it's still an exhaustible supply.

"The other is, I was absolutely convinced we could turn out our electricity with nuclear power cheaper than with coal-fired plants. No question, we could do it cheaper than oil. That was 20 or 15 years ago. As things turned out, as nuclear power was developed, we could not turn them out that cheaply. . . . The cost of hydrocarbons has not gone up with the general inflation rate, but has declined."

Critical thinker that he was, Corey also saw the ways nuclear power went wrong.

Was he wrong in advocating for nuclear power? He responded: "You ask if I thought it was wrong. Yeah, sure, wrong in the sense of being too early, given all the problems that have cropped up. Particularly the fact that President Carter put the kibosh on reprocessing spent fuel," which in the beginning was considered an important way to deal with radioactive waste, by recycling it.

Yes, arguably wrong in that the U.S. really was not prepared for fully managing all stages of this new technology, although that probably would be too much to expect. Humans rarely understand the full, long-range implications and unintended consequences of what they do. Critics of nuclear power and other forms of emerging science and technology say: just because you can do something, is that a good reason to do it? The late David Comey, a nuclear critic, made that point in a colorful way at a nuclear energy conference in a downtown Chicago hotel auditorium: "Just because you *could* fill this room with cow plops, *should* you?"

But abandoning the promise of scientific discovery probably is something humans cannot tolerate. John von Neumann, a Hungarian-American scientist less known than—but as brilliant as—his colleague Albert Einstein, might have spoken for many scientists when he said, "It would be unethical for scientists not to do what they know is feasible, no matter what terrible consequences it may have." That's how scientific inquiry works, and what led to atomic power.

Technical brilliance did not go far enough. Scientists fell short when it came to simple housecleaning. Who's going to take care of this mess? A Department of Nuclear Sanitation would have helped.

On April 7, 1977, President Jimmy Carter banned reprocessing commercial reactor spent fuel because of the risk that plutonium extracted in the process might be diverted or stolen to make nuclear weapons, and Carter wanted other nations to follow the U.S. example.

Instead of recycling used reactor fuel, as once intended, it is piling up at nuclear power generating stations across the nation until the U.S. Department of Energy finds a permanent disposal site for it. Congress in 1982 directed the president and the secretary of energy to recommend a disposal site by March 31, 1987, and a second site by March 31, 1990.

"The federal government is still sitting there doing nothing," said Corey, "and by law, all spent fuel is supposed to be under the aegis and control of the federal government, including all the weapons, all the spent fuel resulting from the weapons program."

After working at Commonwealth Edison for 40 years, from 1939 to 1979, Corey was hired by Bechtel Corporation, the largest construction company in the United States, to calculate the economic performance of typical nuclear utilities from 1979 to 1985. The project allowed him to track the growth and development of the U.S. nuclear power industry, with a critical eye.

"I could scarcely believe that operating costs were escalating as they were escalating," said Corey. The industry was growing, but in a quirky and erratic way. "It was a whole array of things going on during the 80s that increased the manpower at nuclear plants. I'll be conservative—five or 10 times. Not only the manpower requirements, but cost of the equipment has escalated tremendously."

Corey discovered that the security guard staff at a particular nuclear station in 1985 outnumbered the entire operating staff 10 years earlier. In another case, more people were working on construction paperwork than the number of construction workers.

Working as a presiding arbitrator, Corey got involved with the Washington Public Power Supply System (WPPSS), in the state of Washington. It became known as "Whoops" after attempting to build five nuclear power plants in the late 1970s and early 1980s. Two of the projects went bankrupt and two more were abandoned as unaffordable, producing the largest municipal bond default in U.S. history.

Another project that caught Corey's attention was the Marble Hill Nuclear Power Station near Hanover, Indiana. The Public Service Company of Indiana in 1984 announced it was abandoning the half-finished plant, on which $2.5 billion had been spent.

These blunders reminded me of Ayers's comment that the utility business attracts its share of dumbbells. The path to nuclear power in the United States is littered with casualties, although it should be said that public opposition to nuclear energy also plays a role in this, known as the NIMBY or "not in my backyard" effect. Still, Corey said, "We're going to have nuclear power in the world."

President Trump in 2017 pulled out of the Paris climate accord meant to curb emissions that cause climate change. On January 20, 2021, President Biden signed a 24-page "Executive order on protecting public health and the environment and restoring science to tackle the climate crisis," which reversed what Trump did and brought the United States back to the global agreement.

"I don't see any way to implement that treaty but to turn to nuclear power for a good chunk for the reduction of [emission-producing] power," said Corey. "If we want something to run our cars, trucks, and trains that's not going to emit CO_2, it's got to be electricity, as far as I know, which is generated with either hydroelectric, wind, sunlight, or nuclear power. If we accept the premise that CO_2 emissions are bad, nuclear power has got to be one of the important ways to reduce it."

Always looking toward the future, at new technology, Corey mentioned working for a company in the San Diego area. "They were developing plans for the federal government for a fast breeder, which was fail safe, a very, very simple piece of equipment mechanically, which would be small enough to be delivered by freight car. They call it the safer fast breeder reactor. The concept was developed here at Argonne. I would guess that something like that, which in and of itself could avoid the spent fuel problem, will within 50 or 100 years become fairly widely used. But I think before it would be acceptable, it'd probably have to be adopted, built, and operated by the federal government in order to avoid unnecessary fear."

To avoid placing a stigma on breeder reactor technology, he believed, "you have to say our elected officials are seeing to it that nothing will be done that isn't carefully looked at."

The Commonwealth Edison executives turned to nuclear energy in the belief that it was a wise business decision, forced upon them by coal miners and the uncertainty of coal supplies. Confronted with an existential climate dilemma, the world might be forced to do the same thing, with coal once again being part of the dilemma.

Roughly 60 years after ComEd embarked on its nuclear adventure, the world's leading climate scientists reported in 2021 that human activity unquestionably unleashed a warming trend in the earth's atmosphere, oceans, and land that is believed to be irreversible for centuries.

"Humans have already heated the planet by roughly 1.1 degrees Celsius, or 2 degrees Fahrenheit, since the 19th century, largely by burning coal, oil, and gas for energy," said the report from the United Nations' Intergovernmental Panel on Climate Change. "And the consequences can be felt across the globe: this summer alone, blistering heat waves have killed hundreds of people in the United States and Canada, floods have devastated Germany and China, and wildfires have raged out of control in Siberia, Turkey, and Greece." To that could be added seven tornadoes touching down in four northern Illinois counties near Chicago in a single summer evening.

"Bottom line is that we have zero years left to avoid dangerous climate change, because it's here," said Michael E. Mann, a lead author of the IPCC's report, the panel's sixth and most conclusive proof that humans are responsible for widespread and rapid climate change from carbon dioxide and other greenhouse gas emissions from burning coal, oil, natural gas, and auto pollution. The report was approved by 195 member governments of the IPCC.

The 12-chapter climate report had 234 authors focusing on global climate trends proving "multiple lines of evidence" that climate change is shockingly real, stating that more starkly and bluntly than ever before.

To stave off worse outcomes, the report said, humans must reach net-zero emissions by 2050. With this "code red" flashing, rich Western countries responsible for most legacy emissions must "effect deep cuts, transfer technology without strings to emerging economies and heavily fund mitigation and adaptation."

The climate scientists did not say what technological solutions they favor, leaving that up to their applied sciences colleges to figure out.

Chapter 45: Nuclear Reactors of the Future

WHITE DEER ROAMING THE GROUNDS GIVE THE IMPRESSION THAT ARGONNE National Laboratory near Chicago is a scientific wonderland.

It's more than that. It's a scientific powerhouse that took part in creating the world's first human-made nuclear chain reaction in 1942 and today is helping to design what tomorrow's nuclear facilities might look like.

No place better nails down the proposition that human-made nuclear energy was invented in Illinois, and keeps making global impacts even now. It's what Argonne has done and continues to do.

"Nearly all the deployed nuclear technology we see in the world today—different reactor ideas, different reactor concepts—almost all of it initially was developed here by scientists and engineers at Argonne," says Jeffrey Binder, associate laboratory director for energy and global security at Argonne.

"We unravel some of nature's deepest mysteries," he says, by offering Argonne's 3,500 employees and 8,000 visiting researchers from around the world "some of the most advanced tools in modern science," including five of the fastest computers in the world. "We contributed to many evolutionary energy breakthroughs."

Argonne designed the model for the Experimental Boiling Water Reactor, the forerunner of many modern commercial nuclear power plants. And it helped design the reactor for the world's first nuclear-powered submarine, the USS *Nautilus*.

Designated the nation's first national laboratory on July 1, 1946, Argonne's portfolio includes science and engineering research, renewable energy and storage, astrophysics, environmental sustainability, supercomputing, and national security—much of it called applied research.

Argonne's earliest incarnation, though, was as the University of Chicago's metallurgical laboratory. During World War II, metallurgists from that laboratory were recruited in the 1940s for the super-secret Manhattan Project racing to beat Germany and Japan in building an atomic bomb.

Under extreme wartime pressure and urgency, those metallurgists joined a team of physicists to assemble Chicago Pile-1, using 45,000 blocks of ultra-pure graphite weighing 360 tons and 19,000 slugs of uranium about the size of hockey pucks weighing 50 tons inserted into holes drilled into the graphite. It worked, on December 2, 1942, creating the first human-made self-sustaining nuclear chain reaction. Some say the deed ranks with the discovery of fire.

It all happened in a very unlikely place: in a squash court under the bleachers of the University of Chicago's football field. A more remote site originally was chosen, but it was not ready by the time the scientists wanted to launch their historic experiment. So they opted for the squash court.

Upon inventing the atomic age, their work had hardly begun. More atomic piles were needed to produce plutonium for weapons. But scientists recognized that conducting potentially explosive experiments on a university campus on Chicago's densely populated south side was too dangerous.

The next reactors came together in a more remote area, known as Site A in a Cook County forest preserve, Argonne Forest Preserve, near suburban Palos Hills. But this was a temporary location, lasting from 1943 to 1946. The Forest Preserve District of Cook County granted occupancy for the duration of the war, plus six months. Remnants of the CP-1, the world's first nuclear reactor, are buried in nearby Red Gate Woods forest preserve with warnings against digging into the ground.

The nascent research facility moved a short distance to its present location near Lemont, Illinois, 35 miles southwest of Chicago—3,667 acres of open grasslands, forests, winding roads, and red brick buildings, each housing a single scientific purpose in a rural area with plenty of open space to evacuate the facility if necessary.

Like most things nuclear, it was born of controversy, a nightmarish combination of national and local politics, academic rivalries with midwestern universities, intergovernmental disagreements, staffing difficulties, and protests from some 200 landowners whose property was used to create the new research center. That explains those white deer. They belonged to Gustav Freund, inventor of the skinless casings for hot dogs, whose country estate was taken for Argonne, along with his herd of white deer.

Though the center moved out of the forest preserve, the name stuck. The forest preserve is named for the Forest of Argonne in France, where American troops fought for the first time in World War I.

In 1949, Argonne opened a satellite research center known as Argonne-West in the Idaho desert 30 miles west of the city of Idaho Falls. There, 14 reactors were built, including the controversial Experimental Breeder Reactor II (EBRII), which was sodium-cooled and included a nuclear fuel recycling facility. A breeder reactor is designed to create more nuclear fuel than it consumes. In the process of burning uranium and thorium fuel, it produces Plutonium-239.

Plutonium makes some American politicians nervous. They feared nuclear fuel recycling and breeder reactors opened the door to stealing or diverting plutonium for nuclear weapons, potentially leading to nuclear weapons proliferation. Politicians prefer nonproliferation. As a result, breeder reactors and reactor fuel recycling have had a rocky and tangled history in the United States.

President Nixon in 1971 made liquid metal fast breeder reactor technology the nation's highest research and development priority, thinking it could make supplies of uranium last longer. Advocates contended that extracting uranium from seawater would produce enough fuel for breeder reactors to satisfy U.S. energy needs for 5 billion years.

In 1977, President Carter banned reactor fuel recycling and called for indefinite deferral of constructing commercial breeder reactors, calling it a "technological dinosaur" and a waste of taxpayers' money. During the Clinton administration, Argonne physicists converted the EBRII to a new concept, known as the Integral Fast Reactor (IFR). It was a revolutionary design that reprocessed its own fuel, reduced its radioactive waste, and withstood safety tests including the kinds of failures that rocked the Three Mile Island and Chernobyl plants. In 1994, the U.S. Congress terminated funding for the bulk of Argonne's nuclear programs, including the IFR.

But that is not the end of the EBRII story. It is the nature of research that scientists borrow ideas from each other, building upon the technological successes of others.

GE Hitachi (GEH) Nuclear Energy of Wilmington, North Carolina, did that by designing the next-generation PRISM (Power Reactor Innovative Small Module) reactor. It is a liquid sodium-cooled modular fast reactor that builds on what was learned from the EBRII reactor, which operated successfully for 30 years.

GEH is a nuclear alliance created by General Electric Company and Hitachi Ltd. of Japan. The advanced PRISM reactor would generate electricity by recycling used reactor fuel, helping to solve the problem of storing radioactive waste.

Others feast on ideas developed by Argonne National Laboratory, but Congress sidelined Argonne in reactor-building in 1994. Today, the laboratory does not build or operate any nuclear reactors on its Lemont site, although that could be changing.

"We do a lot of reactor design work," explains Roger Blomquist, a principal nuclear engineer at Argonne and a former naval captain on nuclear-powered submarines. "We work with a lot of these small innovative reactor companies, some not so small now. There is a lot of ferment going on in reactor design."

For Argonne, "the biggest single project is the Versatile Test Reactor [VTR]," says Blomquist, who has the weathered face and squint of a sea captain, bald with a fringe of silver hair.

In February 2019, the U.S. Department of Energy announced plans to build the VTR and complete it by 2026 at one of the national laboratories, not named.

"VTR will be a fast neutron reactor, in which the neutrons are not slowed down," explains Blomquist. Neutrons are slowed in conventional power reactors by water, boron, or control rods that absorb high-energy neutrons that cause atoms to split apart. In a VTR reactor, neutrons fly full force.

"Fast reactors can be designed as breeder reactors because they produce more neutrons per fission than water-cooled reactors," the nuclear expert continues. "These extra neutrons can be used for purposes other than sustaining the neutron chain reaction."

Uranium fuel bundles called breeding blanket assemblies can absorb neutrons and produce plutonium. "This is a breeder reactor," says Blomquist.

The project is called "versatile" because the reactors can be reconfigured to do different things.

"The test assemblies are sometimes contained in their own separate cooling system so that, if they fail, only the test loop is contaminated, but the test reactor is unaffected," explains Blomquist. "This is what VTR is designed to do. So VTR embodies all the main features of EBRII and PRISM [sodium coolant and fast neutrons] without the blankets, without breeding. This is called a test reactor."

To complete the story, Blomquist says, "fast neutron reactors can use their excess neutrons to destroy by fission some of the [highly radioactive] high-heat-generating transuranic isotopes in used nuclear [reactor] fuel. These radionuclides contribute substantially to the challenge of designing a used-fuel geologic repository, so converting them to electricity makes the overall nuclear energy system more fuel efficient, while at the same time simplifying waste disposal. This concept is called a burner reactor."

Burner reactors could be used to get rid of nuclear wastes piling up at commercial power plants across the United States.

A fast burner reactor is not a fast breeder reactor, the expert explains further, "because it has no breeding blanket assemblies. It is the fast neutrons that destroy the transuranics [highly radioactive wastes produced in commercial nuclear power plants]. Roughly speaking, in a burner, the breeder blanket fuel assemblies would be replaced with the 'waste' fuel assemblies, so the burner would not be a breeder."

The design details for each of these missions would alter the VTR reactor core somewhat, Blomquist says, but they would use liquid sodium coolant and the same reactor fuel design.

It's like walking a technological tightrope, always making sure that these new reactors are strictly for peaceful uses of atomic power.

Fast-neutron reactors are expected to be among some of the next-generation reactors of the future.

The United States found itself in an embarrassing reactor research bind when the government in 1992 and 1994 canceled funding for the EBRII and a so-called Fast Flux Test Facility in Hanford, Washington, where fast-neutron tests could have been possible. Another was the Clinch River Breeder Reactor project in Tennessee, authorized in 1970 and terminated by Congress in 1983, seen as unnecessary and wasteful.

The Versatile Test Reactor is intended to fill the research gap left by abandoning those two projects. It boils down to testing materials and fuels that will be used in the next generation of nuclear reactors, called Generation IV reactors. The majority of reactors operating around the world today are considered second-generation reactor systems; the vast majority of the first-generation systems retired long ago. Only a few Generation III reactors are operating as of 2022.

After the 1979 Three Mile Island accident, nuclear power plant construction stopped in the United States, killing the nuclear industry for 33 years through lack of public or governmental support. The nation lost a generation of nuclear discovery—a blessing or curse depending on your views on nuclear energy. Not until

2012 did the NRC approve construction of new nuclear power stations in Georgia and South Carolina. But more about them later. Now the U.S. nuclear industry must redefine itself after that 30-year paralysis.

"At Argonne, we figure out safety, but it does not prove to a regulator it is safe," says Blomquist. "There are some gaps in many different advanced reactor technologies, mostly involving experiments. All the reactor design work we do is computer designed against experiments. Those experiments are relevant. It gives us a great deal of confidence in the analytical design.

"A lot of new reactor designs of these 50 or so smaller advanced reactors in the United States and Canada use new coolants, like molten salts, liquid lead, sodium, and water. They are all over the place. Some of these reactors use new materials, and we need to qualify those materials in a new reactor, and subject them to conditions in the reactor core proposed, at the same temperatures and pressures. That means you need to stick samples of these materials in a real reactor and bombard them in a reactor as they would experience in real life. It's important because the environment in a reactor, like neutron radiation, is different from all other industrial applications."

In the VTR's fast neutron storm, flaws in test materials show up faster in those punishing conditions. Those new reactor coolants also pose some harsh demands. Liquid sodium, for example, burns on contact with air and explodes on contact with water, so it must be tightly contained.

VTR research is expected to lead to the development of new nuclear technologies to produce affordable electricity in remote areas, high-temperature heat for industry, and clean water from brackish water, salt water, or wastewater.

Argonne is managed by UChicago Argonne LLC for the U.S. Department of Energy's Office of Science. In 2005 Argonne-West merged with another research facility to become the Idaho National Laboratory. The VTR project is led by the Idaho National Laboratory in partnership with five national laboratories, including Argonne.

In July 2022, the U.S. Department of Energy announced that it selected the GE Hitachi small modular design as the basis for building a multibillion-dollar VTR at the Idaho laboratory to "help develop fuels for advanced nuclear reactors." If Congress approves funding, the agency added, the VTR "would be the first fast nuclear test reactor to operate in the United States in nearly three decades."

"We're engineers," says Blomquist. "We know what we create is beautiful and operational, but we have to ask what happens when something goes wrong. It probably won't go wrong and if it does, the consequences will be insignificant and acceptable. Especially reactors. We want them to be inherently safe. Possibly safe, naturally safe. That's the design philosophy for all experiments, including reactors. Nothing possibly could go wrong, but if it does go wrong, there is no serious problem. Defense in depth."

Defense in depth. Hearing that shopworn phrase is disturbing. Nuclear power advocates used that comforting phrase for decades. It meant nuclear plants were built with redundant safety systems, layers of protection, so if one fails, another kicks in so nothing can go wrong. Then we had Three Mile Island, Chernobyl, and Fukushima. A lot went wrong. Were nuclear advocates blind to that?

"Additional clarification is needed here," answers Blomquist. "I just described defense in depth as response to the possibility of failure. We ask how could it fail, what if it did fail? It accepts the possibility to fail. It doesn't require a perfect system, but a system that is robust. Defense in depth includes quality of the components you build the plant with. Making sure the software used in design calculations is correct. A lot of quality assurance. It includes operator training. One of the outgrowths [of TMI] was a simulator at every [reactor] site so the operators could experience events that are unusual."

It appears they changed the definition of defense in depth. Instead of assuming nothing will go wrong, the assumption now is that something will go wrong, so design and get ready for it.

"One of the revolutionary aspects of the inherently safe designs is you don't need as many safety systems to guarantee they are safe," explains Blomquist. "If you make something harder to break, you don't need as many safety systems to keep them from breaking."

Beyond that, one group of Argonne scientists is working on computer-assisted diagnostics to detect early faint sounds of potential future failures that humans cannot hear. Others work on advanced fuels that do not overheat, and reactors that require refueling less often.

They are not the only ones. Like the race to create the first controlled chain reaction at the University of Chicago, another race is underway to build reactors of the future for peaceful uses of nuclear energy. Like the Manhattan Project, timing is critical.

But let's pause here for a moment and explain why advanced reactors are sought and needed. Almost all nuclear power reactors operating or under construction around the world are light water reactors, so-called because they use ordinary water to cool their hot, highly radioactive cores. It is the industry workhorse, but its flaws are inhibiting nuclear power's growth. These include high cost and long construction times, and susceptibility to severe accidents like those at Chernobyl and Fukushima.

The new generation of nuclear reactors is intended to fix those flaws while addressing future energy challenges.

Global energy demand is expected to grow nearly 50 percent by 2040, with most of that from fossil fuels. This does not bode well for a world threatened by climate change. Nuclear power advocates say the atom is the only carbon-free power that can satisfy an electricity-hungry world at the levels that most towns and factories need. They say new advanced reactors, such as fast reactors, offer advantages that can address environmental, fuel availability, and spent radioactive fuel storage issues.

This is where the big league of future nuclear reactor development, called Generation IV reactors, comes in. Today, the platforms for building new reactors are far bigger than a squash court.

The Generation IV International Forum (GIF) was started by the U.S. Department of Energy in 2000 and formally chartered in 2001 as an international collective representing 13 countries where nuclear energy is significant and seen as vital to the future.

They are Argentina, Australia, Brazil, Canada, China, France, Japan, South Korea, Russia, South Africa, Switzerland, the U.K., and the U.S., plus Euratom, representing members of the European Union. A Canadian developer of a molten salt reactor became the first private company to join GIF in 2019.

The purpose of GIF is to share research and development information rather than build reactors. But the forum's target date for deployment of GenIV reactor technology is 2030, which is closing in fast.

"There is a window of opportunity open now, but it may close in the 2030s," warned Diane Cameron, head of the Nuclear Energy Agency's Nuclear Technology Development and Economics Division, at a GIF panel discussion in April 2021. "There's a clear possibility that 2030 technology decisions in both the public and private sectors will be locked in within the next five to 15 years. So there's an urgency to the Generation IV International Forum now."

Technological innovation comes fast and could leave Generation IV reactors standing at the starting gate. The technological graveyard is full of dead-end hopefuls.

In the earliest days of the automobile, the Stanley Steamer with a steam-powered engine sold briskly until Henry Ford's cheaper Model T captured the market with its internal combustion engine. Thomas Edison fought to make direct current the electrical standard, but lost to George Westinghouse's alternating current. A Kodak engineer invented the first digital camera in 1975. Kodak executives thought it was cute, but decided to stay with tried-and-true film until the company filed for bankruptcy protection in 2012. Airships known as blimps looked promising until the Hindenburg spectacularly crashed in boiling flames. It was a tragically explosive dud. The great hope of the 1990s and 2000s was that the internet would be a force for openness and freedom; it became a swamp of bad information.

The Generation IV International Forum has identified six promising reactor technologies for development, four of which are fast neutron reactors.

GIF set four goals for advanced reactor systems: sustainable energy with minimum waste; clear cost advantages over other energy systems and a level of financial risk comparable to other energy projects; excellent safety and reliability, eliminating the need for off-site emergency response; and nuclear weapons proliferation resistance and physical protection.

The six models are expected to have a variety of uses. They are:

Sodium-Cooled Fast Reactor (SFR)—Initially the main technology of interest to GIF, with more than five decades of test experience in eight countries, including 400 reactor-years of operating experience logged on demonstration or prototype SFRs. Needs development of a sealed coolant system and more information on performance and safety, such as a loss of coolant accident. SFRs can be used to generate commercial electricity and to manage highly radioactive wastes, including plutonium. It operates at near-atmospheric pressure and is a leader in the race to find a reactor of the future.

Molten Salt Reactor (MSR)—This concept is believed to have the greatest inherent safety features of the six models. The main MSR concept is to have fuel dissolved in the coolant as fuel salt. Thorium, uranium, and plutonium all form suitable fluoride salts. Molten salt reactors operated in the 1960s. They are seen as a promising technology for using spent fuel from light water reactors as fuel for MSRs. Global research is led by China. Scientists are working on two versions of the MSR.

Very-High-Temperature Reactor (VHTR)—With a graphite core, the VHTR is designed to operate at 1,832 degrees Fahrenheit. It could produce heat for industrial purposes, with an option for high-efficiency electricity production with desirable safety characteristics.

Supercritical Water-Cooled Reactor (SCWR)—The main mission of the SCWR is generating low-cost electricity. It is built upon two proven technologies, the current generation of light water reactors used around the world, and the superheated fossil fuel-burning boilers also used worldwide. It is being investigated by 32 organizations in 13 countries. Water in the reactor core becomes a supercritical fluid more like superheated steam upon reaching the temperature of 705.2 degrees Fahrenheit and can be used directly in a steam turbine.

Gas-Cooled Fast Reactor—The helium-cooled reactor operates at about 1,562 degrees Fahrenheit and is similar to the VHTR. It is suitable for power generation, thermochemical hydrogen production, and producing heat for industry. A fast neutron reactor, it would have a self-generating breeding fuel core. Like the SFR, used fuel would be reprocessed and all highly radioactive elements recycled repeatedly to minimize production of long-lived radioactive waste.

Lead-Cooled Fast Reactor—The flexible fast neutron reactor can use depleted uranium and thorium as fuel and burn highly radioactive elements in spent fuel from conventional power reactors. The reactor is cooled by natural convection with lead coolant temperatures ranging from 1,022 to 1,472 degrees Fahrenheit. One of its advantages is very long intervals between refueling.

Developers contend fourth generation reactors offer several major advantages, including producing nuclear wastes that are radioactive for a few centuries instead of thousands of years; producing 100 to 300 times more energy yield from the same amount of nuclear fuel; using a broader range of fuels; in some reactors, consuming existing nuclear wastes in the production of electricity; having improved safety features; producing no climate change emissions; and making nuclear power a renewable energy.

America's road back from 30 years of oblivion will be treacherous, as shown with its first steps back in 2012, when the NRC approved construction of new power plants in Georgia and South Carolina. They did not go well, and serve as a warning for GenIV reactor developers.

After three decades of engineering and regulatory reviews, Westinghouse Electric Company planned to make a big rebound with its Advanced Passive 1000 (AP1000) pressurized water reactor, arguably the world's most advanced

Generation III commercial reactor. Westinghouse sold four to China and four in the United States—two reactors each at the Vogtle Power Station in Georgia and at the Virgil C. Summer plant in South Carolina. They were touted as safer with better reliability and cost-competitiveness.

Swamped by construction delays and cost overruns, the construction campaign drove Westinghouse into bankruptcy, sending shockwaves through the nuclear power community. Westinghouse filed for bankruptcy protection in March 2017, because of $9 billion in losses from its two U.S. nuclear construction projects. It emerged from bankruptcy protection in August 2018, an economically battered company with a new owner.

In January 2018, Toshiba Corp. of Japan announced it was selling 100 percent of Westinghouse for $4.6 billion to a subsidiary of Canada's Brookfield Asset Management Inc. Toshiba bought Westinghouse in 2006 for $5.4 billion from British Nuclear Fuels and paid almost $6 billion to the owners of the two U.S. projects in guarantees. It was seen as a disastrous foray into the U.S. nuclear energy business.

Construction on Georgia Power's Vogtle nuclear power station's Units Three and Four near Waynesboro, Georgia, was almost finished and they were scheduled to operate by 2021 and 2022. But they hit a snag.

In August 2021, the NRC refused to authorize Georgia Power to load fuel and operate Vogtle Unit Three until safety-related changes were made. NRC inspectors found that builders failed to separate safety and non-safety-related cables for reactor coolant pumps and equipment designed to safely shut down the reactor, a violation of safety standards. This could involve more than 600 cable separations.

Georgia Power said this would add $460 million in costs to the utility's share of the project cost, which is $9.2 billion, and delay startup until 2022 and 2023 for the two units. In 2018, total costs of the Vogtle project were estimated at about $25 billion. When completed, the Vogtle station will become the largest nuclear power station in the United States.

Argonne's Blomquist added some insight to the Vogtle delays and cost overruns by saying: "They began it before they had a complete design." The nuclear industry went to a new concrete reinforcement bar system that was new to the NRC. "NRC had to circle back and make sure the new concrete rebar system was as good as the old one. It had nothing to do with the design and construction, but it caused a construction delay. When you have construction delays, you have to pay the workers not to work, or fire the workers and hire them. The costs are obvious for a mess like that."

South Carolina Electric and Gas was building two Westinghouse AP1000 reactors at its Virgil C. Summer Nuclear Power station near Jenkinsville, South Carolina, and expected both to be in service in 2020. In July 2017, after an extensive review of the costs of constructing the two reactors, the utility decided to stop construction and abandon the project.

By 2019, construction on all four Chinese AP1000 reactor plants was completed and connected to the power grid, after going through delays and cost overruns of their own. Westinghouse hoped to profit in the Chinese market, a country that builds nuclear reactors by the dozen. But it was another disappointing illusion.

In 2020, the *South China Morning Post* reported that China ditched the AP1000 in favor of a homegrown alternative, the domestically developed Hualong One reactor, because of worries over energy security and geopolitical uncertainties.

Twelve of the Chinese Generation III pressurized water reactors were either under construction or in the approval process. No new AP1000 reactors were approved for more than a decade. The last U.S. reactors went into commercial operation in 2018.

When Westinghouse filed for bankruptcy protection, the world had 450 commercial power reactors. At the end of December 2020, the United States had 94 operating commercial nuclear reactors at 56 nuclear power plants in 28 states. U.S. nuclear electricity generation peaked in 2012, when it had 104 operating nuclear reactors. Nuclear power plants have generated about 20 percent of U.S. electricity since 1990. They account for 25 percent of electricity generation in the European Union and 10 percent around the world.

But let's not forget Watts Bar reactor Unit Two, owned by the Tennessee Valley Authority, and its bizarre history.

Technically, it shed a little rain in the 30-year U.S. nuclear drought. Officially, the newest reactor to enter service during that time was Watts Bar Unit Two, which came online in 2016—43 years after construction started in 1973 near Spring City, Tennessee. It has the longest construction history in the world.

Reports on original cost estimates for both reactors vary. Some say $825 million, others say $2.5 billion. In either case, it ballooned to $4.7 billion. Both reactors were expected to start up in 1977.

Watts Bar Unit Two began operating 20 years after Watts Bar Unit One started in 1996. Both are Westinghouse pressurized water reactors. Construction on Unit One suffered many delays. Unit Two was 80 percent complete when construction on both units stopped in 1985 in part because of a projected decrease in power demand.

It is one more example in the helter-skelter, unpredictable history of nuclear power in the United States.

Private enterprise also is hustling to develop nuclear reactors of the future, and one of the actors is Bill Gates, the Microsoft Corporation co-founder and one of the world's richest and most influential men. He's putting his considerable wealth behind TerraPower, a nuclear power company he founded in 2006. At first a nuclear energy skeptic, Gates became convinced it could be a source of safe, clean energy to combat climate change.

TerraPower, based in Bellevue, Washington, and GE Hitachi Nuclear Energy developed the Natrium reactor, a class of sodium-cooled fast reactor. It's called a traveling wave reactor because the concept uses a small core of enriched fuel in the center of a much larger mass of depleted uranium. Neutrons from fission in the core "breeds" new fissile material in the surrounding mass, producing plutonium. Over time, enough fuel is bred in the area surrounding the core that it begins to fission, sending neutrons further into the mass and continuing the process while the original core burns out. Over decades, the reaction moves from the core of the reactor to the outside, giving the name "traveling wave."

In October 2020, the United States Department of Energy awarded Terra-Power a matching grant totaling between $400 million and $4 billion over the next five to seven years to build a demonstration reactor that transfers heat to molten salt which can be stored in tanks and used to generate steam for electricity production whenever needed. The reactor runs continuously at constant power.

The Natrium reactor is expected to be available in the late 2020s, making it one of the first commercial advanced nuclear technologies. TerraPower executives hope the new reactor will make the United States a dominant force in nuclear power and an exporter of reactors.

NuScale Power, headquartered in Portland, Oregon, designs and markets small modular reactors (SMRs) that can be manufactured in a factory and trucked to an installation. The NuScale small modular nuclear reactor gained design approval in 2020 from the U.S. Nuclear Regulatory Commission, meaning it met safety requirements. In January 2023, the NRC issued its final rule, making NuScale the first SMR design certified by the agency and cleared for use in the United States.

Reactor vessels are about 65 feet tall and nine feet in diameter, using conventional technology like uranium fuel rods to heat water. It uses natural cooling water circulation without pumps or equipment to keep the reactor operating safely. They can be used to produce electricity, for heating, desalination, and industrial heat. A power plant could house up to 12 of the reactors.

"You just add on pieces and all the structures are there. Dig a hole, put a reactor there, connect it up, and away you go," said an advocate.

As part of a carbon-free power project launched in 2015, the U.S. Department of Energy issued permits for the Utah Associated Municipal Power Systems and NuScale Power to build 12 SMRs at the Idaho National Laboratory in Butte County, Idaho. They are expected to be in full commercial operation in 2027.

The NuScale reactor design is based on research funded by the Department of Energy and conducted by Oregon State University, the Idaho National Laboratory, and other colleges beginning in 2000. When federal funding was cut, scientists obtained the patents in 2007 and started NuScale to commercialize the technology.

Small modular reactors are not among the six Generation IV International Forum designs. But some developers are interested because size matters. They represent a retreat from the massive, complicated nuclear reactor plants of the past that were plagued by delays and cost overruns.

SMRs are intended to be easily transportable and flexible. Units could be added to meet growing power demands. Greater safety comes from passive safety features that operate without human assistance. SMR concepts range from scaled-down versions of existing reactors to GenIV designs using fast neutron reactors, along with molten salt and gas-cooled reactor models. Drawbacks include the Nuclear Regulatory Commission's lack of history in licensing such unconventional reactors, including time, costs, and risks. They also pose concerns over preventing nuclear weapons proliferation.

Floating nuclear plants are attracting some attention, pushed by the rush toward decarbonization. The Electric Power Research Institute offered a proposal to produce electricity from a floating nuclear power plant anchored at sea. In South Korea, Samsung Heavy Industries formed a partnership with a Danish firm to develop floating nuclear power plants on offshore barges. In some scenarios, a small modular reactor could be built on a platform assembled in a shipyard, then floated away. In the U.S., that would be subject to existing regulatory rules.

One issue not discussed in this scientific frenzy over the next big thing in nuclear technology is: where will the workforce to operate all this new technology come from? Typically, an industry grows and matures as it gains experience, developing a skilled workforce. That did not happen with atomic power.

The Three Mile Island accident brought a 30-year halt to nuclear power development in the United States, and the number of operating power reactors declined. More than 100 orders for nuclear power reactors, many already under construction, were canceled in the 1970s and 1980s, bankrupting some companies.

It seemed reasonable to turn to experts for an answer. I had two key questions:

First, are career opportunities in nuclear energy considered good now?

And second, the Generation IV International Forum identified six types of advanced nuclear reactors likely to be available for commercial use by 2030—but where will the workforce needed to operate those reactors come from, given the 30-year hiatus in orders for new atomic power plants and the decline in the number of operating commercial nuclear plants?

"Ask us anything," says the website for American Nuclear Society (ANS) in LaGrange Park, Illinois, which describes itself as the "premier organization" fostering the development of nuclear science, engineering, and technology to benefit society. So, I asked.

"Nuclear power is a good career prospect," answered Steven P. Nesbit, ANS president. "There is increasing recognition in the U.S. and abroad that the carbon-free, always-on energy from nuclear power plants will be an essential part of our clean energy future. The U.S. operates more than 90 nuclear power plants, many of which are expected to continue operating for decades. New reactors at the Vogtle nuclear power plant are completing construction in Georgia, and new and innovative designs—including small modular reactors and advanced non-light water reactors—are in the pipeline for the late 2020s and beyond. The current nuclear workforce is mature and robust. As the world decarbonizes and electrifies our global economy, there will always be a need for nuclear scientists, engineers, and technicians to support existing and future nuclear power plants."

Daniel Carleton chairs the ANS Education, Training & Workforce Development Division and responded by email, saying enrollment in nuclear engineering courses across the nation indicates an enthusiasm for careers in nuclear power.

"One of the merits of several Generation IV reactor designs is that they are simpler to operate compared to the current fleet of reactors. Because of this, several designs will require fewer staff members to operate per unit of power output. Second, nuclear engineering schools have adapted their programs to cover Generation IV designs."

Carleton pointed to a 2021 report by the Oak Ridge Institute for Science and Education saying the number of nuclear engineering degrees awarded in 2019 were at the highest level since 2016, and the number of doctorate degrees awarded in 2019 were at the second highest level since 1966. The institute surveyed 34 U.S. universities with nuclear engineering programs.

Beginning in 2012, more than 600 bachelor's degrees in nuclear engineering were awarded annually, said the institute, a trend significantly higher than reported in the previous decade. The number of master's degrees grew by 21 percent from 2018 with 316 degrees and doctorate degrees were nearly identical with 194 degrees.

"Keep in mind that operating nuclear power plants also requires expertise in other engineering disciplines such as mechanical and electrical engineering," said Carleton. "A study by the National Center for Education Statistics indicates a general increase in interest in engineering degrees."

Judging from those remarks, the nuclear power industry's workforce development was not crippled by the 30-year lag in growth as much as I suspected. A respectable number of students appear to be betting they can find careers in the next generation of atomic energy.

The Nuclear Energy Institute (NEI), based in Washington, DC, is a nuclear industry trade association, calling itself a "unified industry voice." I asked them the same questions. After a week, the institute emailed to me an economic impact report on the NuScale small modular reactor construction project at the Idaho National Laboratory site. I politely pointed out that this was not responsive to the questions I asked, but heard no more from NEI.

Looking far into the future, we're likely to glimpse the twinkling hope of nuclear fusion power. But that's where it's always been—far in the future, almost like a mirage. A long-standing quip about fusion since the 1970s is that commercial fusion power is about 40 years away, even after the passage of 40 years.

Like a police reporter checking daily police reports for news, an energy reporter routinely checks for scientifically significant developments. It's been a long time since anything newsworthy about fusion power appeared on the scientific radar screen.

That changed on August 8, 2021, when scientists at the Lawrence Livermore National Laboratory near San Francisco announced in almost giddy tones that they had reached a milestone, maybe a breakthrough, in laser fusion that brought them closer to the Holy Grail known as "ignition."

This created a flash of excitement in the world of energy scientists.

The California research laboratory houses the National Ignition Facility (NIF), 10 stories high and spanning the area of three football fields. Built at a cost of $3.5 billion, it contains an array of optics and mirrors that amplify and split a pulse of photons into 192 ultraviolet laser beams.

Focused on a frozen pellet of deuterium and tritium fuel inside a gold cylinder smaller than a pencil eraser, the laser beams produced a hot spot the diameter of a human hair. Hitting the target with around 1.9 megajoules of energy in less than 4 billionths of a second, the laser beams generated more than 10 quadrillion watts of fusion power for 100 trillionths of a second, creating temperatures and pressures seen only in stars and thermonuclear bombs. It was roughly 700 times the generating capacity of the entire U.S. electrical grid at any given moment.

Most importantly, it appeared to yield more energy than it took to produce it, putting researchers at the threshold of fusion ignition.

Ignition is a tipping point in the fusion process where the fusion heats itself and overwhelms all the cooling losses that occur. Once this happens, a feedback process is generated where heating creates more fusion, which creates more heating, which creates more fusion, and so on.

"To me, this is a Wright Brothers moment," said Omar A. Hurricane, chief scientist for the Inertial Confinement Fusion program at Livermore. "It's not practical, but we got off the ground for a moment," he told CNBC. The fusion energy generated was about five times the energy absorbed by the hydrogen target and about 70 percent of the laser energy shot at the target. Hurricane called that significant.

"This result is a historic advance for inertial confinement fusion research," said Kim Budil, director of the Livermore laboratory, which operates the National Ignition Facility. It was a mega blast of energy, eight times more than they had ever done in the past. Another test on December 5, 2022, achieved similar results, with the *New York Times* reporting that the Livermore National Laboratory "had crossed a major milestone in reproducing the power of the sun in a laboratory."

"The fact that we were able to get more energy out than we put in provides an existence proof that this is possible," said Mark Herrmann, program director for weapon physics and design at the Livermore lab. "It can be built on and improved upon and made better and could potentially be a source of energy in the future."

Energy companies want to commercialize fusion and sell it as a clean energy source, if the fusion reaction makes more energy than it uses. Whether the Livermore scientists achieved "net energy" depends on further study by them and scientists who review the experiment.

Fusion is the opposite of fission, where a neutron smacks into an atom and breaks it apart. Fusion happens when two atoms slam together to form a heavier atom, fusing or bonding them. It is the same process that powers the sun and creates huge amounts of energy.

"Fusion power offers the prospect of an almost inexhaustible source of energy for future generations," says the World Nuclear Association, a claim similar to one made in the early days of nuclear fission. But the association admits fusion poses unresolved engineering challenges.

"The fundamental challenge is to achieve a rate of heat emitted by a fusion plasma that exceeds the rate of energy injected into the plasma," the association explains. Scientists have been working on creating fusion energy since the 1950s, producing what the World Nuclear Association considered the "main hope" for fusion—Tokamak reactors and Stellarators. Both use powerful electromagnetic fields to confine superheated deuterium-tritium plasma in a torus or doughnut-shaped reaction chamber to produce controlled thermonuclear fusion power.

Inertial confinement using lasers is a relative newcomer to fusion research, since the laser was not invented until 1960 and not applied to fusion research until the 1970s. Construction of the National Ignition Facility began in 1997 and was completed in 2009. This technology involves compressing a small pellet containing fusion fuel to extremely high densities using strong lasers or particle beams.

We'd have fusion power by now if it were not so difficult to copy the pyrotechnics of the sun and the stars, using massive gravitational forces to fuse hydrogen atoms to form helium and convert matter into energy.

On earth, the trick is to force hydrogen atoms to do what the sun does. Normally, fusion is not possible because electrostatic forces between positively charged atomic nuclei prevent them from getting close enough to collide and fuse.

Using fusion fuel consisting of different types of hydrogen, nuclei can be forced together with extreme heat—90 million degrees Fahrenheit—breaking down the repulsive electrostatic forces holding them apart. Atoms move faster and reach speeds high enough to bring them close enough so the nuclei fuse, causing an explosion of energy. The reaction must be kept stable under intense pressure, dense enough and confined long enough to allow fusion.

The aim of the controlled fusion research program is to achieve "ignition," which occurs when enough fusion reactions happen for the process to become self-sustaining, with fresh fuel added to continue it. Once ignition takes place, net energy yield is about four times more than with nuclear fission. The amount of power produced also increases with pressure, so doubling the pressure leads to a fourfold increase in energy.

Once again, Argonne's Blomquist offers his insights, comparing nuclear fission with fusion. A fission reactor containing bundles of uranium fuel is "a pile of hot rocks. You have pumps, but the reactor sits there and stays hot. In a fusion reactor, you have to force materials into a small area with high temperature and high density and hold them there. They repel each other and you can hold them there. Fusion is something you have to continue holding there. The actual fusion reaction we use every day is the sun. It's a huge fusion reactor. The gravitation reaction is enormous. That's what holds the materials at high temperatures when they fuse. We don't have the mass of the sun working for us. We must use magnetic fields or lots of powerful lasers. These are big, complicated systems. With fission, once it's there, it runs."

Blomquist calls these fusion reactors "really the state of the art."

In this high-tech, high-cost sweepstakes, winners and losers are likely. Some scientists will win accolades, while others might spend a lifetime on research that proves worthless. How do scientists deal with that? What is their mindset?

At least a partial answer comes from Dave Grabaskas, an Argonne nuclear engineer and risk analyst speaking to an audience about nuclear energy and technology from 1942 to the third millennium.

To avoid bias, said Grabaskas, "I never fall in love with technology."

Chapter 46: Two Prestigious Universities
View Nuclear

A BRONZE DOME-SHAPED MONUMENT RESEMBLING THE MUSHROOM-SHAPED cloud of an atomic bomb explosion, or a human skull, stands on the University of Chicago campus, marking the spot where Chicago Pile-1 once stood.

Near that bronze monument symbolizing nuclear energy at Ellis Avenue and 56th Street is a wall covered with five bronze plaques marking the University of Chicago as an historic site, accolades expressing society's appreciation, if not love, for the new technology that was spawned there.

Try to imagine yourself in that squash court where Enrico Fermi and 48 other scientists gathered that freezing winter day in Chicago—December 2, 1942. Most of them were clustered on a balcony overlooking the 25-foot-wide and 20-foot-tall graphite and uranium pile with cadmium control rods running through it sideways. Those graphite blocks each weighed about 20 pounds and covered everyone who handled them with black dust.

Accounts differ on exactly what Fermi and his colleagues did and said the moment the pile reached criticality and the first controlled nuclear chain reaction. You'd think the record would be perfectly clear on such a momentous scientific achievement. But it isn't. The shock of being present at such times appears to cause a numbing amnesia. Or too much to process.

In one account, Fermi is portrayed like a scientific impresario, raising a hand like he was conducting a symphony and saying dramatically, "the pile has gone critical." In another version, far less dramatic, an eyewitness to the event, scientist

Harold Agnew, recalled becoming "antsy" while hearing the growing growl of a Geiger counter until it couldn't keep up with the neutron emissions from fissioning uranium atoms, which was the point of the experiment. To everyone's relief, said Agnew, Fermi said, "okay, zip in," a decidedly unscientific way of telling George L. Weil, the physicist handling the control rod, to stop the first human-made, self-sustaining chain reaction after 28 minutes. "Everyone cheered," said Agnew, who was puzzled by where that "zip in" term came from.

Let's pause here to consider something that seems lost in this historic moment of scientific discovery, or just shrugged off. These scientists were midwives to a new kind of power that could flatten a city, and they were doing it inside the Chicago city limits. That moment is celebrated without much comment on the risks, other than Agnew saying he was getting "antsy" from hearing that unsettling neutron howl coming from the nuclear genie as it emerged from its atomic confinement. You could also say the atomic age began with a certain amount of bravado, even brazen recklessness.

Theodore Petry Jr. was a teenaged laboratory assistant when he witnessed the historic event, and was the last surviving eyewitness until he died in 2018 at the age of 94. He agreed there was clapping from observers on the balcony who understood what had just happened.

"But nobody said anything, which I thought was a fairly remarkable thing," said Petry. "There was no big hullabaloo. This was a truly momentous occasion. And it was quiet and all I've been able to come up with, it was because everybody knew it was momentous and what's there to say?"

Every version of the event agrees on one point: Fermi, an Italian, was presented with a classic straw-covered bottle of Bertolli chianti, which was shared in small paper cups by everyone there as a toast to the achievement, including Petry.

The result—sustainable nuclear energy—led to the creation of the atomic bomb and nuclear power plants, two of the 20th century's most powerful and controversial achievements.

Seventy-five years later, Petry was recognized as "the last man standing" at a December 1–2, 2017, University of Chicago conference on the experiment that unleashed the nuclear age, titled "Reactions: New perspectives on our nuclear legacy." The hazards of cozying up to technology were on full display at the two-day conference in the ornate Leon Mandel Hall.

Amid the campus Gothic architecture, intended to convey a spirit of inquiry and a sense of history, the conference started with a video describing nuclear energy as a way to provide electricity for the world, and a tool for ending World War II. The Manhattan Project, said the narrator, was "the beginning of Big Science" and nuclear power could be "the largest supplier of carbon-free energy in the country."

Nuclear power was a "city-killing technology," said the narrator, a "world-making and a world-breaking technology." Military matters became more important because it could be used as a weapon.

After the Hiroshima and Nagasaki bombings, said the video voice, "some scientists were appalled by what they saw" and considered the bomb a weapon of genocide. "For the first time, physicists really had to think about what they were doing and why they were doing it. What would be the implications of further research on nuclear energy?" This suggests some scientists had not considered the consequences of their work.

The bombings terrified the world. But as an upside, it brought together experts in science, law, humanities, philosophy, and the social sciences interested in forging a peaceful future with atomic energy.

Electric power production became a prominent peaceful use of nuclear energy, beginning with high hopes of electricity too cheap to meter and with the potential of turning deserts green. Over time, it also showed potential as a business-killing technology that failed at least in some ways to live up to expectations.

That became clear at a panel discussion on December 1, 2017, at the University of Chicago event exploring the role of nuclear energy in a climate constrained world, featuring Christopher M. Crane, Exelon's president and CEO, and Michael Greenstone, a University of Chicago economics professor.

"Currently, the economics of nuclear are terrible," said Greenstone, because of another new technology: hydraulic fracturing, also known as fracking, the high-pressure injection of fluids into deep rock formations, forcing natural gas and petroleum to flow freely for easy removal. By 2019, fracking made the United States a major crude oil exporter.

"In 2007, we could not have imaged how we are awash in natural gas," said Greenstone. "Petroleum markets around the world have been completely upended."

Nuclear energy lost its competitive edge. Natural gas produces a kilowatt-hour of electricity for 5.6 to 6 cents, said Greenstone. Nuclear power for 10.5 cents. Atomic power became the most expensive way to boil water for steam.

"The big, blue elephant in the room is natural gas," said the economist, "and that makes it complicated for other sources of energy."

Asked what's keeping nuclear from being a bigger part of the U.S. energy picture, Crane answered, "economics . . . economics is the albatross," adding that "competing with low-cost natural gas is a major issue." He also believed governmental subsidies for wind and solar power give them an unfair advantage.

The passage of time did not heal Exelon's economic wounds. They only got worse. But by 2021, Exelon found a way to get subsidies of its own. The company threatened to close the pioneering Dresden power plant and the Byron station unless the Illinois State Legislature approved millions of dollars in electric bill hikes to keep the uneconomical nuclear plants running for at least another five years.

On September 13, 2021, the Illinois senate gave Exelon a $694 million bailout. It was an unsavory request considering officials at Exelon's Commonwealth Edison subsidiary in 2020 admitted to paying about $1.3 million in bribes to state legislators in exchange for two laws that benefited the company. On July 17, 2020, Edison agreed to pay a $200 million fine after federal investigators probed into a nearly decade-long campaign of bribery and influence peddling.

Admitting "wrongful conduct," Commonwealth Edison's CEO, Joe Dominguez, told the Illinois Commerce Commission late that same month, on July 29, 2020: "On behalf of ComEd, I want to tell you that I am sorry for that conduct. It violated a trust with you. There are no excuses for our conduct, and I will offer none today." The company offered to refund $21 million to ratepayers, amounting to about $5 per customer, but critics said the refund should be about $40 million.

Crain's Chicago Business pointed out that Exelon's CEO Crane paid no financial price for the scandal, while Exelon shareholders paid the $200 million fine. As if the Edison scandal had never happened, reported the newspaper, Crane's 2020 executive compensation totaled $15.2 million in cash, stocks, and other benefits.

At the University of Chicago conference, Crane pleaded for a sustainable market base "with viable market rules" until Exelon transitions to whatever comes next, depending on public support and "what the country wants. We'll get behind it." Meanwhile, the nation's interconnected electric power grid "is not going away any time soon. It is crucial to our way of life today."

In 2016, Commonwealth Edison made a $2.6 billion investment in what is known as its "Smart Grid," modernizing the electric transmission grid in Illinois by improving power lines, installing smart meters, and upgrading the company's badly outdated infrastructure.

Economic professor Greenstone sees the problem, and solution, largely in economic terms. "Is nuclear any different from any other fuel source? Let the market sort it out," he said, but added, "Something should be done to level the playing field. If nuclear can make it, fine. If not, fine.

"Something would be lost if we did not have a nuclear industry in the United States. I don't know how to price it or set up a market for it. It's sort of the wild card."

In a separate panel discussion on the role of nuclear weapons in the modern world, Madelyn Creedon, a former U.S. Department of Energy nuclear security official, said: "Nuclear weapons are not for fighting. They are used for maintaining peace. I can keep you at bay because I have a nuclear weapon."

Yet another separate session was devoted to one remarkable man, the enigmatic Hungarian physicist and biologist Leo Szilard. Upon hearing of his exploits at the very heart of nuclear invention, one is tempted to wonder why he is not more prominent in world history. In a report on nuclear fission, the Library of Congress states flatly that "Szilard was responsible for the establishment of the Manhattan Project."

Szilard also was among those 49 eyewitnesses to the birth of the atomic age at the University of Chicago in 1942, standing with Fermi on the balcony overlooking the atomic pile, which he co-designed and co-patented with Fermi. The successful chain reaction was as much Szilard's achievement as it was Fermi's. Historians have a tendency to name those cheering great moments in history, but Szilard was not cheering.

Fermi and Szilard were standing alone on that balcony after all the others left the squash court, according to Szilard's biographer, William Lanouette. Fermi and Szilard shook hands, then Szilard said: "This day will go down as a black day in history."

Szilard had worked with German nuclear scientists and knew they also were working along similar lines to create an atomic bomb, and he feared the Germans. So much, that he helped to convince Albert Einstein to write a letter to President Roosevelt, and helped to compose it. It said the Germans were closing in on building an atomic bomb and American scientists needed to get there first. This led to the Manhattan Project.

Lanouette is the author of *Genius in the Shadows: A Biography of Leo Szilard, the Man Behind the Bomb*. It describes how the baby-faced Hungarian, standing alone on a street corner in London in 1933, was the first to conceive of the idea of a nuclear chain reaction. He imagined the possibility of splitting an atom, but at first didn't know which one. Szilard filed for a patent on the concept of the neutron-induced nuclear chain reaction in 1933, which was granted in 1936.

Considered "a flake" in his time, Szilard expressed his ideas in paradoxes, and his paradoxes were not always understood. Szilard exasperated Fermi, said Lanouette, because Fermi "was not sure when he was being serious in his flurry of ideas." He had an air of mystery, but was cheeky and outspoken. You could say he was quirky.

After helping to create the atomic bomb, Szilard worked to outlaw nuclear weapons. In 1962, he created the first political action committee for arms control. He was a professor at the University of Chicago from 1946 to 1960, but Lanouette described Szilard as an "intellectual bumblebee, wandering from one university to another, working as a lab assistant. He earned money through patents he filed. He got the idea of the atomic bomb from H.G. Wells, a book he wrote." Szilard blended one field of science with another, "he just couldn't help himself." He was suspected of being a Soviet spy, or at least a troublemaker. He loved biology because experiments could take hours, rather than weeks or more in nuclear physics, and biological discoveries could help people.

Szilard, said Lanouette, was "on the right side of history." He spent the rest of his life trying to control nuclear weapons, believing that scientists were more rational than politicians. Szilard's wife Gertrud said, "he was not smarter, but one day earlier." The scientist knew others were capable of coming up with ideas he had.

Ideas are the lifeblood of places like the University of Chicago. Twenty miles to the north is another great university, Northwestern University. Both are research universities committed to research as a central part of their missions. They are similar in other ways. Both reside in diversified communities along the Lake Michigan shoreline—the University of Chicago in Chicago's Hyde Park neighborhood and Northwestern University in Evanston, a suburb bordering Chicago on the north.

Both have Latin mottos lauding the value of learning. For Northwestern, it's "Whatsoever things are true" (*Quaecomque sunt vera*). For the University of Chicago, it's "Let knowledge grow from more to more; and so be human life enriched" (*Crescat scientia; vita excolatur*). Their basic, founding truths say something about them and their goals.

Nuclear power was a monumental idea that took root at the University of Chicago and changed world history.

Two years after the 2017 University of Chicago conference celebrating that achievement, Northwestern floated another idea about nuclear power, hoping this new idea would also take its place in world history as a palliative to what the University of Chicago did.

It's called victim compensation.

On a sunny April day in 2019, Northwestern University opened its classroom doors to the community for its annual Day With Northwestern event, demonstrating some of the university's thought-provoking activities. One of the eye-catching sessions was "Global Collaboration in the Twenty-First Century: Lessons from the Fukushima Nuclear Disaster."

A leader of that discussion was Annelise Riles, executive director of the Northwestern Roberta Buffett Institute for Global Affairs, and founder of Meridian 180, a multilingual international forum of thought leaders seeking to generate ideas and guidance on the most important problems of our time. Meridian 180 aims to develop strategies for crisis preparedness, but needed to know what key questions needed to be asked.

Trying to identify the world's most pressing problems, the two groups practically stumbled on the realization that compensating victims of nuclear power plant accidents "is the question of the century." When considering costs of nuclear energy, it is largely unrecognized and unaddressed, an issue hiding in plain sight.

That observation rang a bell and reminded me that Governor Richard Thornburgh of Pennsylvania, the nation's only governor to face a nuclear crisis during the 1979 Three Mile Island nuclear accident, said something very similar that year. The governor asked the Nuclear Regulatory Commission to establish "a national nuclear compensation fund, supported by the industry, to offset economic losses to consumers, communities, and state and local governments caused by accidents like the one on Three Mile Island." Nothing came of it.

"There is a widespread myth that nuclear energy is safe," said Hirokazu Miyazaki, a Northwestern University professor of anthropology speaking at the 2019 university event. "We can assume there will be another accident." In those few words, Miyazaki introduced the question of the century, and a counterpoint to the University of Chicago's achievement. He coordinates the two Northwestern working groups.

Riles and Miyazaki were living in Tokyo with their families on March 11, 2011, and felt the ground convulse when a 9.0 magnitude temblor struck northeast Japan, unleashing a 50-foot tsunami wave that swept entire cities into the ocean, claiming 15,893 lives and leaving hundreds of thousands homeless.

The wave smashed into the Fukushima Daiichi Nuclear Power plant, knocking out four of six nuclear reactors. The fuel cores of three of them melted down, spewing enormous amounts of radioactive debris and water into the air and sea. Japan declared its first-ever nuclear emergency and 140,000 residents were evacuated from a 12-mile zone around the stricken power plant. The catastrophe displaced 50,000 households in the evacuation zone because of radioactivity released into the air, soil, and sea.

Stunned, Riles and Miyazaki followed news reports of the devastation and aftermath, which informed their thoughts about disaster preparedness.

After their presentations at Northwestern, I asked Miyazaki for time to talk with him in more detail about the accident, its consequences, and his ideas about victim compensation. He was living in a stately, two-story red brick house on leafy Garrett Place, just two blocks from the Northwestern campus, a mix of traditional university Gothic and modern steel and glass buildings in Evanston.

A distinguished-looking man with silver hair, Miyazaki focuses on the history of citizen diplomacy for peace and a world without nuclear weapons, and is driven by a very simple question: how do we keep hope alive?

Watching the Fukushima accident unfold from Tokyo, the professor realized that more than lives and cities were lost in the accident. The toll included trust in public officials and operators of the Fukushima Daiichi power plant.

"When the accident happened," he said, "even a few hours after the earthquake, all the experts at the universities specializing in nuclear energy and people who served in government said a meltdown would never happen. They were saying that on television. They really discredited themselves. There was a collapse of authority about the expertise of nuclear energy and some citizens took on the task of doing some of their own scientific work."

Victims of major disasters are not just victims, he observes. They are acting and creatively thinking about their own livelihoods. "When we design a compensation mechanism, we first actually think we should talk to the victims. They are not just passive actors waiting for the government and industry to process their claims." Though public officials who cobble together ad hoc disaster recovery plans believe they are acting in the victims' interests, Miyazaki points out they seldom talk to the victims.

The world saw two nuclear accidents that the International Atomic Energy Agency (IAEA) classified as level 7 major accidents—Fukushima and Chernobyl.

Three Mile Island was classified a level 5 accident, far less severe than the other two but with wide consequences. The Fukushima and Chernobyl mega-disasters are the best known. By some accounts, 33 serious incidents have struck nuclear power stations worldwide since 1952, six of them in the U.S.

On April 26, 1986, a sudden power surge during a poorly managed test at the Chernobyl Nuclear Power Plant near the city of Pripyat in northern Ukraine caused an explosion and fire that destroyed reactor Unit Four, spreading massive amounts of radioactive debris across western Soviet Union and Europe. It killed 31 workers outright and another 29 firemen died from acute radiation exposures. Approximately 220,000 people were relocated from their homes. The World Health Organization estimates another 4,000 long-term deaths will be linked to the accident.

The Chernobyl, Fukushima, and Three Mile Island accidents had this in common: "When something really big happens, government has to come in. That part is strategically left undefined and ambiguous. We assume nuclear accidents are rare, and when something major happens, we have to deal with it on a somewhat ad hoc basis. That seems to be the mindset. We have to think more concretely about the process before a major accident happens again. We tend to come up with a process that is only for that moment, created for political expediency for that moment. So we need to think," said Miyazaki.

After the Chernobyl accident, Germany, in 2000, decided to phase out all of its nuclear power plants. By 2021, six nuclear plants were operating, while 26 were undergoing decommissioning. The remaining six will be phased out by the end of 2022.

Nuclear power plant accidents tend to be regarded as "anomalies," unlikely events caused by human error, weather, or cultural issues, rather than as part of a growing pattern. "Those types of explanations reduce the accidents to some local conditions and factors, but we are simply saying there will be another accident in the future," said Miyakazi.

Some thinkers in Miyakazi's milieu go so far as to believe nuclear power plant accidents are becoming "normal" as the risks grow with the construction of nuclear stations around the world, especially with rapid reactor development in China, South Korea, and India.

Japan ordered all of its 54 nuclear power reactors to stop operating for safety inspections after the Fukushima accident. By March 2021, nine nuclear reactors were operating in Japan and 18 were expected to be back in service by 2030. The Japanese Cabinet adopted a document in 2011 saying "public confidence in nuclear energy was greatly damaged" by the Fukushima accident, and many assumed Japan would phase out nuclear energy. But the Ministry of Economy, Trade, and Industry in 2017 said the nation needs nuclear energy to meet its obligations under the Paris climate accord.

Japan is uniquely sensitive to nuclear power and radiation issues because of the World War II bombings, and the push and pull of the Fukushima disaster adds a certain amount of tension.

"Even now, I think," says Miyakazi, "in Japan, after Fukushima, the government is not really thinking about the next nuclear power accident. The government is thinking and not thinking. The system they put in place has already exceeded its own capacity. Compensation payouts have exceeded the capacity of the system for more injection of funds from taxpayers and ratepayers."

Tokyo Electric Power Company (TEPCO), the largest electric utility in Japan, had paid $80 billion in personal and property compensation by 2019. "This is by far the largest amount of damage compensation ever paid to victims of a nuclear disaster anywhere," observed Miyakazi, "and is possibly the highest amount of compensation for any industrial accident disaster, including the disaster at Union Carbide's pesticide plant in Bhopal, India, and BP's Deepwater Horizon oil spill."

The figure rose to $88 billion by January 2021. Two months after the accident, Bank of America–Merrill Lynch had estimated the sum could go as high as $130 billion. A private Tokyo think tank, the Japan Center for Economic Research, said it could go to $626 billion. In 2012, the Japanese government gave TEPCO $9 billion to prevent its collapse and ensure electricity for Tokyo and surrounding municipalities.

By comparison, the Three Mile Island accident cleanup started in August 1979 and officially ended in December 1993, with a total cleanup cost of about $1 billion. In the United States, the Price-Anderson Nuclear Industries Indemnity Act, first passed in 1957, governs nuclear accident compensation. Effective January 1, 2017, nuclear power utilities are required to buy $450 million of liability insurance for each operating site. Claims beyond that are covered through mandatory participation in a financial protection pool providing an additional $13.1 billion for each incident to pay for public liability claims, of which a stricken nuclear utility would pay $2.8 billion. Beyond that, the federal government or Congress could mandate nuclear utilities to increase their liability levels.

Those liability levels are far below what Japan has seen.

"The difficulty here is that the Price-Anderson regime has not been really, properly tested because we have not experienced a major, major accident," said Miyakazi. U.S. regulators have warned of the possibility of a Fukushima type disaster happening in the United States, especially from flooding. Current reactors need lots of cooling water and are located near oceans, lakes, or rivers. Former energy secretary Rick Perry in 2017 warned that a Fukushima-like disaster could strike at some facilities storing spent nuclear fuel in cooling ponds.

"In the Fukushima case," said Miyakazi, "it is tragically interesting because the amount of compensation paid already exceeds other cases of environmental disasters, and yet there are a couple of mutual issues. In one, people are affected but not included among people who should be compensated for loss and other impacts. And two, even those who have received compensation have not been compensated for the types of loss that cannot be calculated in financial costs. Those are two major issues. And now the government is quickly canceling the mandatory evacuation order."

The government began reopening 11 towns in Fukushima's mandatory evacuation zone after extensive decontamination, hauling away tons of radioactive soil and material, hoping to reunite the refugees with their homes, although many fear to do that because of radioactivity.

"Technically, everyone should go back except for a small area around the power plant," said Miyakazi. "Once these evacuation orders are lifted, the former residents are not eligible for any type of compensation." This creates another kind of hardship.

Simply telling people to go back home is not helpful. The town of Namie, directly downwind of the Fukushima power plant, had 17,114 residents before the accident. It was evacuated, but the evacuation order was lifted for the business district and town hall. Access to more heavily contaminated parts of town is restricted. By 2020, almost 10 years later, the town had a population of 1,238.

"The nuclear power plant accident robbed us of our livelihood and lifestyle," said Kazuhiro Yoshida, mayor of Namie. "We lost our land, homes, occupations, jobs, friends, and acquaintances and even our families were dispersed. Many residents have suffered both mentally and physically from their prolonged evacuation." Essentially, the mayor's community fell apart. Some of these losses are not considered in Japan's compensation calculations, nor anyone else's.

Nuclear accidents have complex social effects. "We have to think about victimhood in that context," said Miyakazi. "We cannot just think only about this kind of simplistic causal linkage between an accident and its effects. There are layers of things related to the accident and layers of things that create the accident. It will take a long time for Japan to reconcile those events."

Governments like to set boundaries as narrowly as possible, like the 12-mile evacuation zone around the wrecked power plant, which also marked an area that qualified residents for compensation. But the accident released radioactive debris that was carried by wind and waves to other parts of the planet.

"The impacts of nuclear radioactivity can appear anywhere," Miyakazi pointed out, making it impossible to have a completely fair system of victim compensation. But he believes society should agree that "we cannot preclude the possibility of compensating for losses that occur outside the boundaries of victimhood that were set in an artificial fashion after the accident. That's why it is very important to suggest we speak to a broad range of people affected and think about philosophical and consequential issues like categories of victim harms. . . . The problem is clear. There are people outside Fukushima who were affected too. We need to create a process where we need to think about the murkiness of these quandaries."

In 2018, the Japanese government reported that one Fukushima worker had died from lung cancer as a result of exposure in the event. Radiation exposure was not the only threat to human health. The official death toll is 573 who died in the evacuation and from stress. Miyakazi contends that 1,500 lives were lost through physical and mental stress brought on by the emergency evacuation.

For that reason, Miyakazi and others advocate for using a social framework, rather than a legal framework, for compensation in nuclear power accidents. They argue that because of the long-lasting radioactive contamination in such accidents, people are robbed of their past, present, and future by being dislodged from their social bearings and their culture, temporarily or permanently.

All domestic and international frameworks for compensation originally were set up with the nuclear power industry's interests in mind, said Miyakazi. The underlying concern, he said, was that private enterprises would not enter that business if they were entirely responsible for losses and damage.

The Fukushima accident demonstrated how complicated that can be. Ninety percent of corporate bonds in Japan consisted of TEPCO bonds, in which companies and pension funds were invested, according to Miyakazi.

"That was a major consideration for the government," said Miyakazi. "Basically, they deemed TEPCO too big to fail. That created a condition for setting up a compensation system which would not jeopardize that company's future. They had to save this company for greater interests. They were not thinking of victims of the accident. They were thinking of the future of the Japanese economy and the future of this company."

But Miyakazi also believes the Three Mile Island and Chernobyl accidents brought a shift from the idea of protecting industry to protecting victims. That is driving an effort to focus on impediments to victim compensation. These include cross-border radioactive contamination and a legal system that largely dismisses the likelihood that exposure to low levels of radiation can cause health effects.

"That's why we also need to bring together experiences of atomic bomb survivors, U.S. soldiers, and other people exposed to radiation at testing sites, and hear their experiences," said Miyakazi. "They share the same kind of problems of proving their harm. This issue really points to the limitations of the legal framework. The report drafted by our Japan-based team suggests that we need to go beyond the legal framework, we need a social framework for addressing their issues."

Miyakazi and his followers are working to phase out nuclear power, but they also recognize that nuclear power plants are expected to play a role in combating climate change. They argue that nuclear energy is not a sustainable source of energy.

For the future, Miyakazi and those working with him believe in the use of international forums that bring together citizens, specialists, experts, political leaders, victims of past disaster, potential victims, taxpayers, and ratepayers to discuss nuclear disaster compensation before the next disaster occurs. It will allow ongoing learning between victims and potential victims.

"Given the gravity of an accident, we are suggesting we need to have a level of agreement in society so that victims know what to expect and what the government ought to do and what industry ought to do when an accident occurs. We should have a more open conversation about where this whole issue stands, and what roles each stakeholder ought to play. The conversation often is abstract, probably because the design for compensation systems do not speak to victims."

Miyakazi believes it should be a deliberate process, not one that seeks rapid "closure."

"We like to put a time limit on things," says the Northwestern professor. "We like to put closure on something like an accident, particularly. My argument is that we should not be in a rush to close. A rush to put closure to the accident, the idea of lessons learned and the accident is over. Actually, the accident is still going on for various factions. And given the long-term effects of radioactive exposure, we need to keep things open and the conversation open. We ought to have a process for keeping things open given the ongoingness of disasters in the past."

That idea of "ongoingness" links the two events on nuclear power at the University of Chicago and at Northwestern University. The University of Chicago offered an update since the dawn of the nuclear age, and the passage of 75 years. An economist said market forces should decide if the commercial nuclear energy industry survives, "if nuclear can make it." He was talking about the United States.

At Northwestern, the Fukushima disaster inspired the question of the century—compensation for victims of nuclear power plant accidents, with the assumption that more disasters are going to happen.

The University of Chicago economist called nuclear energy "a wild card." One of the factors that may, or may not, change the assumptions reached at the two university conferences is nuclear reactors of the future, now in the research stages. They are supposed to be safer, more economical, more reliable than present-day nuclear reactors.

They also are wild cards that may change assumptions of the past, opening a new era of nuclear energy, or bringing it to an end as an interesting technological experiment, like the Stanley Steamer operating from 1902 to 1924.

Chapter 47: A Conspicuous Failure

IN ONE PARTICULAR ASPECT OF WRITING THIS BOOK, I WAS AN UTTER FAILURE.

That was in my attempts to get information directly from ZionSolutions, the company that supervised decommissioning of the Zion nuclear power station. My attempts began on January 25, 2016, after I explored the ZionSolutions website, which contained nine fact sheets about decommissioning and minutes of the Zion Station Community Advisory Panel meetings. Those were helpful, but needed some updating and further explanation.

"Contact us," said the ZionSolutions website. "We'd love to hear from you." That appeared welcoming enough. Press kits about the Zion decommissioning project could be obtained by contacting Mark Walker, it said, and gave a phone number.

Walker is vice president of marketing and media relations for Salt Lake City–based EnergySolutions, the international nuclear services company and parent company of ZionSolutions.

To this day (I'm writing this in 2023), I've never gotten a word of information from Walker. Do I blame myself for being less adroit than I should have been? Do I blame Walker? Or both of us for being headstrong?

My ill-fated attempt began on that January Monday morning with a call to Walker's phone number. I got a recording and hung up. After waiting a few minutes, I dialed again and Walker picked up. I identified myself, asked for a press kit about the Zion project, and told him that I had a few questions. "Is this a good time?" I asked. I mentioned that I was interested in "what comes next" for the people of the City of Zion.

Walker commented, "The people of Zion should have been thinking about this long ago," and thank goodness they've got the cancer center for tax relief.

Walker told me to call back at 4 p.m. Chicago time, which I did and got a recorded message. I left a message, saying I called at 4 p.m. as instructed. I tried his cell phone and got a recorded message.

The next day, January 26, I tried again, but got recorded messages. On Wednesday, January 27, I called Larry Booth, who was identified on the ZionSolutions website as the community outreach manager, and explained that I was trying to get a press kit and answers to some questions. He told me to call Mark Walker. I explained that I tried that, with no luck. Booth told me Walker was a very busy man, but he would call Walker and tell him that I was trying to reach him. Booth told me he would call me by the end of Thursday if I didn't hear from Walker. On Thursday, January 28, Booth called me and said he'd reached Walker and told him that I was trying to talk with him. Booth said he would call me next Tuesday if I didn't hear from Walker. During the conversation, Booth said he was the former Zion police chief.

On February 2, Walker agreed to send a press kit about the Zion project to me, and asked me to send a list of questions. During a brief discussion, he expressed concerns about being "burned" by media people. That day, I emailed a list of 23 questions to him, including a separate message saying, "It seems everyone involved in the coverage, whether you are in the media or in the nuclear industry, ends up feeling bruised or burned one way or another. But I agree with you that it is justifiable to stop working with someone who has treated you unfairly or unprofessionally."

On March 25, 52 days after that exchange, I emailed Walker, saying I was still waiting for that press kit and answers to the questions I submitted at his request. "If you don't intend to respond," I said, "please man-up and say so." It appeared that he was stringing me along. I mentioned that professionals do not act that way.

The same day, Walker emailed: "My apology for my lack of response. Can you please give me a clear understanding of your intentions, and I will have a response for you next week. If you would like to give me a call to discuss, I can be reached on my cell phone."

I called his cell phone immediately, and got a recorded message saying he was not available. I emailed, saying I tried calling him and "can't help but wonder if this is a game you are playing." But I explained that my intentions were to be fair, honest, and accurate. I described the outline of the book I was writing about the

City of Zion, and that I had personal knowledge of the Zion station from my days as a reporter for the *Chicago Tribune*. My purpose, I said, was "in fairness, to get your side of the story."

Walker responded: "There is no reason for your [*sic*] to respond the way you have. I did apologize for not getting back to you. No games being played. I will have answers for you next week. I did apologize for my plane take off 15 minutes early. I will call when I land to discuss."

Later that evening, Walker called, saying he had never been so insulted by a reporter. He objected to my questioning his manhood, referring to my "man-up" comment. Walker said that he had gotten knee surgery, and his mother had gotten knee surgery. He apologized for failing to contact me sooner, but said there was no reason for the insulting tone of my email. I said I believed he did not like reporters. He said he loves reporters, but had never been accused of what I accused him of. I said I took responsibility for my words, and apologized.

Walker said he did not know if I would be fair to him. I said I was more interested in the issues, not in him personally. What I was doing had nothing to do with him. Walker said he had been in Chicago about three weeks earlier and an environmentalist had told him that his eyes were brown, and that meant he was full of shit. He said it was his fault for not answering sooner, but he was insulted by my remarks. He said he was surprised that somebody from the *Chicago Tribune* would act as I did. My email, he said, questioned his professionalism. He said that was a personal attack on him. I apologized again.

It seemed Walker was reacting not only to what I did, but to that insulting environmentalist. And I was long retired from the *Chicago Tribune* by that time, so it was unfair to make that connection, and to link me with the insulting environmentalist. It sounded like I was wrapped up in a long-standing grudge.

During this conversation, Walker said something like, "I could take you on a tour of the Zion site." Without hesitation, I answered, "I accept your invitation of a tour of the Zion site." In a rather sly manner, he said, "I thought you would."

On Monday, April 11, Walker emailed: "I will finally have your answers by Wednesday. The review process has taken forever because of other work. Can you look at your schedule and let me know when a good time for your [*sic*] to tour the facility will work. I also am collecting a group of high-resolution photos that I think will be very helpful of the facility as we have been decommissioning it."

I responded that day, thanking Walker for the update, and telling him my schedule was pretty much open, except for some vacation time in July.

And that's where everything came to a screeching halt. That was the last I heard from Walker, who, according to his biography, was a former ABC affiliate executive producer in Salt Lake City. No answers to my questions materialized, or that promised tour of the Zion plant while it was being decommissioned. And no photos.

I could have simply dropped it there and decided that I made the attempt and struck out. But lacking fresh and insightful information from ZionSolutions seemed like a glaring omission. So I decided to press on. On November 17, 2017, I wrote a letter to David Lockwood, who was chairman and CEO of EnergySolutions at the time, asking if he could use his influence to encourage Mr. Walker to provide the press kit and answers to my questions that he had promised. I explained some of the background, but it made no difference. Lockwood did not respond, either.

Taking another tack, I contacted Viktoria Mitlyng, regional senior public affairs officer for the U.S. Nuclear Regulatory Commission, asking if NRC requires ZionSolutions to provide information to members of the public, since ZionSolutions was dismantling the Zion station under NRC rules and regulations. I explained the stonewalling problem I had with Walker, and wondered if that was a federal violation?

"While the NRC's goal for our interaction with members of the public is to be transparent and responsive, the NRC does not regulate licensees' communications with the public," she answered. I figured it was worth a try.

Still not entirely ready to drop the matter, and thinking of it as an interesting challenge, in June 2020, I asked the Exelon communications department to send some of my questions to Mark Walker, thinking if he would not respond to me, he might respond to Exelon—a $1 billion paying customer.

The Exelon spokeswoman, Elizabeth Archer, said she contacted Walker several times and would persist. On November 30, 2020, Archer emailed me, saying: "I'm so sorry that it is November and we are still having this conversation [about Walker]. I'm not sure what the issue is, and as you know, we aren't able to answer questions on behalf of ZionSolutions. I do expect that we will speak with them this month and we plan to follow up, again." But to no avail.

This quest for some basic information did not go well, and I take some responsibility for that. I labor under the belief that corporate public relations and communications people are paid to provide information and should give it to those who ask for it, even to ordinary citizens like me. The ZionSolutions website said the company welcomes calls and questions. Perhaps naively, I thought ZionSolutions representatives should take that as seriously as I did, no matter who is asking for information, especially in matters as sensitive to the public as nuclear safety. It is their responsibility—not a matter of only doling out information to people they like. That's not professional.

One of the distinguishing characteristics of people in the nuclear industry is an abiding belief that everyone is out to get them. To be fair, it should be said that anti-nuclear activists *are* out to get them. And public opinion about nuclear energy shifted in the United States after the Three Mile Island accident. Their persecution complex is understandable.

Nuclear folks saw the media as biased long before President Donald Trump declared journalists the "enemy of the people." I saw this firsthand at the Zion nuclear station, but I was hoping to leave all that behind when covering the final stages of the Zion plant's history, its destruction.

I don't know if it would have gone better if I waited, silently, patiently, without comment, hoping meekly that Walker would deliver on his promises. There comes a time when a reporter must depend on gut feelings. After the passage of weeks without a response, I am inclined to suspect that a spokesperson is not acting in good faith or not taking the inquiry seriously. And if they act as though they have something to hide, I get suspicious and curious. That's a good time to ask about their intentions to respond, delicately stated or not.

It seemed like I was being hustled.

From my perspective, this is likely colored by my past experiences with people in the nuclear business, who tend to be suspicious and defensive of the motives of journalists asking questions. This "grudge" factor appeared to be at work even at the demolition and decommissioning stage.

Willingness to communicate should be listed as an asset for anyone interested in nuclear power's future, or its past. And leaving the grudges behind.

Chapter 48: Life After Nuclear Power

GOING TO WORK EVERY DAY AT THE ZION NUCLEAR STATION, RECALLS CORRINE Simon, was like going to the Land of Oz, a magical place.

"You walk through those gates. The world changes. You're badged. You go through a metal detector. Nobody understands that. You just walked into Oz Land, over the rainbow. When you step out, you're in the real world. You just left the rainbow."

That's the way Simon remembers it, 44 years after she first set foot into what was known as a male-dominated world of commercial nuclear power. She was chosen, she acknowledges, because she had good-looking legs. (More about that later.) In her 21 years working in the station, she believes she paved the way for more women in nuclear energy by setting a good, professional example while working on risky and radioactive projects.

It's 23 years after the lakefront power plant powered down for the last time, and three years since it was reduced to a pile of tangled, dusty, slightly radioactive rubble.

But like nuclear fusion, the camaraderie that bound many of the power plant's workers keeps them together for reunions or volleyball games. When they heard that I was looking for volunteers to talk about life after nuclear power, nine of them emerged like those actors in the movie *Field of Dreams* coming out of a corn field.

They are not ghostly wraiths, though. They are real, with lively stories to tell about romance, their accomplishments, new careers, retirements, and newfound freedom to travel with spouses, or simply walk their dog. All remember their Zion years with pride, filled with hard work and long hours.

Some of them believe the Zion station should still be spinning out electricity, while others recognize reasons for closing it. None of them recall seeing anything that justified tagging the Zion station with a reputation for sex, drugs, and alcohol during working hours.

Their memories are colored by their friendships and the rigid rules of nuclear safety that come with working in a place that is radioactive. When you talk to nuclear people, you don't get small talk, you get exact details. It's a habit inculcated by years of measurements and writing technical reports.

For a woman, especially in the 1970s, it took a certain amount of determination to break into that man's world. If you thought Corrine Simon had a flaming desire to be part of the shining new world of nuclear power, you'd be wrong. It was a calculated decision based on advancement and job opportunities.

Simon, who hails from Burlington, Wisconsin, was working as a Commonwealth Edison customer service representative in the Crystal Lake office, and "getting bored with it." Because of seniority, "someone would have had to retire or pass away for an opening to present itself. So I was not going anywhere for a long time."

But at the nearby Zion nuclear power plant, she noticed, "there were many management job opportunities and job postings." The drive to Zion from her home in Burlington was not far, "so when a job posting for Zion station would get tacked onto the job posting bulletin board, I would apply."

Told she was not qualified, she was turned down repeatedly. "Frankly, I get it," she says. "There were not many women who qualified because you would not get the experience to qualify." Then Simon was told to report to the Zion station for an interview. Because of federal law, "they had to have women and minorities on the workforce."

She recalls the interview with a Zion supervisor. "Basically, he was honest with me. He said, 'I had to hire a woman, so I figure I had to pick the one with the best legs. In the [clothing] change room, might as well give the guys a good look at good legs.' He was sort of joking. Back then, there was no sexual harassment training."

Upon working inside the Zion station, Simon encountered other aspects of a man's world: tool company calendars bearing photos of semi-nude models hanging on the machine shop walls and elsewhere.

"I was not offended by that," she says. "My family worked in construction. My brothers were bricklayers and my father built homes. I was aware of the male attitude, you're talking late 1970s. . . . I didn't care that they had girlie calendars. When I went into an office, we were not talking about calendars. We were talking about the job.

"There was nothing crude or vulgar, in my opinion. It was like family banter. I was respectful to everyone, and everyone was respectful to me. I behaved like a lady and was treated like a lady. I never felt uncomfortable, except around the boiler-makers. The boilermakers were the scum of the world. Boilermakers were creepy, they always made you feel uncomfortable. But I wouldn't take any guff."

Her first day at work, Simon arrived at a room equipped with pens, pencils, and a stapler. Upon becoming modifications (mods) coordinator during a reactor refueling outage, she was tasked with keeping track of 433 modifications for each reactor.

"So here's these ledgers for Unit One and Unit Two," she recalls. "You had to go through books and ledgers" as each modification was completed and reviewed, with a line for each. "I said this is ridiculous. This is too cumbersome." With the help of a "computer geek," she created a database for modifications that could be updated and printed out, rather than entering data in ledgers. "I didn't see any guy do that, or think about doing that. It worked so well, the other sites, Byron, Braid-wood, Dresden, and LaSalle adopted it."

More harrowing were her challenges upon becoming a quality control inspector.

"I was one of the few women who went into containment and had to suit up" in anti-contamination clothing. She recalls an early assignment, "walking the plank." After refueling the nuclear reactor, a quality control inspector had to walk upon two temporary 10-inch-wide boards to reach the top of a reactor to take measure-ments and make sure the top was bolted securely.

"That was the scariest part of my job," she says. "You walked the plank, it was wide enough. I thought if I fell down there, I'm dead. To heck with radiation expo-sures. The first time I got out there, I'm like, are you kidding me? Okay, here we go."

She walked the plank twice before a safety coordinator decided it was too haz-ardous and replaced the planks with a walkway with handrails.

A quality control inspector examines completed work to be sure it was done properly. When a gasket on a relief valve on the pressurizer had to be replaced, "I would have to verify it was the right gasket by serial number, it was placed correctly and the flanging surface that holds it in place was torqued right to the foot-pounds."

A slender woman, Simon stood 5 feet, 4 inches tall and weighed 112 pounds. That qualified her to do things no bulky man could do. "Sometimes I had to climb down steam pipes to make sure debris was not left behind when they overhauled the turbines. I was small enough to crawl down a pipe." Turbine steam pipes are about as big around as a manhole cover.

Asked to retrieve a dropped hammer, "I had to crawl in face first, head down the pipe, find the hammer. I just fit in. Bring the hammer with me and crawl back out. It was freestyle. I think about that now; you're young and foolish." She decided she'd refuse to do anything like that again.

A woman in a radioactive workplace must make choices, not only for her own safety, but for the safety of the unborn. Friends and family often asked Simon if she feared radiation exposures.

"When you are young, you are still living for the moment and feel nothing bad can happen to you," she answers. And for her, radiation exposures were a career choice. During a refueling outage, she needed approval to take an extra 300 millirem radiation exposure for the day to work in a highly radioactive area, which was allowed under federal regulations. A radiation protection technician politely suggested that Simon refuse the additional radiation exposure.

"But there was a science behind their concern for me," explains Simon. "They wanted me to understand that a female only has so many embryos in their ovaries that will run out as we age. An egg is the same egg forever and gets radiated each time I get exposed. I was 26 years old and very capable of still being able to have children, so this needed to be considered in my dose approval request. I would just inform them that I was a career woman and have no intentions of having children.

"After all, being a woman in a man's world and being given the opportunity to have this job, I felt like this is my job and I should not be treated any differently than anyone else who has to do their job. I shouldn't have any special treatment. I would never give someone the opportunity to say she couldn't handle the job. After all, I felt I was leading the way for more women to have the opportunity to work in these types of job positions. I was being watched for any mistakes all the time."

Simon became a training instructor and a technical staff training leader, a title she held until the Zion plant closed.

"You basically took graduate students who came into the technical staff as new engineers and taught them the ropes of the plant and the outages. When they come out of school, they are textbook smart and can do calculations. Now they have to learn where the valves are and be tested to be sure they know what is involved in a nuclear plant."

Looking back at her time in the Zion plant, Simon says, "The employees were some of the smartest people I have ever met. From mechanics, instrumentation personnel, electrical personnel, rad protection to the engineering staff. I don't want to leave out the construction and training staff as well. It was one big family that looked out for each other's safety and skill sets. If you wanted to learn anything, there was someone there that could teach you. Everyone was held to a high standard and very serious about the jobs that were performed."

It was not all work. She remembers the after-work gatherings at the Pit Stop restaurant in Zion, where the specialty was double-decker pizzas. And the baseball and sand volleyball games at Illinois Beach State Park.

Describing herself as a "tomboy," Simon has four brothers and four sisters. "My mother probably would have liked me to be a girlie girl." She has no regrets about her career decisions. "I'm happy kidless with six godchildren. I'd rather be a god-mother, because you can send them back to Mom." If she had chosen to have children, she would have been a stay-at-home mom, not a working mom.

Although she suffers from arthritis, "I'm fairly healthy for a 68-year-old," says Simon, who sews and dabbles in oil painting. She mows the grass and trims the trees at her home. For a time, she cared for a sister who suffered from an autoimmune illness, which paralyzed her for several months. She recovered enough to use a wheelchair.

"Healthcare is even more complicated than the nuclear industry," quips Simon.

Reactions differ on why the Zion station closed, or if it should have been closed. "My personal opinion about why they shut Zion down is it is too close to Chicago and Milwaukee," says Simon. "We're only an hour from Chicago and an hour from Milwaukee. Other plants are nothing like that."

For those who think otherwise, it's a sore point.

"It bothers the hell out of me to see the Zion station torn down," huffs Ken Graesser, the station superintendent who invited me to see the station from the inside back in 1983. "I know when it was torn down, it was three months of having everything done to operate that plant successfully."

Edison executives said economics doomed the Zion plant, and its need for expensive new steam generators.

"We changed steam generators at Byron," counters Graesser, resisting the idea that the feat was insurmountable. "I was there when we changed steam generators at Byron." Maybe his loyalty to that once-magnificent power plant causes him to come to its defense. But he admits to complications if new steam generators were installed.

"You'd have had to go through a licensing upgrade," he resumes. "They were afraid that close to the Chicago area, it would be difficult to have the license renewed."

The momentary flash of anger passes, and Graesser settles down to describing life in retirement with a home in Antioch, Illinois, and a condo in Indialantic, Florida, which is situated on a barrier reef facing the Atlantic Ocean. He spends six months in each location.

Graesser is active in the Lions Club and the Methodist church in both locations. "I like to stay active in the community, trying to give back to the community that gives me so much. Right now, I'm pretty much locked up in the house, staying out of the way and minimize time wearing the mask." This was in February 2021, when Covid-19 was raging, and Graesser was waiting to get his second Covid vaccination while staying in his Florida condo.

"My doggie and I take a mile walk in the park," he says. "That is his responsibility, taking me for a walk. I have a Bichon Frise, a little white dog. The second one we had. My wife and I had another one 18 and a half years. We finally put him down. We were going to wait before getting another dog, but waited three months. He will be 11 years in July. He's my little buddy." His name is Cody.

In retirement, Graesser and his wife traveled extensively to all parts of the world on cruise lines. "Prior to her passing, we had nearly 500 days on Holiday Cruise Line. Four months before she passed away, we completed a 64-day trip on the Mediterranean with Holland America." Graesser and his wife, Shirley, were married 53 and a half years and had three grown daughters.

Grief-stricken, Graesser joined a church Grief Share program for those who lost spouses, children, and relatives and struggle with grief. "I'm still involved in that, after my wife passed away in 2013. I think that's important."

Fishing is one of Graesser's passions. Besides periodic shore-fishing excursions to local spots like nearby Cape Canaveral, the retiree always looked forward to annual fishing trips to Canada with friends for a week. "I fish four times a year, mainly it's my Canada fishing. Last year, we could not cross the border because of Covid. We're hoping to go this year. We'll see what the rules are."

Speaking of rules, I always wondered if Graesser suffered any consequences for inviting me into the Zion station for a close look at nuclear power, resulting in articles about an accident that contaminated workers. For that reason, I was happy to hear that Graesser was among those who volunteered to speak with me many years later.

"Nope, not a bit," Graesser answers. "Nothing was ever done. Cordell Reed [Edison's chief nuclear officer at the time] said just make sure you are honest and if you can't be honest, tell him that. I never got any heat from you being up there, and you being in there. Reporters, what the hell. Reporters, you never know what their message will be, what they are trying to accomplish or how they interpret what you said. I felt that you were honest and forthright and gave us a fair shake. I read a number of your articles in the *Chicago Tribune*. You were trying to be pretty honest, and you were honest with us also."

I'm glad to hear that. It was strictly a Commonwealth Edison decision to allow me into the nuclear station.

As in the past, he still sees the news media as hostile to nuclear energy. "The media are now so hell-bent on solar power and wind turbines, and nuclear waste became a big issue. It's too damn bad nuclear power didn't come first before the bomb. People think of the bomb, and not of the positive uses of nuclear power. I think the media are not fair. If it wasn't for all the government assistance to solar panels and wind turbines, nuclear power would be further down the road. My brother in Texas has solar panels on his house. He said they are covered with snow. They can't get the necessary sun rays for them to operate properly. I talked to him two days ago. It's a huge mess."

When Bill T'Niemi heard I was going to talk with Ken Graesser, he coached me on how to get his attention: "Ken was a big fisherman. Ask him about his fishing days." A Commonwealth Edison/Exelon employee for 34 years, with a senior reactor operator's license, T'Niemi remembered that Graesser kept a set of fishing gear in the closet of a trailer parked behind the Zion station to fish in Lake Michigan.

"He wasn't supposed to tell you that," quips Graesser. "We had a water intake bay, cooling for condensers. Some fish would get past the screen and get into the intake bay. I'd cast into the intake bay and pick out salmon and lake trout, smoke them, and have fish for dinner. Yeah, I did that."

T'Niemi stands six feet, weighs about 200 pounds, and has a round, friendly face.

While Graesser was big on fishing, the T'Niemi family loved sports. "We are a big volleyball family," he says. When his three daughters were young, T'Niemi's wife of 45 years, Nora, began coaching one of the daughters in Little League. That was not enough.

"We jumped to the Mundelein Thunder," a traveling girls fast pitch softball base-ball organization. T'Niemi was coach and president of the Mundelein Thunder for 16 years until 2010. Nora also coached girls in volleyball, leading to the family's involve-ment in the Sky High Volleyball organization based in Crystal Lake, Illinois. It hosts volleyball camps and clinics for girls and boys from the Northwest suburbs of Illinois.

"We went all over the country in volleyball, and all over the Midwest in soft-ball. We took the team to the nationals in Minnesota."

In Finnish, T'Niemi means "peninsula," shortened by his father and an uncle from a longer, more complicated version of the name. T'Niemi is descended from Finns who settled in the upper peninsula of northern Michigan, generally near Toivola, Michigan, where he still has relatives. Finnish names sometimes have multiple translations, and one of those for Toivola is "Place of Hope."

"In summertime, I always have projects up north," he says. "I'm pretty handy with tools. I help relatives up north. They are older and need help with things around the house."

As a retirement project in 2013, T'Niemi and Nora expanded their Michigan cottage into a four-bedroom house on an inland lake six miles from Lake Supe-rior, in an area where both of them grew up. He was finishing interior work on the house when approached to participate in the bittersweet task of demolishing the Zion station because of his intimate knowledge of how the plant worked.

At Zion, T'Niemi headed the project management office and managed infor-mation technology projects.

"We were telling them how to decommission while maintaining certain systems we had to have, like fire protection and ventilation, and air, water, and cooling. Dem-olition guys would not know one pipe from another. We came into an area and say, take this, this, and this. Don't touch this. It has 100 pounds of pressure in the line.

"It was strange. I'd walk into an area and know this used to contain this pump. It was a pump room, and there was nothing there. It was weird, strange, and sad. By then, we were resigned to it. It was what it was. A handful of guys, a small group, five of us, helped to coordinate demolition. We'd start talking and a lot of memories came back. Old stories came out. In a way it was good; in a lot of ways it was sad."

Life after nuclear power is good. Ask George Pliml and Barbara Newsom.

"If you want a love story, we have a good one," says Newsom. "We're mainly a couple for 30 years."

Pliml was the Zion nuclear station superintendent after Graesser, then moved to Commonwealth Edison's Chicago corporate office in 1989 to become nuclear quality manager of the six Illinois nuclear power facilities.

"He wanted to redo the safety training," says Newsom. "That's when I met George. We worked together." She was an independent contractor specializing in nuclear safety and security in nuclear power plants, starting her own business after working for the U.S. Department of Defense. "I was a curriculum director and instructor for nuclear power plants."

They started dating in August 1991. "I was divorced, George was divorced," says Newsom. "A training director at Commonwealth Edison kind of put us together. We followed all the rules about not dating at work. We broke no rules."

As an example of how tightly knit the nuclear community is, Pliml would go to another Edison nuclear plant and hear "there goes Barb's boyfriend."

"About 24 hours after George and I had our first date, we were asked if something is going on," says Newsom. "You'd think they would not know what we were doing. Everyone was very happy about it."

After retirement, Pliml attended classes for professionals at the University of Chicago from 1998 to 2002, and earned a degree in classic literature. He had a degree in mechanical engineering from the University of Illinois and an MBA from the University of Chicago, but "we spent all our time in technical issues, not on classics. I wanted to get that."

Newsom points out that Pliml spoke no English until he was five years old. He was born in Cicero and raised in a Bohemian neighborhood speaking the Czechoslovakian language at home and with neighbors until he started school. It was a common phenomenon in ethnic Chicago neighborhoods and suburbs, including mine in Chicago's Humboldt Park area. My family spoke Polish.

"I got a motorcycle when I retired," says Pliml. "I had a Harley, of course. Had a ponytail. My buddy and I rode to the West Coast, Florida, and Alaska. Did a lot of traveling." He lives in Oak Brook, Illinois.

He went to Czechoslovakia about 15 years earlier, walking around a town where his mother once lived, looking for her home. A woman approached and asked what he was doing there. He explained, identified his mother, and the woman answered: "Agnes! I knew your mother. I went to school with her." His father's town was destroyed in the war.

"I'm involved in working with homeless people," says Pliml. "We have a group that is called Shelter for All." Newsom volunteers for work in nuclear disarmament. Both have their causes. "I don't think we can progress with nuclear power until we progress with nuclear issues. It's too precarious. We're in a completely different era."

For Peter LeBlond, it was fear, and love for his special needs sons, that drove him to the next chapter in his life after the Zion nuclear plant.

"I was pretty scared," admits LeBlond, when he quit Unicom in 1999, a year before the utility became Exelon. "We have two special needs kids. There's a school for special needs kids in Northbrook [close to his home]." The company changed, and LeBlond reasoned it was only a matter of time before he would be tapped on the shoulder and told to relocate, which is common in management ranks.

"I ended up quitting," he says. "I just didn't fit. I was not the person they wanted or needed. And if I was that person, I'd have to move to Quad Cities, or Harrisburg, or New Jersey [where Exelon has nuclear stations]. I wanted to be in the Chicago area and get what the kids needed. That prompted a crisis." Living apart from his family would have been "a quiet life of frustration."

Describing those moments of intense soul-searching, LeBlond characteristically breaks the tension with a moment of humor: "I had a vision!" he says grandly, and laughs. "I could say I had a vision. I was more scared of staying with the company than leaving. Trust to God and God provided. Here I am."

With glasses and a patch of white whiskers on his chin, LeBlond resembles a smiling, mischievous wizard. He talks fast, poses questions, and answers them, sometimes repeating the last words of a sentence as though pondering them. I first met LeBlond in 1983 when he was training manager at the Zion nuclear plant. He accused me of asking a frivolous question when I visited the station's reactor training simulator and asked to see what the control board looked like at the height of the Three Mile Island accident. Neither of us mention that incident while talking about life after nuclear power, but it reminds me that he could be cutting and blunt.

LeBlond was the Zion station operations manager when Exelon decided to stop it 15 years earlier than scheduled. "Nobody asked guys like me if they should shut the plant down," he comments. "I watched it happen. First off, we were never an excellent performer. It's a hard thing to grasp."

Power plants are judged by capacity factor, a measure of how often a plant is running at maximum power, expressed in percentages. The Zion station's capacity factor was around 75 percent, low compared with other nuclear stations. "To compete," says LeBlond, "we had to make changes."

Changes came, but too late for the Zion station. Refueling outages went from 70 days to 15 or 18 days. The Nuclear Regulatory Commission allowed maintenance on pumps and diesel engines while plants were operating, rather than wait until an outage shutdown. This resulted in longer run-times for nuclear plants, gradually rising industry wide from the 60 percent levels in the early 1980s to the 90 percent mark by the 2000s. In 2020, nuclear energy's average capacity factor was 92.5 percent, according to the U.S. Energy Information Administration. It was 40.2 percent for coal generating stations, 35.4 percent for wind and 24.9 percent for solar. Because the wind does not blow and the sun does not shine all the time, solar and wind power sources are intermittent.

It's worth mentioning that nuclear technology has moved on since the Zion station closed, just as those who worked at the station have moved on, providing a different picture of how the nuclear industry operates currently compared with other energy sources.

After leaving the Zion station around 1995, LeBlond went downtown to corporate headquarters and headed the nuclear rules and regulations department, a move that paved the way to his next chapter. He began his own company, LeBlond & Associates LLC, which he operates out of the Libertyville home he shares with his wife, Jean.

"When I left Exelon, I was not planning on doing this," says LeBlond, who specializes in explaining the Nuclear Regulatory Commission rules and regulations that govern the operation of nuclear power plants, known as 10CFR50. Around 2000, those rules were revised in major ways. The Nuclear Energy Institute (NEI) hires LeBlond to conduct classes at conferences exploring the regulations and how they are applied.

"My mother tells me I shouldn't talk this way, but I've become the Jedi Master of these things," explains LeBlond. Basically, it boils down to "how to ask the Nuclear Regulatory Commission for permission" to comply with or deviate from the regulations.

"What I learned at Zion gave me the background to say I know how this works," says LeBlond. "The thing that is unusual was the NEI group. There was a niche and I fit in it. There was confusion about this regulation, and they wanted an answer. There is an answer, and I can give it. Nobody could tell you what to do, so

I had parts of the puzzle. That's it. I've worked on my own for as long as I worked for Exelon, 22 years." LeBlond's website lists 16 available training courses.

His sons, aged 29 and 27, are living in Illinois and in Colorado. "In the world of special needs, both are living on their own," says LeBlond. "In special needs, that counts as a win."

"You're talking to members of the volleyball team, although you don't know it," LeBlond informs me. Word that I wanted to interview former Zion workers went out by email to the Zion station volleyball team.

"We played volleyball 38 years running, Wednesday nights, which was an excuse to drink beer and eat pizza. Last time we went to a place in Antioch. There's a pizza place bar, guys, and their brides. We're all a big happy family. If Covid breaks, we'll do it again and call it a team meeting. T'Niemi was the captain."

The fate of nuclear power is always on their minds.

"You're really worried about climate change?" asks LeBlond. "Cool. Really? Why are you shutting down nuclear plants? It's silly. Can't get it out of my mind." He sees nuclear plants as a necessary "bridge" to whatever comes next in producing electric power. "Why not say we should not shut down another nuclear power plant until we get fusion? That should be the nation's power goal. Why don't they do that? They don't. They don't."

Another sore spot is their perception that the American news media are anti-nuclear. LeBlond offers his own opinion about that by describing how he believes journalists work: doing a story on the shape of the Earth, a reporter has a lot of data to consider, including "some people saying the Earth is flat and carried on the back of a turtle." Aiming to be neutral or nonjudgmental, he says, a reporter might include those viewpoints in his report.

"If you give those guys equal credence, that the world is flat and carried on the back of a turtle, then bit by bit, industry dies." He's clearly talking about the credence journalists give to anti-nuclear critics who attack the nuclear industry.

A small point of rebuttal: As a reporter, I think of all the times in human history when those who challenged conventional wisdom proved to be right. They are "oddballs," like Nicolaus Copernicus, the Renaissance astronomer who was attacked for saying the Earth circles the Sun. It's hard to recognize the wisest people in the world, until some passage of time. Reporters are not in the business of ignoring points of view, even odd ones. We need to keep an open mind.

"Are media biased?" LeBlond asks, before delivering the punch line. "I don't know. No, they are not biased. They are just ignorant. If nuclear power is bad, how come I'm not dead?"

His mood brightens, as it sometimes does when he's getting too serious. "My wife bought me a Lladro [porcelain figurine] of a lamplighter, because I'm a lamplighter. Where do you get power? A nuclear power plant. I thought I was doing a service. My kids thought it was cool dad worked in a nuclear power plant."

The admiration of family is not exactly what John Brandes is enjoying in retirement, which can be a sad time for some. "I was divorced in the last year and a half," he says. "I'm trying to get my life together."

It's a life that sort of wandered into nuclear energy in a serendipitous way when he joined Commonwealth Edison in 1974. Zion station workers often bragged about "growing their own," training people from the bottom up, and Brandes is a prime example of that.

"I got out of the service and got a job reading meters," he recalls. "I was trying to decide what I wanted to do with my life. I didn't want to climb poles and work on high voltage, 34,000 volts some of them. I didn't care for heights and I didn't care for high voltage. Sometimes in life we get directed to what we want to do by things we don't want to do. I heard about nuclear plants, the jobs there were very technical in nature, working with chemistry, radiation protection, instruments, electrical, and mechanical maintenance. All looked like good career paths.

"I was going to be in radiation protection. Two of us bid on the job, the senior man took it. I got a job as a station laborer, a stationman, sweeping floors, doing radioactive laundry, general house maintenance in a nuclear plant around the equipment."

Brandes took courses after work on the basics of nuclear power and boasts that he "started there sweeping floors and worked my way up to reactor operator. From 1978 to 1982, I was a reactor operator. . . . I loved my job as a reactor operator."

The trouble, says Brandes, is "I didn't develop many hobbies during the time at the power plant. I ran a dairy farm in the 1980s. I spent a lot of my time off [after work] running the farm. My daughter loved it. We had about 90 head of calves, heifers, and cows. We had Holsteins," dairy cows identified by their black and white colors. He still owns the farm and rents it out. "The big companies do everything efficient. For small people to farm, the economies of scale make your return negative."

Brandes lives in Wheatland, in the far southeast corner of Wisconsin, near the Illinois border. Before retiring from the utility in 1999, like T'Niemi, he took part in dismantling the Zion station. "When we took it down, that plant was like Fort Knox. Very well constructed. When you tear it apart, you realize all the civil engineering and structural engineering that went into that place. It was very hard for me to be involved in tearing it down. It could still produce power."

The upside, he says, is that decommissioning "turned [the Zion site] back to green space. It shows the rest of the world it would not be a dump site for eternity."

"I was blessed to work with a bunch of great people," says Brandes. "A lot of people made a good living working in that power plant. We got to be very close. Every once in a while, someone passes away and you go to the funeral, and you reflect on that and know they were a blessing in your life. Some maybe less."

According to Brandes, "the original plan was to have six [reactor] units at Zion."

"I don't remember that," says Charles E. Schumann, who has an equally long history at the Zion station. As an operating engineer, Schumann helped start both Zion reactors in 1973 and 1974, one of them on Christmas Day. But he can vouch for construction fatalities at Zion.

"There were two I remember," said Schumann, who left the Zion station as assistant superintendent of operations. "One was in containment, don't know which unit. The guy fell through grating. They had a temporary sheet of plywood, the plywood shifted, and the guy fell 100 feet.

"The other guy was burned to death, working on a stainless steel fuel transfer channel. There was kraft paper on the stainless steel, and they were using solvent to take it off." A dense mist of solvent fumes accumulated inside the channel. It might have been somebody smoking a cigarette, Schumann's not sure, but something ignited the fumes, turning the place into a scorching inferno and killing the worker.

When I mention that Schumann qualifies as a pioneer at the Zion station, considered the biggest and best of its time, he is rigidly humble, and adds, "with other people. I didn't start it myself."

The Zion station demolition "hurt me a lot," admits Schumann. "A year ago, we drove by the plant to see what was left. There was only the lakeshore. I don't want to get into the politics of it," he says cautiously, only later adding, "I don't think it should have been shut down."

A resident of Bristol, Wisconsin, Schumann left the navy as a boiler technician in 1959 and applied for a job at Commonwealth Edison to work in a boiler room.

He became a senior reactor operator, eventually specializing in starting reactors. Before retiring in 1992, he became a certified financial planner.

Jim LaFontaine has the clear, resonant voice of a broadcast announcer. Ask him a question, and he gives details, not casual answers, but details. Times, dates, places, particulars.

It's not just his time at Commonwealth Edison, which he joined in 1966 and left in 2000, that accounts for that. After taking a buyout, he joined the Kenny Construction Co. as project manager for its power group from 2001 to 2013. He owned his own business, LaFontaine Construction Consulting LLC, for almost five years beginning in 2016.

Now retired, LaFontaine and his wife, Karen, play golf and bike. They have a home in Spring Grove, Illinois, and another in Naples, Florida. Biking is better in Florida, he says. "We can ride six miles up and six miles back. Here in Illinois, we have to load the bikes on the car."

Like Graesser, he's a member of the Lions Club in Antioch, Illinois, and Naples, Florida, and is involved in eye screening for the underprivileged and eyeglass donations.

For LaFontaine, life after nuclear power is "my wife of 50 years, my two daughters, and my six grandkids. And our ability to share our lives together. Our family is Lutheran, and we have churches here and in Naples."

In his career at the Zion station, LaFontaine carried out a variety of high-level technical positions, including senior reactor operator for fuel handling, outage coordinator, shift and staff supervisor, and maintenance staff supervisor.

Graesser remembers LaFontaine as "probably one of the best project managers in my life. He'd take a project and lay it out, move it forward. Very positive results. He was the project manager for replacing the turbines at Zion."

When I talked with him, LaFontaine wasted no time getting to his central points. "I'm a pronuclear guy," he says. "If you ask me, I'd say Zion should still be there. We had spent the last year cleaning, painting, fixing, and repairing. It was probably in the best shape operationally than it had been in a while, based on leaks and cleanliness."

But that is behind him now. LaFontaine spends parts of his leisure hours reading murder mysteries, action fiction, and historical fiction. "We watch the news until we get sick of it and turn it off. We watch the Chicago TV, public television, and movies on Netflix." Among his observations: "Most people's knowledge of electricity is, they turn the switch on, and the lights go on."

Like most of the others, LaFontaine knows exactly how much radiation exposure he got in the station.

"Radiation exposure is there," he says. "It's controlled and monitored. Over my lifetime in nuclear power, I've been exposed to 19.1 REMs. It is something that goes with the career, it goes with the job. Is there any risk with radiation exposures? Yes, there is." He points out that radiation exposures also come from traveling by air, and from medical and dental X-rays. "You probably get more from them than in a nuclear plant. The risk is minimalized."

Radiation exposures do not concern him, but he admits to a health issue. "Do you know why I'm still in Illinois?" he asks. "We opted not to go to Florida. We want to help our kids. We didn't want to take an airplane. They told me I have aortic regurgitation," in which his heart's aortic valve doesn't close tightly, allowing some blood to leak backward into the main pumping chamber. "The reason I'm here is my health. I'm very comfortable with my health and not concerned about the radiation exposure I took in the power plant. A friend died of lung cancer and she never smoked in her life. When it comes down to it, God knows when people pass and what they pass from, and you go forward with the best decisions you can make in life."

Judging by this small sample of former Zion workers, they were dedicated to work, faith, and family. And they liked each other, even now, and they're looking forward to a strenuous game of volleyball together again, followed by rehydration with beer and pizza at one of their favorite pizza joints.

Their retirements are marked by the enthusiasm they showed for their work. They are a lively bunch. They still have stories to tell and to look forward to.

The Zion nuclear power plant, though, is nothing but a memory now, says Corrine Simon. "Everything now is like a ghost town with only the ghost stories left to tell of those who are still alive," she says. "Rest in peace for all those souls who have passed away and taken their stories with them."

Chapter 49: Energy in a Time of Plague

2020 BROUGHT A GLOBAL CORONAVIRUS PANDEMIC AND JOE BIDEN TO THE United States presidency—both reminders that the world as we know it can change in a flash.

As I press the keys on this keyboard in 2023, the World Health Organization estimates 6.6 million people have died globally in the Covid pandemic. It brought more than a million deaths to the United States, more than the 1918–19 Spanish Influenza epidemic—at that time considered the worst case of contagious calamity to strike the nation.

Both developments are playing out in ways unknown because it's early, but conflict is baked in as a main ingredient.

Declaring climate change the "number one issue facing humanity," the new president launched a Clean Energy Revolution aimed at achieving a 100 percent clean energy economy with net-zero emissions no later than 2050 and using nuclear power reactors as a centerpiece to quell greenhouse gas contamination causing global warming.

After months of haggling, the Biden administration advanced the $3.5 trillion Build Back Better plan, later reduced to $2.2 trillion, to invest in climate, energy, and public health—the most significant investments of their kind in the nation's history. It was blocked in late 2021 by Senator Joe Manchin, a West Virginia democrat whose family owns a coal company, saying "I can't get there" because the program was too expensive.

Another setback to the Biden climate change agenda came when the U.S. Supreme Court sharply cut back the Environmental Protection Agency's authority to reduce carbon emissions from existing power plants.

But with astonishing speed, the administration roared back in just days with the $738 billion Inflation Reduction Act of 2022, authorizing $391 billion for energy and climate change and $238 billion for deficit reduction, prescription drug reform, and tax reform, while raising taxes mainly on the rich. With all Republicans opposed, President Biden signed the act into law in mid-2022, a stunning reversal in what appeared to be a lost cause a short time earlier.

The U.S. Environmental Protection Agency called the act "the most significant climate legislation in U.S. history." It was the nation's biggest boost to the climate change effort since the U.S. Senate ratified the world's first climate treaty 30 years ago. The sweeping bill allotted nearly $30 billion in tax credits over 10 years for nuclear power plants, $700 million to develop fuel for advanced nuclear reactors, and $30 billion for a production tax credit for wind turbines and solar panels.

A fraction of the size of the Build Back Better proposal, the Inflation Reduction Act is full of items unappealing to environmentalists, such as guarantees for new oil and gas leases, tax credits for carbon capture, and an absence of any requirements to reduce emissions that cause climate change.

In his first two days in office, President Biden signed 10 executive orders to combat the Covid-19 pandemic, including mandating wearing masks on public transportation. In a nation that prides itself on self-determination, independence, and scientific sophistication, both a mask mandate and an ambitious climate agenda were rebuffed to some degree. Coronavirus deniers refused vaccinations against the pandemic, some for political reasons. And the war between those for and against nuclear energy intensified.

Covid-19 was a relative newcomer in the nation's struggles, while nuclear reactors had been producing electricity in the United States for 60 years when both seized the Biden administration's attention. The coronavirus pandemic blazing across the world came at a time of heightened global awareness of another existential threat, the atmospheric damage from 200 years of the industrial revolution's scorching smoke and gases.

The American nuclear power industry recognized this as an opportune moment. It promotes itself as the source of power that does not produce the kind of damaging atmospheric greenhouse carbon emissions that cause climate change, like burning coal, oil, or natural gas.

Acting on that belief, the Biden administration's U.S. Department of Energy in

June 2021 announced more than $61 million in funding awards for 99 advanced nuclear energy technology projects in 30 states and a U.S. territory. Of that amount, $58 million goes to U.S. universities, including $400,000 for Northwestern University and $800,000 for the University of Illinois in Champaign-Urbana, both in a state with a long history of nuclear power achievements.

"Nuclear power is critical to America's clean energy future and we are committed to making it a more accessible, affordable, and resilient energy solution for communities across the country," said energy secretary Jennifer Granholm. "At DOE we're not only investing in the country's current nuclear fleet, but we're also investing in the scientists and engineers who are developing and deploying the next generation of advanced nuclear technologies that will slash the amount of carbon pollution, create good-paying energy jobs, and realize our carbon-free goals."

The Biden administration clearly saw a need to boost the nuclear workforce after that 30-year reactor construction paralysis in the United States. It was a sign of welcome relief for a nuclear industry that saw six commercial nuclear plants shut down in the U.S. since 2013, with 12 more scheduled to retire within seven years.

"We are thrilled to see the inclusion of America's largest carbon-free energy technology, nuclear energy, in Biden's infrastructure plan to reenergize and decarbonize our economy," said the American Nuclear Society in a statement. "We applaud the plan for recognizing that both existing and advanced nuclear energy need to be supported and expanded if our power grid is to successfully decarbonize by 2035."

Biden came into office vowing to undo much of what his predecessor, Donald J. Trump, did, including rejoining the 2015 Paris Climate Accords. Biden did not share Trump's interest in reviving the American coal industry, but both presidents wanted to help the nuclear industry. It was one of the rare instances of energy policy continuity between incoming and outgoing administrations.

In 2018, President Trump ordered then secretary of energy Rick Perry to "prepare immediate steps" to stop the closings of unprofitable coal-burning and nuclear power stations, which would have been an extraordinary intrusion into America's energy markets.

Under the Trump plan, the Energy Department would order power grid operators to buy electricity from the struggling power-producing stations for two years. That never happened. A broad alliance of energy companies, consumer groups, and environmentalists opposed it, saying it was legally indefensible and likely to force consumers to pay more for electricity.

If Biden believed the climate change issue would soften attacks upon nuclear power, he was mistaken. The Sierra Club, the nation's biggest and oldest environment group, remains adamant.

"Tragically, it took a horrific disaster in Japan to remind the world that none of the fundamental problems with nuclear power have ever been addressed," said the Sierra Club. "Besides reactor safety, both nuclear proliferation and the required long-term storage of nuclear waste [which remains lethal for more than 100,000 years] make nuclear power a uniquely dangerous energy technology for humanity. Nuclear is no solution to climate change and every dollar spent on nuclear is one less dollar spent on truly safe, affordable, and renewable energy resources."

Nuclear critics have crafted their arguments to account for climate change, saying nuclear energy is a feeble force against an immediate threat. Though Peter LeBlond might believe this lends credence to unwelcome naysayers, democracy depends on a free marketplace of informed thought and ideas. Everyone has a right to be heard, keeping in mind that outcomes seldom are predictable.

Amory Lovins, an environmental engineering scholar long considered an American energy guru, wrote in *CounterPunch* magazine that nuclear power is not the silver bullet for climate because of cost and time. "Stopping climate change requires scaling the fastest and cheapest solutions," he wrote, adding that renewable sources of energy like solar and wind displace three to 13 times more fossil-fuel generation per dollar than nuclear, and far sooner. By using nuclear energy against climate change, "we shoot ourselves in the foot."

Such opposition from longtime opponents to nuclear power can be expected. Less expected is surprising opposition from the likes of Gregory Jaczko, chairman of the U.S. Nuclear Regulatory Commission from 2005 to 2012, once at the very epicenter of the government's dominance over nuclear energy.

"Funding nuclear energy is a waste of government resources, the industry is dying," Jaczko said on September 17, 2021, during an appearance on *Rising*, a daily news and opinion web series produced by the Washington, DC, political newspaper *The Hill*. Among all the advanced reactor designs, he said, the small modular, portable, reactors are closest to gaining federal approval. If demonstrated to be workable by 2030, SMRs could go into commercial operation in another 15 to 20 years, according to Jaczko's timeline. That adds up to something like 25 to 30 years into the future.

Meanwhile, renewable technologies are racing ahead by leaps and bounds.

Because of a tremendous revolution in the cost of renewable energy technologies, Jaczko said, "the alternatives are there, they are cheaper, and we can deploy them."

These new energy technologies already are making an impact. The U.S. Energy Information Administration reported on July 28, 2021, that in 2020, renewable energy sources generated a record 824 billion kilowatt-hours of electricity, or about 21 percent of all electricity generated in the United States, surpassing nuclear and coal for the first time on record because of the steady increased use of wind and solar technology. Nuclear clocked in at 20 percent and coal at 19 percent. Only natural gas produced more electricity in 2020 than renewables. Nuclear power dropped 2 percent from 2019 to 2020 because several nuclear power plants retired and others had slightly more maintenance-related outages. Renewable generation was projected to grow 10 percent in 2022.

2020 saw the biggest annual drop in U.S. energy production on record, most of it driven by economic responses to the Covid-19 pandemic. Those energy production figures will shift again once the world is beyond the pandemic's spell.

Nuclear power plants have generated about 20 percent of U.S. electricity since 1990, although industry figures show they provided 52 percent of America's carbon-free electricity in 2020.

Solar and wind technologies are effective and cheaper because of successful innovation, said Jaczko. "Nuclear has had 60 years to innovate and they really haven't done a good job at that," and, he contended, "it is not something you can use for climate change." He acknowledged, though, that "very powerful interests in Washington push for nuclear subsidies." After leaving the NRC, Jaczko called for a global ban on nuclear power.

Jaczko's successor was Allison Macfarlane, who chaired NRC from 2012 to 2014. It seems unlikely that two consecutive NRC leaders would cast doubt on nuclear power's usefulness for fighting climate change, but that's what happened.

Writing in *Foreign Affairs* on July 8, 2021, Macfarlane said: "Innovations in reactor designs and nuclear fuels are still worthy of significant research and government support. Despite its limitations, nuclear power still has some potential to reduce carbon emissions—and that is a good thing. But rather than placing unfounded faith in the ability of nuclear power to save the planet, we need to focus on the real threat: the changing climate. And we need strong government support of noncarbon-emitting energy technologies that are ready to be deployed today, not 10 or 20 years from now, because we have run out of time. We cannot wait a minute longer."

Macfarlane agreed with Jaczko that small modular reactor developers are clos-est to getting NRC approval, but she disagreed with nuclear advocates who believe SMRs could replace retiring coal-burning power generating stations, saying those chances are "very low in the near future, like zero."

The Union of Concerned Scientists was founded in 1969 by faculty and stu-dents at the Massachusetts Institute of Technology. A science advocacy organiza-tion, it has a long history of opposition to atomic power.

In March 2021, the organization released a report saying an analysis of the designs of a number of the advanced, non-light water nuclear reactors currently in development were found to be "no better—and in some respects significantly worse—than the light water reactors in operation today."

Those advanced models were rated for safety and security, nuclear proliferation, and terrorism risks and sustainability, meaning efficient use of uranium and how much radioactive waste they produce.

"Despite the hype surrounding them, none of the non-light water reactors on the drawing board that we reviewed meet all of those requirements," said the report's author, Dr. Edwin Lyman, director of UCS's nuclear power safety. He con-tends those advanced sodium-cooled, molten salt, high-temperature, gas-cooled models "are largely based on unproven concepts from more than 50 years ago."

The Nuclear Energy Information Service based in Chicago describes itself as "Illinois' Nuclear Power Watchdog since 1981." Its director, David Kraft, said the nuclear industry "is a dead end," but "as long as the feds have an unlimited money spigot, they can keep careers going. The market said no to that."

"Not only is the nuclear age over," said Kraft, "it's into the age of nuclear decom-missioning." Pointing to the Zion power station, Kraft said decommissioning brought a new set of problems, including poor auditing of the hundreds of millions of dollars in the decommissioning trust fund raised from consumer electric bills.

"The NRC does not require auditing of the accounts," contended Kraft. "All they require is a utility have enough funds." Kraft said NEIS tracked the Zion station demolition project and went to the Illinois Attorney General's office with concerns about questionable costs. "They literally dumped it back on us. They forced a ridicu-lous audit out of [ZionSolutions]. It was done badly and never done again."

Viktoria Mitlyng, an NRC spokeswoman, appeared to support what Kraft said about auditing: "The NRC reviews decommissioning funding reports to make sure the funds meet the NRC's requirements. It is done biannually while plants are operating and annually while decommissioning. However, the NRC does not conduct line-item reviews of decommissioning fund expenditures. By law, those funds must be spent for radiological cleanup and the NRC's inspectors review the plant's submittals with that in mind."

Kraft called the Zion station demolition budget "an $830 million account" that resulted in "a four-page audit, including the title page. A total farce after a year of meeting with the Illinois A.G.'s office."

Transparency was a problem while tracking the Zion station decommissioning. "We were on this for 12 years, going to the [public] meetings. Along the way, we had a lot of run-ins in terms of the transparency. We can't say it was done well or badly. They bled that [decommissioning] fund into the ground. Some of those costs would have been questionable."

Decommissioning trust funds totaling some $68 billion nationally are a huge temptation to companies with little to no experience in decommissioning, reported *Fortune* magazine on April 27, 2021, pointing to two examples in New York state and Vermont. The heap of funds amount to corporate catnip.

Following an example set by Exelon and ZionSolutions, Entergy Corporation, based in New Orleans, sold its Indian Point three-reactor power station to a Holtec International subsidiary in 2021, and its Vermont Yankee nuclear plant in 2019 to subsidiaries of NorthStar Group Services for demolition. Holtec, based in Florida, supplies equipment and systems for the energy industry, and NorthStar, based in New York City, is a demolition company in the business of razing buildings, cleaning up the site, and salvaging scrap metal.

"The U.S. nuclear energy era that began in the 1960s is ending," declared *Fortune* magazine, an American multinational business magazine founded in 1929, with close ties and sympathies with industry. Handing off money-losing nuclear power plants to Holtec and NorthStar for quick disposal would allow them "to pocket as profit the remaining money Entergy set aside to raze" the plants.

"This type of arrangement is new," said *Fortune*. "Simply put, the dismantling of the U.S. nuclear power industry, by any measure the largest environmental cleanup in American history, is being handed off to two private companies with low public profiles and histories of bribery or financial failure. . . . These companies are drawn to the business by an enormous prize: tens of billions of dollars in cleanup trust funds held by the utilities and collected from customers over decades."

Like the Zion power station decommissioning project, Holtec and NorthStar intend to finish those demolition projects in just 12 years and return the land to productive use. They expect to dispose of the radioactive wastes in Texas and New Mexico.

"As federal regulators sign off," the *Fortune* magazine report went on, "a chorus of attorneys general and nuclear industry experts is warning that the decommissioning companies have seriously underestimated costs and dangers, are ill-equipped to handle the jobs safely, and have troubling track records that bear closer examination before Holtec and NorthStar are given free rein on such sensitive projects. They worry the companies will bleed the cleanup trust funds dry, go bankrupt, and leave behind unfinished teardowns, dangerous radioactive pollution, and billions of dollars in extra cleanup costs."

Those projects are not like the Zion station decommissioning in one key aspect: Exelon expects ZionSolutions to return the station's operating licenses once the Nuclear Regulatory Commission finishes its final review. As of this writing, the NRC was reviewing Zion's final radiological survey results to ensure the site can meet the commission's release criteria before ZionSolutions may return the station's operating licenses to Exelon, which owns the power plant property.

Fortune magazine strikes a more negative tone today than it did in its August 6, 2007, edition, which struck a cautiously optimistic tone. The front page featured three nuclear power workers clad in yellow anti-contamination garb, and the words: "America's Nuke Revival. Can nuclear power help solve global warming? Soaring energy costs? Only if we can be convinced it's safe."

Fourteen years later, *Fortune* magazine sees threats from companies entering the reactor decommissioning business. NEIS's Kraft raised a related point about the reputations of utilities that will benefit and profit from state and federal plans to revive the nuclear power industry. "The overwhelming amount of these subsidies and state-level nuclear bailout schemes would go to nuclear utilities which have demonstrated a consistent penchant for corruption and criminal behavior in their business model," naming Exelon as an example, along with others in Ohio and South Carolina.

"All have been subject to FBI investigations, federal fraud, bribery, and improper lobbying charges, and outright admissions of guilt, paying hundreds of millions of dollars in fines. These are neither the business partners nor the industry America can rely on to successfully fight and win against the climate crisis."

Speaking of Americans, they blow hot and cold about nuclear energy. A 2019 Gallup poll points out that 40 years after Three Mile Island, they were evenly split on the use of nuclear as a U.S. energy source. Forty-nine percent of U.S. adults either strongly favor or somewhat favor the use of nuclear energy to generate electricity, while 49 percent either strongly oppose or somewhat oppose its use. In 2016, for the first time since Gallup first asked the question in 1994, a majority of Americans said they opposed nuclear energy—the 54 percent opposing it was up significantly from 43 percent a year before. A 2013 poll showed Americans favoring solar power, wind, and natural gas over nuclear.

The American Nuclear Society saw the 2019 Gallup poll results as an indication that public perception of nuclear energy was turning a corner, seeing value in nuclear power for electricity, clean drinking water, and heat.

Public opinion is easily swayed by clever slogans devised by Madison Avenue spinsters, like "clean energy," meaning one thing to environmentalists and another to the nuclear power industry, making it ideal for political sleight of hand.

The nuclear power industry calls itself America's largest carbon-free technology, but seldom mentions the large pollution footprint of the uranium mining and milling industry that produces the fuel that powers atomic reactors, and the transportation network that goes with it.

Worldwide production of uranium in 2019 was 53,656 tons, with Kazakhstan, Canada, and Australia accounting for 68 percent of it. The United States has a relatively small uranium mining industry employing 225 full-time workers in Utah and Wyoming. By some estimates, the American desert Southwest has 15,000 abandoned uranium mining sites, their scattered radioactive wastes, called tailings, a health hazard. For example, from 1944 to 1986, four million tons of uranium ore were extracted in Navajo territory in Arizona. The *Navajo-Hopi Observer* newspaper in 2015 reported that uranium mining "left more than 500 abandoned uranium mines, mills and processing plants—many near homes and livestock grazing areas, which has led to a unspecified number of health issues."

The biggest unsolved problem with the nuclear power industry, dating from its very beginning, is the extremely dangerous radioactive spent nuclear fuel scattered across the country because the U.S. Department of Energy has failed to find a permanent disposal site for it.

In what the U.S. Government Accountability Office (GAO) calls "an ad hoc system for managing commercial spent nuclear fuel," about 86,000 metric tons of burned-out fuel rods are stored using a variety of technologies at 75 operating or shutdown nuclear power plants in 33 states, an amount that grows by about 2,000 metric tons each year, said the GAO in September 2021.

DOE was aiming to put a spent fuel burial ground, called a geologic repository, at Yucca Mountain, in Nevada. But intense public opposition derailed that in 2010. DOE terminated efforts to license the site as a repository and congress stopped funding efforts to do so.

"Since then," said GAO, "policymakers have been at an impasse on how to meet the federal disposal obligation, with significant financial consequences for taxpayers." Unable to meet its disposal commitment, the U.S. government has paid reactor owners about $9 billion for storing the spent fuel.

Often called the congressional watchdog, GAO is an independent, nonpartisan agency that examines how taxpayer dollars are spent and how well governmental departments are performing. Its assessment of DOE's performance in finding a final disposal site for reactor wastes is dismal.

One surprising factoid surfaced from a search for information about spent nuclear fuel disposal. It was that the nation's only permanent radioactive waste disposal site exists near Morris, Illinois, 61 miles southwest of Chicago and close to the Dresden nuclear power generating plant. Called "the Morris Operation," the site is owned by GE Hitachi Nuclear Energy and contains 772 tons of spent nuclear fuel from five nuclear power plants, including the Dresden station. The spent fuel storage pool there is full and stopped taking shipments in 1989.

In what appears to be a breakthrough in the longstanding nuclear waste disposal dilemma, Finland in June 2022 completed building underground storage caverns half a kilometer deep into the granite bedrock of Olkiluoto Island off the Finnish coast. Built to take wastes from Finland's two nuclear power stations, the disposal site makes Finland the only country to have a deep geological storage facility.

If nuclear critics are to be believed, the future of nuclear power does not appear very bright. But then, nuclear power had been counted out long before *Fortune* magazine did so.

The *Economist*, a respected British publication of world events since 1843, featured a 14-page special report in 2012 on nuclear energy: "The Dream That Failed." A year after Fukushima, "the future for nuclear power is not bright—for reasons of cost as much as safety," said the publication. But the *Economist* was telling a different story by 2022. The United States had expected to export its nuclear savvy around the globe. "The loss of Western competence," said the *Economist*, "helps to explain a loss of market share." And why Chinese or Russian designs were used in 27 of the 31 reactors that have started construction since 2017.

The pro-business *Wall Street Journal*, which often advocated for nuclear energy on its editorial pages, in 2017 published a story headlined "Nuclear power sounds the retreat." It reported U.S. utilities were closing nuclear power plants at a rapid clip while facing competition from cheaper sources of electricity and political pressure from critics.

Even the pro-business, pro-technology *Chicago Tribune* expressed doubts about nuclear power coming to the rescue a month after the Arab Oil Embargo in December 1973. That's long before the 1979 Three Mile Island accident! It was in an editorial titled "Nuclear Power Dilemma."

"At a time of energy crisis," said the *Tribune* editorial, "when the backers of nuclear power hoped to be riding confidently on a momentum of past successes and future promise, the industry is plagued by bottlenecks and frustrations. Even worse for nuclear advocates, only a small part of the shortfalls can be attributed to the favorite bogeys, government regulation and interference by environmentalists. Serious problems in nuclear power generation arise from the economics and technologies of the industry itself."

It concluded: "The arguments are becoming more persuasive that the nation should lean more toward coal, solar and geothermal processes and nuclear fusion to meet its energy needs a decade or more hence and plan today's research and investment accordingly. Little is gained by replacing a politically unreliable fuel—oil—with technically unreliable and potentially dangerous power from nuclear fission."

It was a remarkably prescient statement.

Granted, vast portions of the public do not trust the media, but at least give the media credit for keeping track of important trends and reporting them. In time, we'll see if those media heavy hitters got it wrong as a major new influence gained world attention: climate change. Another one of those sudden shifts in direction.

But the nuclear power industry battles on, insisting it can do what its wind and solar rivals cannot. And that is summed up in a word, "baseload," the ability of an atomic reactor to run nonstop for a year or more, dependably pouring out power. Exelon's LaSalle Two reactor ran continuously for 739 days before stopping in February 2007. A Canadian reactor claimed a world record in 2016, with 940 days.

Solar and wind power sources are intermittent and can't come close to matching nuclear power's nonstop marathon power runs.

So I asked nuclear watchdog David Kraft to answer that argument, and he referred me to comments by Jon B. Wellinghoff. From 2009 to 2013, Wellinghoff was chairman of the Federal Energy Regulatory Commission (FERC), which regulates the interstate transmission of electricity, natural gas, and oil.

Only a month after President Obama formally designated him as FERC chairman, Wellinghoff astonished the U.S. energy community by saying the nation may never need new nuclear or coal-fired power plants because renewable energy from various parts of the country could be transported by the national power grid to wherever it's needed, meeting future power demands.

"I think baseload capacity is going to become an anachronism," Wellinghoff told reporters at a U.S. Energy Association press conference in Washington. Renewables like wind, solar, and biomass combined, he said, provide enough energy to meet baseload needs. "You have to be able to shape it," using the geographical spread of the power grid, tapping into wherever the wind is blowing or the sun is shining.

"So if you can shape your renewables, you don't need fossil fuel or nuclear plants to run all the time," said Wellinghoff. "And, in fact, most plants running all the time in your system are an impediment because they're very inflexible. You can't ramp up and ramp down a nuclear plant. And if you have instead the ability to ramp up and ramp down loads in ways that can shape the entire system, then the old concept of baseload becomes an anachronism."

Digital grid technology already is moving in the direction to accomplish this, said the FERC chairman. "We are going to have to go to a smart grid to get to this point I'm talking about. But if we don't go to that digital grid, we're not going to be able to move these renewables, anyway. So it's all going to be an integral part of operating that grid efficiently."

Consternation and jubilation met his ideas, but little came of them, even by the Obama administration.

Nuclear critic Kraft likes the idea of being able to computerize the transfer of renewable power around the country where it is needed. But he fears the national power grid is "probably in worse shape than all the rest of the energy systems. It is decrepit and born out by Texas."

In February 2021, the state of Texas had a major power crisis because of winter storms sweeping across the United States. They caused a massive electricity generation failure in Texas, leading to shortages of water, food, and heat. More than 4.5 million homes and businesses were without power, some for several days. At least 210 people were killed directly or indirectly by the crisis; some estimates say as high as 702. Damages were estimated at about $20.4 billion.

State officials initially blamed the outages on frozen wind turbines and solar panels, although they later discovered that poorly winterized natural gas equipment contributed to the power grid failure. Making it worse, in 2002, Texas had isolated its power grid from two major national grids in a successful effort to reduce power costs in the state and deregulate energy. That disconnection made it difficult for the state to import electricity from other states during the crisis.

The Lone Star State decided to go it alone where the power grid was concerned, and paid a big price for that. A more unified power approach would have been helpful in that crisis. But Americans who refused to wear masks against the coronavirus pandemic showed that cooperation is not always a national trait. They do what they believe is in their own best interest.

"The bottom line," said Kraft, "is we have the technical capability for making baseload something that is an anachronism, if we choose that. We're not choosing to get rid of coal plants," which account for about 20 percent of the nation's electrical power. But the U.S. coal industry is shrinking fast, losing 141,500 domestic coal jobs between 1985 and 2016. The Bureau of Labor Statistics said coal mining employed 53,000 workers in 2016. By comparison, the renewable energy industry employed 777,000 people, according to the Environmental Defense Fund in 2018.

Along came Covid-19, and that reshuffled the deck in workplace trends and how work is performed, which adds another dimension to the picture. In a matter of weeks, the percentage of employees working from home doubled, from 31 percent to 62 percent, establishing a new trend of people preferring to work from home. Workers furloughed or laid off found other jobs, and did not return to their former jobs, causing worker shortages when the economy began emerging from the pandemic. It's a picture that is shifting and likely to impact the energy sector in unpredictable ways, except that unpredictability has long been part of the United States energy scene.

Chapter 50: Three Mile Island and Me

THE THREE MILE ISLAND ACCIDENT WAS A HARD SLAP TO THE AMERICAN CONsciousness, and put the entire nation on edge. Nobody expected anything like that to happen, and everyone was totally unprepared for it. Everybody. We never saw anything like it, and it scared the heck out of us.

I had been covering nuclear energy issues for about 10 years, including nuclear disaster drills, but never saw anything like this. The terror, the chaos, the confusion, the anger, the deceit, the kindness. The way people react in a *real* emergency. It stands out as the most memorable, and frightening, story of my career in journalism. America's most severe accident in commercial nuclear power history, it had a profound impact on government policy, the nuclear industry, and the way the public thinks about nuclear energy.

It still reverberates. It's impossible to talk about nuclear power in the United States without thinking about that close brush with nuclear disaster. Here's what it looked like from a reporter's viewpoint at ground zero:

I got into the Three Mile Island story early, the day it began at 4 a.m. on Wednesday, March 28, 1979. Wire service stories already were describing the accident as "probably the most serious nuclear reactor accident to date," forcing evacuation of the generating station and causing radioactive steam to escape. Before leaving Chicago for Pennsylvania, I did a fast story asking a Commonwealth Edison spokesperson and the director of the Illinois Emergency Services and Disaster Agency if anything like that could happen in Illinois?

"In all honesty, it could happen in Zion," said Steve Goldman, an Edison spokesperson, pointing out that the Zion station had two pressurized water reactors similar to those at Three Mile Island. Days later, Edison's chairman, Thomas Ayers, disagreed. Erie Jones, the state's emergency management director, when asked if Chicago could be evacuated in a nuclear emergency, said: "I don't think so. I've tried to get out of there on weekends. I wonder if people could get out in two days. I haven't figured it out."

Flying to Pennsylvania, I got a room in the Hershey Motor Lodge and Convention Center in Hershey, Pennsylvania, known as Chocolate Town U.S.A. This is the home of the Hershey Company, where the air is infused with the mouthwatering aroma of chocolate produced there. Streetlights are shaped like chocolate kisses, and the main intersection is Cocoa and Chocolate streets.

Hershey is eight miles from Middletown, an historic but little-known town dating to 1755 that suddenly became world famous for being closest to the stricken Three Mile Island Nuclear Power Plant. Contrary to popular belief, Three Mile Island is not named for being three miles downriver from Middletown, Pennsylvania. Instead, it is named for the length of the island upon which the power plant sits on the banks of the Susquehanna River.

Setting up my headquarters in Hershey, I used a portable typewriter to take notes over the phone or write my stories before dictating them verbally by phone to a rewriteman at the *Chicago Tribune*. This was before cell phones and before computers, which made their first appearance in the *Tribune* newsroom in 1981. It was a different world.

Soon I experienced my first encounter with the unpredictable and downright bizarre nature of the unfolding Three Mile Island story. I knew the stricken reactor spewed radioactive fumes into the air when the accident started, so I avoided going there until Metropolitan Edison, the company that operated the facility, announced it had stopped the fumes from escaping. I knew, as an experienced nuclear reporter, that it is wise to stay upwind of a sick atomic reactor. Or just stay away.

Assured that it was safe to approach the 200-acre nuclear plant, I drove there on the third day of the accident, Friday, March 30. Standing outside the front gate, I interviewed plant workers as they came and went in the shadow of four tall concrete cooling towers visible for miles around. Later I learn that a heavy burst of radioactive gas gushed from the plant and spread over surrounding communities while I was standing there. It was the strongest radioactive "puff" since the accident had begun. Radioactive gases continued pouring from the plant for several hours. I had not yet received a dosimeter from the *Tribune* to check my radiation exposures.

So I called Jan Strasma, a Nuclear Regulatory Commission media relations specialist who, along with others from NRC, came to the area to take media calls. Strasma normally worked in the NRC's Midwest regional office in Lisle, Illinois. I knew him and we'd worked together many times before. It was a lucky break for me to have an NRC contact at the scene who knew me and was an expert in nuclear details. He had a gift for answering questions without resorting to the mind-numbing jargon often used in the nuclear trade which sometimes confuses even nuclear people. He also had a gift for being calm and patient under extreme pressure.

On the day of the accident, Wednesday, Strasma went to the NRC regional office in King of Prussia, Pennsylvania, a Philadelphia suburb, remaining there until Saturday, when he went to Middletown and set up a media base in the community center.

I was one of about 400 journalists from throughout the world who descended on the Middletown area, some of whom had never or seldom covered a nuclear story, while others were experienced science writers or specialists like me. Strasma had the unenviable task of explaining the accident to some journalists who did not know a reactor vessel from a pressurizer.

I usually called Strasma for a quick clarification about a recent development. I'd leave my name and phone number at the NRC's central message desk, and he usually returned the call soon. But that Friday, I wanted to know what that big puff of radioactivity when I was at the plant front gate meant to my health.

"You were exposed," he affirmed, "but it probably won't hurt you." I was somewhat assured, but that word "probably" stuck in my mind. Uncertainty. Right from the beginning, there was uncertainty. It really never went away at Three Mile Island. But, to this day, I remain grateful to Strasma for all his help on a very vexing assignment that required a lot of explanation.

As a check on my memory, I contacted Strasma, now retired. He, too, was afflicted by the rampant quandaries of that tumultuous time. "I can't be sure of my guidance that Friday, but I recall it was a time of considerable uncertainty. Even at the NRC regional office that afternoon, there was little certainty and, for a time, I was particularly frustrated because I did not have solid information to respond to media inquiries. . . . We knew that relatively low levels of gaseous radioactivity were being released—and your recollection of my guidance seems reasonable."

In those early day at TMI, said Strasma, "there were always more inquiries than I/we could handle. I rarely took calls directly, but put the yellow message slips in piles—my priorities generally were local media, national media and wire services, and reporters I had worked with and trusted—and you, obviously, were in the latter category. Lots of calls just didn't get returned."

Clashing accounts at TMI began almost immediately, and stayed that way.

"It doesn't look like a very serious accident at this point," Metropolitan Edison vice president Jack Herbein told reporters hours after the accident began, minimizing the accident's severity. Met Ed spokesperson Dave Kluscik insisted, "We are not in a China Syndrome type situation."

By coincidence, a movie by that title had come out only 12 days earlier, starring Jane Fonda and Jack Lemmon, about a fictional nuclear power plant meltdown. China Syndrome gets its name from the theoretical possibility that a ball of flaming hot melted uranium fuel could burn its way through the power plant floor, into the ground and head "all the way to China" on the other side of the planet.

Shortly after Herbein's press briefing, Pennsylvania lieutenant governor William Scranton III assured the public "everything is under control." But five hours later, an angry Scranton declared "the situation is more complex than the company first led us to believe. Metropolitan Edison is giving you and us conflicting information."

This led to dueling press conferences. NRC held its daily briefings in the Middletown Borough Hall, a massive gray stone building in Dauphin County's oldest community. The hall is an all-purpose kind of place with a stage at one end, bleachers, an electronic basketball scoreboard, and basketball hoops at each end of the hall. Population about 8,000, Middletown was 10 miles from Harrisburg, the state capital.

MetEd held its press briefings a short distance away in American Legion Post 594 on East High Street. Waves of reporters swept from one location to the other to get the latest updates that were often conflicting. Then they dashed away, looking for a telephone to call their offices. Think of what it's like competing with hundreds of journalists for a telephone. Standing in lines at telephone booths was common, or paying local residents to use their telephones.

But I was blessed by the kindness of Frank Corby, owner of the Corby Beverage Center on East Main Street in Middletown. He allowed me to make collect calls from his liquor store any time I wished. It was a precious resource in dire times. In gratitude, I'd send him copies of the *Tribune* occasionally because he liked to read out-of-town newspapers.

Most of what I recall from Middletown is serious, but anytime you're surrounded by journalists, there's bound to be a few jokers.

While waiting for a press briefing to begin in the crowded borough hall, it was common to see journalists lift their personal dosimeters up to the light to check their radiation exposure levels on the scale inside the dosimeter. During one of those idle moments, a television crew member said, "I think I'm radioactive." The noisy chatter around him immediately quieted down. Asked why he believed that, he answered: "This morning at breakfast, I stuck a spoon in my grapefruit and got a dial tone."

Jokes about glowing in the dark were common. What's the five-day forecast? Two days. The American spirit of free enterprise kicked in only six hours after the accident began. Three Mile Island T-shirts appeared in local shops. One said, "I survived Three Mile Island." Another said, "I survived Three Mile Island, I think." Or, "I didn't die at TMI." Or, "Hell no, I don't glow." And, "Happiness is a cool reactor." And "TMI recovery team." T-shirts sold for $4.50, sweatshirts for $8.50, belt buckles for $8 and coffee cups for $3.50. There were bumper stickers too. "At one time I had 21 different sayings," said shopkeeper Joyce Yinger.

But levity was a rarity in that tense time. The situation got really serious with that big radioactive gas leak on Friday. Governor Richard Thornburgh, in office only two and a half months, needed to decide whether to evacuate up to 950,000 people from a radius of 20 miles around the nuclear plant, an exodus of stupendous scope and complexity.

Such an evacuation was recommended at 9:15 a.m. Friday by the NRC's director of state planning in Washington. "We spent a great deal of time worrying about that recommendation," said Paul Critchlow, the governor's press aide. "We didn't know if he had authority to make such a recommendation. We wasted an hour before calling Chairman Hendrie," referring to NRC's chairman, Dr. Joseph M. Hendrie, who said the mid-level official was not authorized to make such a recommendation. Instead, Hendrie told state officials to "advise people to stay indoors."

Frustrated by such contradictions during a crisis, Thornburgh acted by urging pregnant women and preschool children living within five miles of the power station to leave, since they were most vulnerable to radiation. This was expected to involve about 5,000 people, but it triggered the beginning of a much larger exodus of wary citizens and their children. Twenty-three schools in the area closed and half a dozen evacuation centers opened. Shortly before noon, a siren wailed in Harrisburg, touching off fears that something bad was happening. It was an unauthorized alarm set off by a fire official intended to warn people within 10 miles of the plant to stay indoors.

Growing doubts about the NRC's capabilities and control of the situation caused Thornburgh to ask Hendrie to send somebody who could speak with authority. Hendrie conferred with President Jimmy Carter, who asked: "Who is the best in the country?" Hendrie named Harold Denton, NRC's director of the Office of Nuclear Reactor Regulation, who quickly helicoptered to Middletown and became Carter's personal advisor on the accident. Not only frazzled state officials, but the public also needed somebody to believe and trust. Denton became known as "the Hero of Harrisburg."

Denton was a soothing influence, but that proved to be a lull before the storm. A bubble of highly flammable hydrogen gas was noticed Saturday inside the reactor.

By Saturday, the *Chicago Tribune*'s team covering the TMI accident numbered nine, led by Aldo Beckman, the *Tribune*'s Washington bureau chief, a funny, quick-witted man. We met in Harrisburg for dinner that evening in the popular Santanna's Seafood House, where draped fishing nets adorned the paneled walls along with a stuffed barracuda and a swordfish. Red tablecloths were topped with burning candles covered with amber-colored glass shades.

Our group included two photographers from Chicago, Frank Hanes and Val Mazzenga; Washington bureau reporters Raymond Coffey, Bill Neikirk, and Carol Oppenheim; Chicago reporter Howard Tyner; and Dr. Arthur Wolff, a radiation expert from the University of Illinois School of Public Health, hired by the *Chicago Tribune* to take independent radioactivity readings near TMI and write about his public health assessments. My job was to pay close attention to the technical aspects of the broken nuclear reactor.

The restaurant was crowded with media people. Soon after we were seated, Coffey left the table to answer a phone call from a Chicago news editor and came back, saying, "something is happening." Beckman asked me to check it out.

Using the public telephone on the wall in Santanna's dank basement, I called the governor's office and spoke to Patty (Patricia) McCormick, fresh out of college and helping to take a storm of media calls.

"The gas bubble is explosive," she told me in a tense voice. Unknown to me at the moment, her remarks were based on an Associated Press report at 9:33 p.m.: "Federal officials said Saturday night that the gas bubble inside the crippled nuclear reactor at Three Mile Island is showing signs of becoming potentially explosive, complicating decisions on whether to mount risky operations to remove the gas."

That was the most horrifying moment of the historic TMI accident, when word flashed around the world that a large hydrogen gas bubble that fumed into existence inside the reactor during the accident might explode. My notes call it "the night of terror."

Walter Cronkite, one of the most trusted men in America, intoned: "The world has never known a day quite like today. It faced the considerable uncertainties and dangers of the worst nuclear power plant accident of the atomic age. And the horror tonight is that it could get much worse. The potential is there for the ultimate risk of meltdown at Three Mile Island. . . ."

Knowing only what Patty McCormick told me, I returned to the *Tribune* group and told them what she said. Beckman asked for my opinion. If the report is true, I said, "we ought to get the hell out of here." But I asked to make another call to Patty and headed downstairs, where I found a striking redhead, Ellen Hume, a *Los Angeles Times* reporter, holding the phone to her ear. She turned toward me, her face pale and rigid with terror, and said, with some defiance in her voice: "Does this mean we're all going to die?" I wish to this day I could have said something to ease her fears, but I simply shook my head and said, "I don't know."

In another call, McCormick explained that the Associated Press story was based on an unidentified NRC official. "If I thought it was true, I'd get out of here," she said. But she did not know if it was true. I called three more times. On the third call, McCormick said the governor's office had reached people at the power plant, and they denied the bubble was explosive. Turning from the telephone, I faced a young female radio reporter on the verge of tears with a trance-like expression. "Are you a reporter?" she asked. I nodded, and she said, "Our office just called and told us to get back, the bubble's going to explode. Get out of here while you can."

I updated the *Tribune* group waiting in the restaurant. Beckman made a decision: "Let's go," he said. "If I'm going to be responsible for eight lives, I think we should leave. No story is worth our lives. And if it isn't true, we can cover it from outside town. If it is true, we'd better get out of here."

The restaurant owner, James Santanna, came to our table and told us he was closing, saying the place emptied in three minutes once word of the exploding bubble got around. He said he lost about $700 in business because of the abrupt closure.

The scene outside was surreal. Sirens were howling. People were running in the streets, some jumping into cars and screeching away. Television crews raced down to the power station 10 miles away, eager to get videos in case the place exploded. Absurd. More sensible, a newsman said: "We don't want to be martyrs. Let us know. We'll be in the coffee shop."

Our group drove to the Holiday Inn, where Beckman and others were staying, to collect belongings. Ms. Oppenheim decided to stay in Harrisburg. While at the Holiday Inn, I made more telephone calls as the others urged me to get going. I reached Curt Ashenfeldt, the governor's assistant press secretary. He read a statement from the governor: "The news report that the gas bubble in the nuclear reactor is becoming potentially explosive is not true, according to Harold Denton, director of the Office of Nuclear Reactor Regulation." Ashenfeldt added, "Denton is at the plant and he knows." The erroneous assertion came from NRC officials in Washington. "Anybody is free to go," said Ashenfeldt, "but the governor is not ordering an evacuation. There are no plans to evacuate anywhere."

Informed of this, Beckman decided to err on the side of caution. Our group in several cars headed north, although frankly I don't know if that was upwind or downwind of the reactor. Our goal was to drive about 20 miles beyond the reactor and find a motel for the night, but thousands of others had the same idea. Cars clogged the highways, some with children in the back seats wrapped in blankets and clutching teddy bears or other stuffed toys.

An estimated 150,000 people evacuated the Harrisburg area in one big rush, starting with the warning the day before to pregnant women and preschool children. That was out of a total of about 950,000 within 20 miles of the reactor. It's a minor miracle that nobody was hurt or killed in that mass, frenzied flight. These were America's first nuclear refugees.

We fled into the night, made more eerie by dense fog. While on the road, we listened to a radio broadcast of a continuous news conference in the state capitol about the reactor, taking notes. "Keep listening," said the soothing announcer. "There's no need to panic."

At 11 p.m., Governor Thornburgh made a statement: "There have been a number of erroneous or distorted reports during the day about occurrences or possible difficulties at the facility on Three Mile Island," he said. Denton had assured him "there is no eminent catastrophic event foreseeable at the Three Mile Island facility and I appeal to those who may have reacted or overreacted to reports of the contrary today, to listen carefully to his characterization of the current status of the situation. I appeal to all Pennsylvanians to display an appropriate degree of calm and resolve and patience in dealing with this situation." Thornburgh mentioned that President Carter was coming to Middletown the next day, "a further refutation of the kind of alarmist reaction that has set in, in some quarters." A broadcast question-and-answer period followed.

We stopped for gasoline at a Mobil station near Lickdale, Pennsylvania, about 25 miles north of Harrisburg on U.S. 81. Jim Adams, the station's owner, said one or two cars normally stopped each hour for gas; now he was getting 15 to 20 cars an hour. "If they ever start evacuating," said Adams, "you're going to see mass confusion." I used the gas station telephone to call the *Chicago Tribune* and describe the mass evacuation flooding the highways and the governor's broadcast comments. We stopped several times looking for a place to stay, but motels were full. Early Sunday morning, we found lodgings for the night in Hazelton, Pennsylvania, 80 miles northeast of Harrisburg. Cars carrying members of the *Tribune* team got separated in the dense traffic and the maelstrom of confusion.

"Who would have thought all this could be caused by a bubble?" Neikirk observed from the back seat of our car. He pointed out that on Saturday, we reported three conflicting stories: that the reactor crisis was over, that the bubble was about to explode, and that the bubble was not expected to explode. All within hours of each other.

Ray Coffey was a world-traveling, experienced war correspondent. He was a compact man with an air of quiet ability. I asked him how Harrisburg compared with the risks of covering combat. "I discussed this with my wife," he said. "She was very much against me coming here. I've been to Vietnam and other war zones. She didn't like it, but there was a provision in the insurance. If I was hit, I was covered." But with radioactivity, "you can't see it or hear it or smell it. You can't see it coming. You can't tell where it's coming from. You're helpless. You know, I'd like to see them color this stuff purple."

It was a short night. President Carter was coming on Sunday to calm public fears and make sense of a situation made worse by the Nuclear Regulatory Commission and by wrangling between utility and government officials, and I had to be there.

MetEd vice president Herbein on Saturday said, "I think the crisis is over." But shortly afterward, Denton in a press conference disagreed. "My own view is that the crisis won't be over until we have a cold shutdown" of the reactor.

Adding to the confusion were the warring NRC camps, one in Washington and the other in Middletown, over the bubble. One side said it was potentially explosive, the other side said it was not. They were still squabbling when Carter, dressed in a crisp gray suit and tie, arrived with his wife, Rosalynn, in Middletown. First, the president visited the power station, wearing yellow plastic anti-contamination foot coverings and radiation detection devices, accompanied by Thornburgh and Denton.

Then the presidential party motored to the Middletown Borough Hall, where the two key NRC adversaries, Roger Mattson and Victor Stello, were in a heated debate over the bubble in a back room when the president arrived. Carter listened to them briefly, then told them to "figure it out" as he turned to walk to a podium in the borough hall basketball court to make a statement to the press and to the citizens of Middletown.

A trained nuclear engineer, Carter told the crowd, "My primary concern in coming here to the plant is to learn as much as possible as I can about the problems at the Three Mile Island plant, and to assure the people of the region that everything is and will be done to cope with the problems of the reactor." The president expressed his "admiration and appreciation for the citizens of this area who, under most difficult circumstances, have remained calm. . . . The primary and overriding concern is the health and safety of the people of the area. If we make an error, we want to err for extra caution and extra safety."

Important decisions must be made in the coming days before the reactor is brought under control, he said. "If it does become necessary, your government and the governor of the state of Pennsylvania will ask you to take appropriate action. If he does, I ask you to take them calmly, as you have in the last few days." He added, "This does not mean danger is high." Some interpreted that as a strong hint that an official evacuation might be coming.

Behind the scenes, NRC experts Mattson and Stello still battled. Mattson, an NRC expert from Washington on emergency core cooling, had come to the astonishing conclusion that a hydrogen gas bubble at the top of the 43-foot-tall reactor vessel caused by a chemical reaction between super hot steam and the zirconium metal cladding on the fuel rods posed two hazards: that the hydrogen bubble might explode, or that it might expand as the reactor cooled, blocking cooling water from reaching the reactor fuel core below, causing the fuel to overheat catastrophically.

NRC estimated the bubble measured 1,000 to 1,500 cubic feet.

There's nothing in reactor operating manuals about hydrogen bubbles. The fear was that if the bubble did explode, it would crack open the reactor vessel and the concrete structure surrounding the reactor, spewing clouds of deadly radioactive gas, steam, and debris over the countryside.

Stello, a brawny man, was an NRC expert in nuclear operations. He came to Middletown with Denton as second in command of the TMI crisis. In the three minutes that Carter spoke to the crowd, Stello convinced Mattson that his calculations about an exploding bubble were wrong.

They were quarreling about radiolysis, a phenomenon in which chemical bonds of molecules break apart from exposure to ionizing radiation, the kind produced by nuclear reactors. Mattson believed that would break down water molecules in the reactor, producing enough oxygen and hydrogen to fuel a powerful explosion.

"That assumption was not valid," Stello told me at the scene. "Any oxygen formed would immediately recombine because of the excess hydrogen in the system. Therefore, radiolysis really couldn't occur. Because the hydrogen was there, it prevented any oxygen from being formed. Pure hydrogen can't burn. You need oxygen there, and if there is no oxygen, you can't have an explosion." No free oxygen, no explosion.

I can't remember if it was Stello or somebody else who told me that the Mattson faction apparently forgot something they should have remembered from high school physics: an explosion needs a spark, some kind of ignition to set it off, and nothing like that exists in the heart of a nuclear reactor.

In short, the nation had been terrified by a bogus exploding bubble, a hoax, a fumbling miscalculation by one of NRC's masters of nuclear technology, which even he could not fully comprehend. As the *Chicago Tribune* put it in headlines: "Three Mile Island: The fear was real . . . but chance of an incinerating nuclear blast was not."

The exploding bubble commanded public attention for days. Although Denton was the first to publicly deny imminent danger of an exploding hydrogen bubble, he tended to describe it as a remote possibility, rather than flatly denying it. "I'm sure that in the near-term, there is no hazard of flammability, and that is days," he said on Saturday, the day before Carter's arrival. Denton also had the habit of starting a sentence with one thought and changing direction in the middle, ending with another thought. He could be confusing, too.

But Stello agreed with Mattson on the second potential hazard of the hydrogen bubble, that falling pressure inside the reactor "could cause the bubble to expand and expose the [fuel] core. That was what I viewed as the major concern."

Eventually, they slowly bled off and vented the hydrogen gas.

Temperatures inside the reactor reportedly reached up to 2,500 degrees Fahrenheit, and Stello estimated that the hydrogen bubble formed about 16 hours after the accident started.

The reactor operators made it worse by doing the wrong things. An NRC official put it this way: "If the operators had just sat on their hands [during the accident], we wouldn't be in this predicament." They were turning off automatic safety systems trying to stop the accident by pouring water into the reactor.

Even the president of the United States was not spared a fumble at Three Mile Island. In his book, *Three Mile Island and Beyond, Memories of a Life in Nuclear Safety*, Denton admitted to a blunder. As Carter was leaving Middletown, his dosimeter read 85 millirems when it should have been zero. "I had a panic moment as it's definitely unacceptable to give the president any sort of radiation dose," he wrote. The First Lady's dosimeter read about 75 millirems, while dosimeters carried by Denton and Governor Thornburgh read zero. All of them should have been zero. "Panic began to set in on the bus," Denton wrote. Determined later: the dosimeters carried by Carter and his wife had been used previously in the power plant and should have been reset before they were used again. Denton fretted that "Carter's view of the plant slid a little downhill from that incident." You think?

In my hotel room on Monday, I was jolted by a call from my managing editor, Maxwell McCrohon, who wanted a story on what caused the accident. I said the NRC did not expect to announce that for weeks, maybe months. He answered, "You're our nuclear expect, I want that story now." I was pondering how to do that when the phone rang again, this time from Clayton Kirkpatrick, the *Tribune's* editor. "Try Commonwealth Edison," he said, and hung up. Of course! Any company with the largest fleet of nuclear reactors in the country would urgently want to know what caused that accident. Its operators would likely go through possible scenarios and call to pick the brains of contacts at Metropolitan Edison as soon as possible. Reactor operators are a tight-knit community.

NRC and industry sources also were trying to piece together what caused the accident. With their help, including Commonwealth Edison and its research director, A. David Rossin, a clear picture emerged of the failures that led to the accident.

The reactor was operating at 97 percent of power when a maintenance crew was trying to clear a resin blockage in a water line and the reactor tripped off, causing an electromagnetic relief valve at the top of the pressurizer to pop open. The valve should have closed automatically in 13 seconds, but it stuck open for two hours and 19 minutes. That allowed 250,000 gallons of radioactive water in the form of steam to blow out of the open valve and flood the reactor building to a depth of seven feet, including the adjoining auxiliary building. Instruments in the reactor control room falsely showed the relief valve was closed. Later reports said the containment building flooded to a depth of 40 feet.

It was frightfully clear from the beginning that something was terribly wrong with the warble of alarms in the control room and water hammers sounding like colliding freight trains in the massive cooling system pipes.

A shift foreman reporting for duty at a shift change is credited with realizing the relief valve was stuck, and he closed it, stopping the loss of cooling water. An NRC report said that if that stuck valve had remained open another 30 to 60 minutes, the accident would have been far worse, requiring an evacuation of thousands of residents living around the plant and causing a serious public health threat.

But here's the crucial part: with the loss of all that reactor cooling water, emergency backup cooling water pumps should kick in automatically. Two days before the accident, in a test of three backup cooling water pumps, the valves to those pumps, called block valves, were closed by turning wheels manually. Somebody forgot to open the valve on one of the main emergency cooling water pumps after the test. That pump failed when it was needed, triggering the accident.

My story in the Tuesday, April 3 edition was headlined: "Human fault in atom leak. Shut valve triggered accident." Pinpointing human error, the story explained the accident was triggered because a manual valve in a key backup cooling system was inadvertently left closed after a test several days earlier. As a result, the reactor was denied critical cooling water so that the top five feet of the fuel core went dry while temperatures reached 2,500 degrees Fahrenheit. The entire reactor fuel core is supposed to be covered in cooling water. Analysis showed that the top of the reactor fuel core started going dry about two hours into the accident.

Although the accident is usually described as a partial fuel meltdown, a report by the Nuclear Safety Analysis Center said there was no general fuel melting, although fuel rod metal cladding ruptured and caused parts of the fuel core simply to collapse. Subsequent discoveries revealed that when a torrent of cooling water finally shot full force into the superheated reactor, the top of the fuel core crumbled into a bed of rubble from thermal shock.

The *Tribune* story was a world exclusive, and the events that led to the accident were confirmed by Denton and by subsequent federal agency reports. A series of human errors were made at Three Mile Island, before and during the accident.

It took the reactor operators about eight minutes to recognize that vital emergency cooling water block valve was closed. Ron Fountain, an auxiliary reactor operator, was ordered to open it. Wearing anti-contamination clothing and a respirator mask, hyperventilating in fear, Fountain entered the dangerously radioactive room where the valve was located. "My instinct was to rip off my mask," he said, "but I knew the air was heavy with [radioactive] particles. I said a prayer. I had to gather my wits. I was sweating and breathing heavy. I made my walk to the valve. I opened it. Then I walked toward the door." But by that time the fuel core already was seriously damaged.

The nuclear industry believed the "big one," a major accident, would involve something like a break in one of the huge cooling water pipes, called a LOCA, or loss of coolant accident. But Three Mile Island proved to be a series of small and unrecognized events that included elements of a big one, such as loss of coolant, a supposed meltdown, and release of radioactivity. It was not exactly "the big one" of the kind the nuclear industry expected, not the full-blown event that people talked about, but close.

But the "mindset" in the nuclear industry that its worst fears would be a huge LOCA was one of the human failures of the accident, said the president's commission report on the TMI accident. The reactor operators were trained to expect one thing, and failed to recognize something they were not expecting. "It became clear that the fundamental problems are people-related problems and not equipment problems," said the report. The training of TMI reactor operators was "greatly deficient." Response to the emergency "was dominated by an atmosphere of almost total confusion."

Among the things Three Mile Island proved is that clear communication is crucial in a crisis. That was seriously lacking in Middletown. Denton became prominent during the accident in part because he became the only one speaking about it. An unacknowledged gag order was placed on Metropolitan Edison officials. I learned this when I asked a MetEd spokesperson, Don Curry, a question. "As of this moment, we are under strict orders not to say a thing," he answered. Orders from whom? "I do not know, we have an agreement with NRC."

NRC denied such an order, although Denton made all announcements about the accident from then on. MetEd would not issue statements without NRC approval. NRC said its approval was not needed. Nobody would admit that such an agreement existed.

"I'm not aware of it, or I've forgotten," said Robert Arnold, a MetEd vice president. He and others praised the Denton single voice, or single source, approach as one that cured the Harrisburg garble. It is likely to be the strategy in any future nuclear accident, considered convenient and a way to do away with conflicting statements.

But does it serve the truth and public safety? I thought it was disturbing, from a journalist's point of view. There's constitutional freedom of speech to consider. Granted, MetEd officials were muddling public understanding of the accident with evasions and falsehoods. Their repeated claims that there was nothing to worry about fell short of the truth, but should they be muzzled? Critics might argue that more voices are needed in such perilous times, not fewer. What if the single voice is wrong?

Exelon Corporation bought and operated the undamaged TMI Unit One reactor, but closed it in 2019.

Though McCrohon put me under intense pressure to report immediately the cause of the accident, I was glad to learn I did not disappoint him. In an April 13, 1979, memo to me, he said, "I read virtually everything I could get my hands on during the Harrisburg crisis. The *Chicago Tribune*'s coverage was truly outstanding— accurate, comprehensive and timely. Aldo Beckman and I agree that our coverage owed a great deal to your knowledge, tenacity and reportorial skill. Many thanks."

I returned to Middletown several times to see how residents were coping. On one of those visits, I encountered Frank Corby, the friendly liquor store owner. "Every now and then you get a busload of tourists," he told me. "For a town like this to have any tourism is crazy. The first ones I saw are Japanese." They stood in front of the Middletown post office, holding TMI T-shirts, having their pictures taken.

Before long, Metropolitan Edison was conducting a two-hour tour of the TMI station for thousands of visitors from all over the world. "We're literally trying to answer to the whole world here," said Bill Gross, a MetEd spokesperson. "March 28 literally made us world famous." But by going inside the station, Gross warned visitors, "you may be exposed to 1 or 2 millirems" of radioactivity.

Two voices especially stand out in my mind when I think of Three Mile Island. One of those voices belongs to Dr. Joseph Trautlein, acting chief of internal medicine at the Milton S. Hershey Medical Center, about six miles from TMI, who described how people *really* act in an emergency, not as they are expected to behave as portrayed during calm nuclear emergency drills.

Hospital staffs are central to emergency planning, the doctors and nurses are expected to care for others when they cannot care for themselves. Rock solid. But tensions were rising with the uncertainty over conditions at the nuclear plant.

In the hospital, people were trying to get organized for the unknown. "Nurses would console a patient, then come out of the room and break down in tears herself," recalled Trautlein. "At noon Saturday was the first time I really got scared." A representative from Governor Thornburgh's office came to the hospital, said the doctor, and gave a presentation about conditions at the nuclear plant to hospital administrators and the chief nurse, saying "the situation is out of control and we can't give you more than three hours before it blows, and it probably will." They were told the news was too terrible to inform the public.

"The effect was unbelievable," said Trautlein. "Combat surgeons, guys who were up to their ankles in blood [in wartime] were dysfunctional. They were told they would be dead, that there was nothing they could do for their patients or for themselves. The old military types stuck together. The mission was to find if he knew what he was talking about."

Most of the hospital staff lived near the generating station. "TMI is over the hill, that way. People left and came back, more to get their families out of harm's way." By the next day, "a few hospitals lost so much staff, they were in danger of closing. There were clearly instances where people voted with their feet."

Asked to describe the medical staff's reaction when told the reactor would explode, Trautlein replied: "All but paralyzed, not really able to function at their normal capacity. I've seen it on faces many times. You have some bad news to give them. You see it all the time. It is too much to ask human beings to be volunteer kamikazes and function. You feel your stomach drop. It's a dead look. Color drains out of their faces. The eyes shrink. There's no expression. They can't sort it out. It was a moment we really needed leadership. The [hospital] director grabbed the situation by the ears and really jerked it around. You need one person who takes command in a situation like that."

The hospital was told to be prepared for possibly thousands of victims if the power plant exploded. "Five hundred people was the most we could handle, hosing them down in our docking area," he said. "But there is not much you can do. If you have gamma [radiation], you've got it. At 600 RADs, you're dead. At 300, half the people are dead. You can do as much at home with a bar of soap, other than document your own demise."

Disaster victims contaminated with radioactivity must be isolated to prevent hospitals from becoming contaminated, and doctors and nurses must take special precautions to avoid becoming contaminated while treating patients. Health officials say it would be difficult to set aside large, vacant areas in a hospital for such an eventuality.

The other voice I recall vividly is a mother's cry for the safety of her children. On a sweltering August evening five months after the accident, Mrs. Beverly Gorman, mother of two teenagers, was among 200 TMI residents gathered in the Londonderry Township fire hall near Middletown.

Metropolitan Edison intended to release Krypton 85 radioactive gas into the atmosphere as part of a plan to remove the shattered fuel core in the crippled reactor and resume operating the second reactor on the site, which was idle since the accident. Robert Arnold, a MetEd vice president, was facing the people "sitting on the powder keg" who lived in the shadow of the stricken plant.

"I want a straight answer," said Mr. Gorman, who lived one and a half miles from the power station. "How much radiation were we exposed to the first two days of the accident?"

"The estimate is 83 millirems," answered Arnold, who went on to explain the difference between radioactivity inside the power plant and natural background radiation. Then Mrs. Gorman said something that was on every parent's mind.

"I don't want any flowery speeches," she persisted. "Can you tell me my children have suffered no physical danger?"

"We would have to say your children did receive some radiation from the accident at Three Mile Island," Arnold answered. "Any radiation could result in some cell damage, whether it's from a school building, the ground where they live. The judgment has to be made if the cell damage is at levels that are acceptable. I don't know anybody who can decide."

Mrs. Gorman shot back: "You are proposing to expose them to more. I have to live here. We don't have the money to move away. I don't think they should be subjected to more." The meeting lasted from 8 p.m. until midnight. Voices were raised for and against Three Mile Island, as a menace and as an economic necessity.

After the meeting, a frustrated Mrs. Gorman said: "I didn't get a good answer. I want somebody up there to tell me my children did not suffer physically from the TMI accident, yes or no. That's all I want to know. All I hear is REMs and milli-rems. I just want a straight yes or no. . . . It's always a long, drawn-out answer. . . . If I can't be concerned about my children's health, then I can let them eat jelly beans for breakfast and stop worrying about what they eat and how long they sleep. It's all for nothing if they are sacrificed to nuclear energy."

The nuclear industry was bent on explaining what happened at Three Mile Island, yet the Londonderry township meeting illustrated the pitfalls of giving technical answers to emotion-laden questions. Suspicion grows and both sides often come away feeling betrayed or puzzled. The very poise with which utility officials answer angry questions can infuriate challengers, who view that calmness as indifference.

Another voice, not a voice really, a loud laugh comes to mind. I was invited to speak about my Three Mile Island experiences at Argonne National Laboratory. As I described the fear, and the roads filled with those fleeing Harrisburg that Saturday night of terror, one of the laboratory officials guffawed. Clearly he thought it was ridiculous that thousands people were frightened by what turned out, in part, to be a nuclear mirage. He was smarter than that.

That, I thought, is the problem with nuclear people. No empathy. No sympathy for frightened people who might have been fans of nuclear energy, until that chaotic night (and Walter Cronkite) changed their minds. Winning them back by acting like a sympathetic neighbor talking over the back yard fence in plain English would be a good start at demystifying nuclear energy. Treat them with respect. Have a heart.

Chapter 51: Atomic Roulette

THE MASS EXODUS OF 150,000 PENNSYLVANIANS DURING THE THREE MILE Island accident was a mistake, driven by panic and misinformation. They would have been safer staying home, rather than clogging the highways while driving in a fearful frenzy.

Forty-three years later, another far greater mass exodus was prompted by bombs and bullets—Russia's invasion of Ukraine in an attempted land grab with nuclear weapon overtones, putting the entire world's nerves on edge.

It was the work of one man: Vladimir Vladimirovich Putin, Russia's president and the Kremlin's supreme ruler. Like John Alexander Dowie 122 years earlier, few under Putin's leadership question his decisions. A master of the Russian brutal art of purges, Putin was a foreign intelligence officer for the KGB, the nation's former top security service known for skills in propaganda, psychological warfare, assassination, terrorism, and torture while combatting dissident, religious, and anti-Soviet activities.

On February 24, 2022, Russian forces invaded Ukraine in a major escalation of the Russo-Ukrainian war that began in 2014, when Russia seized Crimea. About an hour after the attack began, Putin announced that he had decided "to conduct a special military operation . . . to protect people who have been subjected to abuse and genocide by the Kyiv regime for eight years," repeating a false claim about the treatment of Russian separatists in Ukraine's Donbas region.

Calling the attack a "peacekeeping" mission, he told the Ukrainian military to "immediately lay down your weapons and go home." Speaking angrily at times, Putin said, "Russia cannot feel safe, develop, and exist with a constant threat emanating from the territory of modern Ukraine," and repeated past complaints about the failure of NATO and the United States to satisfy Russia's security demands.

What unfolded in the following weeks looked like the prelude to nuclear warfare, like a horror movie with chilling dialogue. But this was real.

Why am I writing about war? I never intended that, but found it necessary when nuclear power slid in that direction, and because of what Putin said next:

"Whoever tries to hinder us, and even more so, to create threats to our country, to our people, should know that Russia's response will be immediate. And it will lead you to such consequences that you have never experienced in your history."

It was a thinly veiled threat to resort to nuclear war, invoking atomic energy's evil sister. Russia is one of the world's nuclear powers, and Putin has his finger on the nuclear launch button. Explosions were heard in Kyiv, the Ukrainian capital, minutes after he finished speaking, like a ballistic flourish.

Putin's chilling words mark a thunderbolt moment in world history, opening a new nuclear era when the unthinkable idea of nuclear warfare became thinkable, changing the world outlook on nuclear safety, energy supplies, international relations, and food security while the world still grappled with the ravages of the Covid-19 pandemic, climate change, and economic uncertainty. The Ukraine war and Russian imperialism became one of the most consequential dilemmas of our time. Suddenly, nuclear weapons were more than just a peacekeeping deterrent.

Stepping up the threat days later, Putin put Russia's nuclear forces on "special alert," telling his defense chiefs it was because of "aggressive statements" by the West in the wake of widespread condemnation of his Ukraine invasion and economic sanctions against Moscow.

"As you can see, not only do Western countries take unfriendly measures against our country in the economic dimension—I mean the illegal sanctions that everyone knows about very well—but also the top officials of leading NATO countries allow themselves to make aggressive statements with regards to our country," Putin said on state television.

At that point, U.N. secretary-general António Guterres said nuclear war is "back within the realm of possibility." Russia previously said it could vaporize various locations in the U.S. with new missiles.

The Ukraine conflict is the biggest in Europe since World War II, and world leaders were talking about World War III. Nations punished Russia for its attack in various ways, restricting trade while seizing assets, and sent military aid to Ukraine, but not too much in ways that might infuriate Putin. It was like parents who tread lightly around a petulant child. And for good reason.

Russia is one of the world's largest oil producers—the second biggest crude oil exporter after Saudi Arabia. Europe depends heavily on Russia for oil and natural gas. In 2021, 40 percent of the gas Europeans burned came from Russia. More than 25 percent of the European Union's imported crude oil comes from Russia. The invasion caused the price of crude oil to top $130 a barrel for the first time in 13 years, driving the average price of gasoline in the United States to an all-time high of $5.02.

The dictator could weaponize winter by choking off fossil fuels to Europe in retaliation for Western sanctions.

Starvation could be another weapon. Ukraine is known as the breadbasket of Europe. Together with Russia, they supply 28 percent of globally traded wheat, 29 percent of the barley, 15 percent of the corn, and 74 percent of the sunflower oil. Ukraine's food exports alone feed 400 million people.

The war disrupted these supplies. A lot of Ukraine's wheat leaves the country via the Crimean Peninsula, controlled by Russia. Russia was blockading the port of Odessa on the Black Sea basin. The war and extreme weather could drive tens of millions into potentially deadly hunger.

In a psychological warfare gambit, Putin blamed the victims for causing the trouble, and his actions tipped this book and the world into a new reality and a heightened awareness of nuclear power's role in modern life, both the military and civilian forms that began at the University of Chicago in 1942.

Ukraine already had a tragic history with nuclear energy, when the Chernobyl nuclear power generating station exploded in 1986 and spewed radioactive wreckage across the countryside. The war threatened to repeat that catastrophe.

Ukraine has four operating nuclear power plants with 15 reactors, and two of those stations quickly became embroiled in the conflict. In early March, Russian forces seized the Zaporizhzhia nuclear station, the largest nuclear power plant in Europe, with six reactors. Missiles fell dangerously close to the station, and Russian military accused Ukraine of deliberately targeting the facility, while Ukraine denied the allegations and contended Russian forces were shelling the power plant to frame the Ukrainian military. Russian troops reportedly mined the Zaporizhzhia station, deciding it "will be either Russian or no one's."

Horrified, the Nuclear Energy Institute in the United States urged combatants to create safe zones around all Ukrainian nuclear plants. A second Ukrainian nuclear station, the South Ukraine plant with three reactors, reportedly came under fire when a Russian missile exploded within 350 yards of the plant. The United Nations called attacks on a nuclear power plant "suicidal."

Occupying Russian officials demanded that Ukrainian technicians at the Zaporizhzhia plant declare allegiance to Russia or suffer consequences. In March, dozens of plant employees sympathetic to Ukraine reportedly were tortured, held for ransom, or killed.

The nuclear stations became vulnerable pawns in the struggle.

World history is full of despotic, strongmen bullies. But Putin ranks as the first with a nuclear arsenal and a powerful grudge. With delusions of grandeur, Putin sees himself as a man with a mission to recapture the former glory and territory of the Soviet Union, which collapsed in 1991. The superpower split into 15 independent countries, including Ukraine, which Putin regards as an historic tragedy that needs to be remedied. Nuclear saber rattling became part of his frightening arsenal.

With imperialistic ambitions, Putin expected his assault on Ukraine to last a few days before the smaller neighboring country surrendered. It was the first in a long line of disappointments and unexpected setbacks.

Ukrainians fought back ferociously, an underdog slugging it out with a larger adversary, and paid a high price in casualties and ruination. Ukraine had a population of 43 million in 2021. More than 5.4 million fled the country when Russia invaded, most of them women and children. Men aged 18 to 60 were required to stay and fight. Most of the war refugees fled to neighboring countries, causing repercussions. Nobody knows exactly how many left Russia when the conflict started. Estimates range from 150,000 to 300,000.

For those with long memories, Putin's war in Ukraine seems like a continuation of World War II. Details are similar, and it could be said that he learned nothing from the terrifying losses the Soviet Union suffered in that earlier war.

With Napoleonic impulses, Putin grew increasingly frustrated and angry at a string of humiliating defeats against the Russian army, expected to be an unbeatable juggernaut far outnumbering the Ukrainians. And there were rumbles of recrimination in Moscow. This set the stage for an escalation in the battle and in Russia's bellicose nuclear threats.

In a September 21, 2022, televised address, seven months after the conflict started, Putin announced a "partial mobilization" that put the country's people and economy on a wartime footing, Russia's first mobilization since the Second World War. The call was for reservists with military experience, numbering about 300,000, from the ages of 18 to 60.

The goal of "some Western elites," said Putin, "is to weaken, divide and ultimately destroy our country." He accused the Ukrainian government of reprisals against its own citizens, with a "policy of intimidation, terror, and violence" that was taking on "increasingly mass-scale, horrific, and barbaric forms." People in the historic lands of Russia "do not want to live under the yoke of the neo-Nazi regime." He complained: "More weapons were pumped into Ukraine" to resist Russia.

Washington, London, and Brussels (NATO headquarters) "have even resorted to nuclear blackmail," Putin insisted, by shelling the Zaporizhzhia Nuclear Power Plant, "which poses a threat of nuclear disaster." Leaders of NATO countries, charged Putin, have made statements "on the possibility and admissibility of using weapons of mass destruction—nuclear weapons—against Russia."

Then Putin dropped his verbal bomb, heard around the world: "I would like to remind those who make such statements regarding Russia that our country has different types of weapons as well, and some of them are more modern than the weapons NATO countries have. In the event of a threat to the territorial integrity of our country and to defend Russia and our people, we will certainly make use of all weapon systems available to us. This is not a bluff."

Toward the end of his speech, Putin plays the nuclear card again, adding: "Those who are using nuclear blackmail against us should know that the wind rose can turn around," referring to diagrams showing wind direction.

Putin's brutish partial mobilization speech was deeply unpopular in Russia, sparking rare protests across the country and leading to almost 1,200 arrests. Anti-war sentiment hardened as the bodies of soldiers deployed weeks earlier began returning home for burial. U.S. officials estimated in December that around 100,000 military members had been killed or wounded on each side.

The mobilization speech also touched off a stampede of roughly 300,000 Russian men and their families who wanted no part of combat duty, causing turmoil at the borders and stirring fears in neighboring countries about potential instability.

Increasingly, Putin was described as cornered, no longer consulting with Kremlin ministers about his decisions and refusing to inform them about his plans. There is no respect for the leader, they said, only fear.

The war reached a hair-raising stage in October when President Biden warned of nuclear "Armageddon."

"We have not faced the prospect of Armageddon since Kennedy and the Cuban Missile Crisis" 60 years ago, Biden told a fundraiser in New York City in his bluntest comments about the use of nuclear weapons since the Ukraine invasion began. "[Putin is] not joking when he talks about the use of tactical nuclear weapons or biological or chemical weapons" because "his military is—you might say—significantly underperforming."

Facing domestic discontent and military setbacks, Putin told reporters he did not regret starting the conflict and "did not set out to destroy Ukraine," adding, "What is happening today is unpleasant, to put it mildly. But we would have had all this a little later, only under worse conditions for us, that's all. So my actions are correct and timely."

Commentators on the sidelines argued Putin would not dare drop the atomic bomb and risk becoming an international pariah, while others said he would. Putin was quoted as saying the United States set a precedent in using atomic weapons by bombing Hiroshima and Nagasaki.

After months of nuclear saber rattling, Putin in late October 2022 told a Moscow discussion forum that Russia had no intention of using nuclear weapons, saying, "There is no point in that, neither political, nor military." He accused the U.S. and allies of playing a "dangerous, bloody, and dirty" game in hopes of dominating the globe.

In November, Russian forces withdrew from the Ukrainian city of Kherson, and Ukrainian president Volodymyr Zelenskyy told a jubilant crowd gathered there that the city's liberation marked "the beginning of the end of the war," pledging to drive Russia entirely out of his country. He said investigators had documented more than 400 war crimes committed by Russian troops around the city.

The comments by Putin and Zelenskyy seemed to ease tensions over nuclear warfare in Ukraine, but relief was short-lived, a reminder that Putin is not the world's only despot with a nuclear arsenal. Using the test launch of an international ballistic missile that landed near Japan as a backdrop in November, North Korea's leader, Kim Jong-un, vowed to resort to nuclear weapons if his country is threatened, saying North Korea "will resolutely react to nukes with nuclear weapons and to total confrontation with all-out confrontation." Once unthinkable, nuclear threats now come with ballistic missile demonstrations.

As Biden indicated, these standoffs echo the Cuban Missile Crisis, which is seen as the most dangerous confrontation with Russia during the Cold War. In October 1962, President John F. Kennedy said in a televised address that "unmistakable evidence" revealed that Russia had installed nuclear strike capability in Cuba. Kennedy announced a naval "quarantine" of the island, saying any attack would be considered a provocation by Russia "requiring a full retaliatory response upon the Soviet Union."

Lasting 35 days, the missile scare standoff continued until tense negotiations resulted in an agreement between Kennedy and Soviet First Secretary Nikita Khrushchev. The Soviets agreed to dismantle their offensive weapons in Cuba and return them to the Soviet Union. Observers today wonder if Putin has the same grasp on reality that Khrushchev had in moving back from the brink of the nuclear warfare abyss.

Meanwhile, the Doomsday Clock ticks. Created in 1947, the clock represents how close we are to destroying the world with dangerous technologies. It is set by scientists at the University of Chicago. The furthest the clock has been set was 17 minutes to midnight in 1991, after the collapse of the Soviet Union and the signing of the Strategic Arms Reduction Treaty.

The clock moved to 100 seconds to midnight in 2020 and stayed there through 2022. It moved to 90 seconds to midnight in 2023 to show the situation is becoming more urgent.

"For many years, we and others have warned that the most likely way nuclear weapons might be used is through an unwanted or unintended escalation from a conventional conflict," said the Bulletin of the Atomic Scientists. "Russia's invasion of Ukraine has brought this nightmare scenario to life, with Russian President Vladimir Putin threatening to elevate nuclear alert levels and even first use of nuclear weapons if NATO steps in to help Ukraine."

For a time, it looked like the Ukraine conflict was the prelude of nuclear war. As of this writing, comments by Putin seem to have quelled those flames. But the growing threat caused the world to tremble.

On the atom's peaceful side, the other atomic sister, something astonishing was happening too, in the Commonwealth Edison trailblazing tradition of doing things never seen before.

Exelon Corp., Edison's successor and now parent company, transferred ownership of all 21 nuclear reactors at 12 locations to a spinoff company, Constellation Energy Corp. Not only that, after buying Constellation for $7.9 billion, Exelon also paid Constellation $1.75 billion in the deal to get rid of those reactors.

That's almost $10 billion.

This was the work of Exelon CEO John Rowe. Remember him? Back in 1998, when Rowe was chairman and chief executive officer of Unicom, of which Commonwealth Edison was a subsidiary, he said Edison's monumental achievement of building the nation's largest fleet of nuclear power plants was a mistake. Unicom was selling Edison's coal-burning power plants, and Rowe wished he could sell the nukes too. "I do not know how to sell the nuclear plants in an effective way at this time," he lamented. In 2000, Unicom became Exelon in a $16 billion merger. Exelon took possession of all of Edison's nuclear power plants, putting Edison out of the power-producing business and making Exelon a power producer at a time when Edison was getting poor safety reviews from the Nuclear Regulatory Commission.

It took Rowe 14 years and four attempts, but he found a way to shed himself of those troublesome nuclear plants, like a corporate Captain Ahab obsessed by his search for a great white whale.

Under Rowe's leadership, in 2011 Exelon agreed to buy Constellation Energy Group Inc. for $7.9 billion. Headquartered in Baltimore, Maryland, Constellation was created in 1999 by Baltimore Gas and Electric as a holding company. The longest-serving utility CEO in the country, Rowe had failed three times since 2003 to buy power companies in what appeared to be expansion bids. Exelon and Constellation officially merged in 2012, the year Rowe retired. The Constellation deal was his last. With the merger, Constellation Energy Group, which operated more than 35 power plants in 11 states, was rebranded Constellation Energy Corp., an Exelon company.

Now jump to 2021, when Christopher Crane was Exelon's CEO. On February 23, Exelon announced intentions to spin off Constellation as an independent, publicly traded power generating company, taking all of Exelon's nuclear power reactors with it.

The tax-free spinoff was completed on February 2, 2022, completely separating Constellation from Exelon, which now was totally free of those nuclear power reactors. It was a neat bit of corporate prestidigitation, and nobody saw it coming. They punted the nuclear power plants to the spinoff company, not a buyer but a taker. Rowe had persistently paved the way for that.

And it was the second time in 22 years that ownership of an entire fleet of power generating reactors shifted—from Edison to Exelon to Constellation. Those reactors had been ditched twice, like so many radioactive hot potatoes.

The U.S. nuclear power generating industry had never seen anything like it. Ownership of nuclear reactors changes, but not on this scale.

"Today is an important milestone in Exelon's history," said CEO Crane in a statement about the Constellation spinoff. "With the successful completion of our separation, we step forward in a strong position to serve customer needs, drive growth and social equity in the communities we serve, and deliver sustainable value as our industry continues to evolve. As we look to the future, we will advance our core business strategies to meet unique customer and community priorities."

Exelon calls itself the nation's largest utility company, serving more than 10 million customers through six power transmission and distribution utilities— Atlantic City Electric, Baltimore Gas and Electric, Commonwealth Edison, Delmarva Power and Light, PECO Energy Company, and Potomac Electric Power Company. The Exelon board of directors decided the separation was "in the best interest of Exelon and its shareholders for a number of reasons."

From a strictly economic viewpoint, Wall Street analysts praised the split, making Exelon a power transmission and distribution (T&D) utility and Constellation a standalone power generating utility.

"Exelon Corporation took a bold step to unlock value for shareholders with the spinoff of Constellation," observed one analyst. "This move created a generator and marketer in Constellation and a pure transmission and distribution utility in the remaining Exelon. The idea is that some investors will appreciate that stability and predictability of a fully regulated T&D utility, while others would appreciate the upside of a generation and marketing company with more exposure to power pricing." In other words, the separate companies gave investors a clearer idea of what they were investing in. And it allowed each company to pursue their own distinct operating priorities and strategies.

The industry tracker also pointed out that the Inflation Reduction Act passed by Congress "provides considerable subsidies for nuclear power generation."

To me, it looked like Exelon also was escaping the potential liabilities of owning nuclear power reactors, like severe accidents, maintenance on aging reactors, and decommissioning. So I put that question to Exelon's media relations manager, Nick Alexopulos. He said he'd get back to me.

Alexopulos did get back to me, not in spoken words but in links to text and documents filed with the U.S. Securities and Exchange Commission. It was a ton of information and helpful. But it seemed as though "nuclear reactors" became words that must not be spoken at Exelon, like Lord Voldemort of Harry Potter fame.

The benefits and risks of the Constellation spinoff were spelled out in a December 15, 2021, information statement filed by Exelon to shareholders. Among the risk factors listed, for Constellation, were reactor decommissioning obligations, U.S. environmental and climate policy, renewing permits and operating licenses, emerging low-carbon technologies competing with nuclear, the growth of cost-effective residential self-supply of electricity, the cost of upgrading deteriorating nuclear plants or shutting them down, and consequences of a major nuclear accident causing loss of life and property damage. If the split did not produce the economic benefits expected, said the report, the trading price of Exelon's common stock could suffer.

One of those documents filed with the SEC hints that no matter how hard Exelon tried, it could not entirely dodge nuclear liabilities. The document said even after the split, Exelon and Constellation "will indemnify each other for certain liabilities" that might arise in the future. It looks like a mutual protection agreement if some obligations prove to be enormous.

Constellation calls itself America's largest producer of clean energy, generating 10 percent of the nation's carbon-free energy. Its fleet of nuclear, hydro, wind, and solar facilities powers more than 20 million homes and businesses, with 90 percent of it carbon-free.

Brett Nauman is Constellation's senior manager for generation communications. Curious about Constellation's view of the split, I asked Nauman if Exelon was trying to shield itself from the risks of owning atomic power generating reactors.

"I kind of reject the premise of the question itself," answered Nauman. "The health of this nation, the success of this nation, is linked in our mind to the success of eliminating carbon pollution. That is our focus, helping our customers and communities reach that goal as well."

Calling itself a "customer-facing company," Constellation is a Fortune 200 company with about 13,000 employees.

Constellation technically owns and operates 23 nuclear reactors at 13 locations, although Nauman cautioned that the company usually counts itself owner of 21 reactors at 12 locations because it has only a minority ownership stake in the

Salem station in New Jersey. The others are six stations in Illinois—Braidwood, Byron, Clinton, Dresden, LaSalle County, and Quad Cities; the Calvert Cliffs station in Maryland; three stations in New York—FitzPatrick, Ginna, and Nine Mile Point; and two in Pennsylvania—the Limerick and Delta stations.

Constellation will get the two operating licenses for the Zion station when decommissioning by ZionSolutions is finished and the Nuclear Regulatory Commission authorizes the site suitably clean of radioactive contamination for public use.

Land once occupied by two nuclear reactors is bare now, enticing because it inspires visions of what could be there, so close to the busy waves chuckling up to the sandy and rocky Lake Michigan shoreline.

"It's got grass planted there and natural vegetation growing where the reactors used to be," said Nauman. "You have the switchyard and where the spent fuel is stored, the canisters are 19 feet tall." This is the 7.75-acre waste isolation island containing the most radioactive parts of what remains of the Zion power plant, waiting for the federal government to take it away.

Nauman encouraged me to see the former reactor plant site, so I drove there in September 2022, thinking I might see something inviting. Instead, I found more signs warning that trespassers would be prosecuted, jailed for a year, or fined $5,000. The former plant entrance was blocked by a row of upright concrete slabs and orange traffic cones. Seventeen signs were posted at the barricade.

The double chain-link fences that once encircled the power plant buildings were gone.

But this was new: A green sign saying "Final status survey. Clean area. This area is under the control of health physics. No decommissioning activities are to be performed in this area without prior approval. The use, movement, or storage of radioactive material is prohibited without the authorization of health physics." This was evidence of radioactivity cleanup going on there.

A side road led to the nuclear waste island with 61 of the 19-foot-tall cement storage casks containing the most dangerous wastes left by the power station. I drove partway down the road until confronted with more signs saying "Restricted area. No trespassing. Violators will be prosecuted."

Later, I told Nauman about my failed attempt to get a closer look at the former plant site and the waste disposal island, which is guarded and surrounded by its own fence.

Nauman's focus was on the future of that property.

"We would consider things with the land," he explained. "We haven't made any decision yet. Until we get the site back, there is not much we can do with development, with the land or anything else. That's the first thing that has to happen. The site has to be returned to Constellation. We're working with community leaders, once we have the land back and consider options for the use of that land."

As of January 2023, Constellation was still waiting for ZionSolutions and its parent company, EnergySolutions, to return the two Zion reactor operator licenses, which gave ZionSolutions complete authority to conduct the Zion station demolition project. When Constellation gets the two licenses, it will have authority over that ground.

Viktoria Mitlyng, NRC senior public affairs officer, said final radioactivity surveys of the Zion site were still being wrapped up.

Along with the responsibility of managing the spinoff nuclear reactors, Constellation has the responsibility of demolishing them safely when that time comes. The company has no reactor decommissioning experience.

"Zion is the first plant that was decommissioned from our company, and that was handled by EnergySolutions," the Constellation spokesperson explained. "That is the extent of our experience, in handing off decommissioning to another company."

When Constellation inherited those 23 nuclear reactors from Exelon, it also inherited $15.4 billion in decommissioning trust funds, according to documents filed in 2022 with the SEC. By September 2022, said Nauman, that balance dropped to $13.5 billion. "Most importantly," he added, "our decommissioning obligations are well funded, and we meet all of our NRC decommissioning minimum funding requirements."

"Constellation would be responsible for decommissioning all the nuclear plants we currently operate," confirmed Nauman. "The responsibility lies with the plant operator, which has the [operating] license. It would be like that throughout the industry. There [are] decommissioning funds for each plant, and it's a matter of public record how those are funded." Funding is provided through a trust fund projected to grow throughout the plant's operating lifetime.

To swing the spinoff deal, Exelon raised $2 billion through four bank loans, then gave a $1.75 billion cash payment to Constellation "for general corporate purposes." It looked like a $1.75 billion payoff to take the nukes off Exelon's hands, so I asked Nauman to explain it. "It is common for companies to make adjustments to their respective balance sheets in the context of spin transactions," he said. And it was a priority to be sure both companies maintain an investment grade credit rating. Constellation used some of the $1.75 billion to settle a $258 million intercompany loan from Exelon and $200 million in short-term debt, plus provide a $192 million contribution to pension plans. Nauman's explanation offered some insight into the complexities of spinoffs. The separation agreement also allowed two years to sort out the division of administrative and operational services like information technology, accounting, finance, and human resources. I appreciated Nauman's willingness to answer questions. Nuclear utilities have a special obligation to explain how they operate.

Another asset that came to Constellation from Exelon was Tony Orawiec, senior decommissioning manager. Everything Orawiec said about hopes for productive use of the former Zion power plant site remain in force, verified Nauman. It shows that when a power plant goes to a different owner, the entire staff of experienced employees and managers goes with it.

After a deep-dive examination of Constellation's qualifications, the NRC on November 16, 2021, approved transferring ownership of all of Exelon's nuclear reactors to Constellation, as well as their spent fuel storage installations.

"The purpose of this evaluation is to ensure that the proposed corporate management is involved with, informed of, and dedicated to the safe operation, maintenance, and decommissioning of the facility, and that adequate technical and financial resources will be provided to support these activities," says the 29-page NRC review of Exelon's February 25, 2021, application to transfer the nuclear facilities to Constellation.

Noting that Exelon was "the largest owner and operator of nuclear power plants in the United States," the NRC went on to review Constellation's financial qualifications, decommissioning funding, technical qualifications, management and support organization, operating organization, ties to foreign ownership, and nuclear insurance and indemnity, as well as perform an antitrust review.

NRC concludes "there is reasonable assurance that the health and safety of the public will not be endangered" by the transfer, and that Constellation would abide by NRC rules and regulations.

The order named 28 nuclear facilities, not all of them operating now. Operating licenses for Zion reactor Units One and Two were excluded from NRC's transfer approval because decommissioning operations were not finished.

"The review focused on the technical expertise and financial ability of the new company to operate nuclear plants," explained NRC's Mitlyng. "The NRC did not find any challenges for Constellation in those areas."

I thought there might be something history-making in the rapid turnover in ownership of almost two dozen nuclear reactors, twice. Applications for operating license transfers are not unusual, but I thought the size of the Constellation transfers needed further attention from an historical perspective.

By 2022, the tempo of applications for license transfers had increased. NRC reported that it reviewed more than 115 applications between 1999 and 2021. Another nine were in the works. NRC's Mitlyng said she could not recall more than three plants switching ownership at one time.

Wondering if there was some kind of trend at work here, I contacted the Nuclear Energy Institute (NEI) and the American Nuclear Society (ANS) to ask for their insights.

"We tend to avoid commenting on nuclear utility/operator business matters and usually stick to matters of importance to neutrons (and not so much on electrons or the utility business)," responded Andrew Smith, ANS director of communications. He suggested calling NEI, and I did.

"Unfortunately, we are not aware of a resource that has tracked the transfer of plant ownership over time," said NEI spokesperson Jon Wentzel. "We only have records of current owners . . . but not the history of ownership." NRC also said it does not track historical information of nuclear reactor ownership.

ANS's Smith gave me a list of nuclear experts and consultants. One of them, Dan Yurman, a nuclear news blogger, gave this assessment: "The current trend in the U.S. isn't transfers, it is closures." NRC in 2022 reported that 21 power reactors were undergoing decommissioning, and two power reactors recently ceased operations permanently.

Maybe I have an overactive imagination in seeing license transfers as noteworthy, or as a possible trend in the nuclear industry. Another one of those exploratory paths that led to a dead end.

NRC certified that Constellation is technically and financially fit to operate nuclear power plants. But is the utility prepared to operate in Illinois, governmentally one of the most corrupt states in the nation? Corruption has been a problem from the state's earliest history. Election fraud in Illinois pre-dates the territory's admission to the Union in 1818.

A 2022 study by the University of Illinois in Chicago ranked Illinois as the second-most corrupt state in the nation, with four of the last 11 governors serving time in prison. A 2020 investigation found 891 convictions for public corruption and profiteering since 2000 and determined that "Crooked Illinois" was the nation's most corrupt state.

Doing business with Illinois officials can be hazardous to a company's reputation and executives, who can be lured into webs of deception and pay-to-play practices. Constellation might be unaware or unprepared for the chicanery that is part of the political playbook in Illinois.

Commonwealth Edison and Exelon were tarnished by a bribery scandal in which former ComEd CEO Anne Pramaggiore and three others in 2022 were facing federal conspiracy charges for allegedly granting favors to Illinois lawmakers over the course of nearly a decade for passing multiple laws that were highly lucrative for ComEd and its parent Exelon.

Crain's Chicago Business described it this way: "Transcripts of recorded conversations between former ComEd CEO Anne Pramaggiore and her various lieutenants and lobbyists reveal how deeply committed ComEd was to the back-door process of getting what it needed out of lawmakers, not on the merits but by participating in a corrupt game of granting favors to lawmakers who could kill or pass legislation in the manner ComEd wanted." The scandal, said *Crain's*, "still has the power to shock." That's how it's done in Springfield, the Illinois state capital—a revelation made possible because they got caught that time.

Constellation will be stepping into a viper pit of political corruption.

"Constellation, being a standalone entity, must get help from the state or the feds," said David Kraft, director of the Nuclear Energy Information Service (NEIS) in Chicago, where the recorded message on the telephone answering machine says: "And remember, you can't nuke climate change."

Kraft said that Constellation, like its predecessor Exelon, is "looking forward and wants to make sure as much as possible that if something happens, it will be paid for by ratepayers or taxpayers. Constellation would come to the Illinois legislature. They would go to the unions, who get whipped up. If that did not work, they go to the nuclear hostage scenario. They lean on the legislature and get bailed out. That's the game. We dubbed it the nuclear hostage crisis. Pay us or we will shoot this reactor, something like that." Kraft was describing Exelon's threats in the past to close nuclear power stations if the utility failed to get financial relief.

The fleet of nuclear reactors built by ComEd is aging. Of particular concern, said Kraft, are pressurized water reactors, of the kind at the Byron and Braidwood stations, with eroding steam generator tubes and corrosion at the top of reactor vessels. "It has been a fairly widespread, not only national but international, issue. Those are the kinds of things that may come up in the future," requiring costly repairs or replacement, or otherwise taking a power reactor out of service if the cost of repairs is too high. A single steam generator can cost $200 million.

"These are old reactors," Kraft said. "They will have maintenance issues."

Eying the future, Kraft and his organization believe the NRC and the commercial nuclear power generating industry fail to think ahead, concerned only with present conditions. With climate change, Illinois by midcentury could have the kind of weather currently found in East Texas or Mississippi, Kraft contends.

All power stations draw cooling water from lakes or streams. The Byron nuclear station, for example, takes cooling water from the Rock River, a small tributary of the Mississippi River. Changing weather patterns might affect how much water a power plant uses and how much water is available. Recall that the Zion Nuclear Power Station was described as a big water processing plant, with pumps pushing 250,000 gallons of water a minute; "millions and millions of gallons an hour," said Dave Smith. A future generation might consider that wildly extravagant. Kraft believes NRC should require projections of future water use at nuclear power plants and that a power plant operator should identify alternative power sources that could be used if a power plant becomes inoperable.

Kraft is right to point to future water supplies as an issue that needs attention. Warnings about the limits of clean water resources have been sounded for decades and underscored by recent world droughts. People compete with industry and agriculture for water, and the competition is likely to get more intense. In the Great Lakes Basin alone, for example, the population of Americans and Canadians grew 26.5 percent in the 50 years between 1971 and 2021 to 35.3 million, according to the U.S. Environmental Protection Agency. That growth, said the agency, will be "increasing." A century of tapping groundwater in the Chicago area caused water table levels to drop by as much as 850 feet by 1979, causing some communities to turn to Lake Michigan and local rivers. Nearly 120 million Americans rely on underground aquifers for drinking water, irrigation, and manufacturing.

Kraft also believes power plant operators should draft socioeconomic impact statements that analyze the local impacts of early or unexpected power plant closures of the kind that hit the Zion nuclear station, which proved devastating to the local economy. Power generators typically extol the benefits of having a power plant in town. They don't forecast what happens to the local economy when that power plant closes.

"Their abrupt disappearance would wreak economic havoc on the affected governmental and essential service entities' ability to operate," claims NEIS, which proposes temporary "closure mitigation funds" to soften the economic blow of closure.

All of this is to say both of the atomic sisters are active in the world, the good and the evil. They were born together at the University of Chicago in 1942. The sisters should not be compared with Pandora of Greek mythology, who supposedly released all the evils of humanity from "Pandora's box." Some believe Pandora is a victim of mistranslation. An earlier version called her "she who sends up gifts." But the more popular version blames her for evils. And in that version, after evils fly out of that box, Hope emerges.

Hope is the belief that the future will be better.

Epilogue

My bottom line in nuclear safety is that power generating stations are only as good as the people running them, plus a little luck.

The Three Mile Island and Chernobyl reactor accidents proved that. Both were caused by human error, or got out of control because of human error.

One of the details of the Three Mile Island accident that sticks in my mind after all these years is that a paper "out of service" tag dangled in front of, and hid, a control panel light showing that an important emergency water block valve was closed when it should have been open. That caused an eight-minute delay in restoring emergency cooling water. A dangling paper tag. That was a bit of bad luck. One of the major findings into the causes of the Three Mile Island accident was that human error was a major factor.

Yes, nuclear safety is only as good, or safe, as the people operating the plant.

"That's for damn sure!" said A. David Rossin when I talked with him in 1998. "That's a necessary condition. But there can be outside forces, even if you have good people. You can have problems. Relationships with the community, the press. All figure into the balance. That goes all the way to management of the plant, corporate attitude."

In my mind, Rossin was the quintessential, super-smart nuclear advocate who dedicated his life to the industry and could serve as a proxy for thousands of his generation who answered the call to nuclear power. He had degrees from four universities, including a master's degree in nuclear engineering from the Massachusetts Institute of Technology. His research while at Argonne National Laboratory on embrittlement of steel in nuclear reactors produced reference works still read today. He was Commonwealth Edison's director of research and the Electric

Industry Man of the Year in 1982 for improving public understanding of nuclear power and the environment, and he became the U.S. Department of Energy's assistant secretary for nuclear energy. Those are just some of his highlights.

Rossin remembered what called to him.

"In the mid-1950s," he said, "there was a lot of talk about nuclear power. This was going to be important to the future. I was convinced that was right. It was going to be important to the future. . . . To me, here was something we really had a lot to learn, and if we learned it right, we could make a big difference. You would always read about someday harnessing the atom."

And that day came in 1942, at the University of Chicago with the Manhattan Project. Nuclear power took two branches, weapons and electric generation. Rossin and his brethren were interested in the electrical kind.

Sixty years ago, nuclear people saw themselves as supermen of science, keepers of the technological flame. They were an elite corps dedicated to a new kind of power, hardnose engineers who fell in love with a technology and defended it passionately. They staked their careers on it and spoke of "educating" the public to its importance. They could be haughty sometimes and talk down to the public. A nuclear advocate once told me I would not be able to understand what he was talking about because I did not have a degree in nuclear physics. The history of anti-nuclear activism shows the public can be a formidable force when aroused, and that force included denigrated housewives and mothers who became angry, studied nuclear technology, and became well-informed critics.

But not until 1971 did Alvin Weinberg, a University of Chicago graduate, physicist, and administrator of the Oak Ridge National Laboratory in Tennessee, sound a famous warning to give the public some idea of what it was getting into.

"We nuclear people have made a Faustian bargain with society," he said, alluding to Goethe's tale about a man who made a pact with the devil. The modern Faustian bargain, said Weinberg, is unlimited nuclear energy in return for "the eternal vigilance needed to ensure proper and safe operation of its nuclear energy system."

That includes radioactive reactor wastes that are dangerous and must be guarded for 10,000 years, longer than human civilizations have existed. One of the interesting questions raised by this obligation is how to warn future generations of radioactive waste burial grounds, since we don't know how humans will communicate in 10,000 years. What language will they use? Will we even exist? Or will the human race abandon a poisoned planet by then?

Americans must decide whether the benefits and risks of nuclear power are acceptable, said Weinberg, because "it is a choice that we nuclear people cannot dictate" in a democracy.

To put it in another way, how much risk is acceptable? People accept risks every day when driving, traveling by aircraft, or in the workplace. It seems astonishing now, but accidents took 124 lives in the nation's coal fields in 1984, when 206,541 miners were on the job. Coal mining continues because the risks are considered acceptable. In 40 large commercial airline crashes in 2020, five were fatal with 299 deaths worldwide, despite a sharp decline in flights because of the coronavirus pandemic.

Nuclear advocates say such death tolls in nuclear power probably would doom the industry. A major difference, of course, is that people voluntarily accept the risks of air travel, driving, or working. They can choose other sources of electrical power than nuclear, which can be seen as a risk imposed on them. A national referendum on continued use of nuclear energy would be interesting, although that could be seen as discriminatory.

A contemporary of Weinberg's was physicist George L. Weil. He was the man maneuvering the control rod in Chicago Pile-1 that produced the world's first sustained nuclear chain reaction, a participant in that historical event. Weinberg said Weil "became the first articulate opponent of nuclear energy."

The U.S. Atomic Energy Commission spent more than $3 billion of taxpayer dollars to help develop the commercial nuclear power industry, according to Weil, who participated in the first atomic bomb test and was chief of the AEC's reactor branch before becoming an independent consultant in the nuclear field.

Weil is remembered for his 1967 lecture titled "Nuclear Energy: Promises, Promises." Among his observations was the difficulty early nuclear power opponents had to be heard. Nuclear power had enthusiastic support from AEC, the Congressional Joint Committee on Atomic Energy, and the nuclear industry. "These promoters tend to dismiss their critics as 'kooks,' 'professional alarmists,' 'stirreruppers,' 'misguided zealots,' 'troublemakers for profits or kicks' and argue with remarkable self-assurance that the benefits of our current low-performance nuclear plants far outweigh both the risks and the huge publicly financed development costs," said Weil.

Weil's 1967 remarks reveal that a strategy of dismissing critics was adopted early, and that nuclear power plants were not performing well even then. The physicist went on to challenge the nuclear industry's three most popular arguments:

that nuclear energy is cheap and will more than repay development costs through lower electric rates to the public, that nuclear energy is urgently needed because of "dwindling" coal resources, and that nuclear plants are clean and do not pollute the environment. Weil called nuclear power plants "a dead-end street" that "offer too little in exchange for too many risks."

Technology skeptics, critics, doubters, or whistleblowers often are scorned or vilified, as though they breached some sort of moral code.

Weinberg came to Chicago, his hometown, in 1992 to accept an award from the American Nuclear Society/European Nuclear Society while observing the 50th anniversary of the first atomic chain reaction in Chicago. I asked Weinberg about his famous warning.

"People hold it against me," he said with a grin. He was short and trim and wore a Western string tie. "But Goethe's Faust was redeemed. The angels lifted Faust to the heavens." Then he quotes from Goethe, saying anyone who aspires and struggles on, "for him there is salvation. And I think that is what will happen to nuclear power."

I keep looking for it. The *Chicago Tribune* named me environment editor on February 5, 1970, a title that changed to environment reporter. They called me the nation's first full-time environment writer for a major metropolitan newspaper, and the beat included energy.

In 1985, my story about atomic power's midlife crisis at the age of 40 contained some observations that are apt even 35 years later. I wrote that the nuclear power industry was fighting for its life. That no permanent burial ground for highly radioactive reactor fuel existed. That safer and less expensive standardized reactor designs were needed to replace hugely expensive custom-made models operating in the country. Nuclear power plants should be built where they make the most sense for economics and safety. Illinois, with huge coal deposits, it could be argued, does not need nuclear energy, if coal could be used in a less environmentally harmful way, and mined safely.

The United States has been called "the Saudi Arabia of coal." It's a large energy resource if it can be used cleanly without environmental damage, or turned into safe gaseous or liquid fuels. Those technologies have huge air and water disadvantages now. I failed in 1985 to mention the tremendous toll nuclear power plants took on water resources—rivers, lakes, and oceans. Water is becoming an increasingly scarce natural resource. Designers of some of the nuclear reactors of the future recognize that and propose to use different sorts of coolants than water, which could become a limiting factor for future technology. At the same time, population growth brings growing demands for clean water, demands likely to clash with the industry's massive use of water.

Environmentally clean coal technology might require a large research and development effort, on the scale of developing new nuclear reactors. On the other hand, coal has a long and sorry history and maybe it is a technology that should be abandoned as others take its place for the sake of climate change and health and safety. We know now that coal takes a terrible toll on human life worldwide and is the most deadly form of energy production.

Nuclear reactors don't produce greenhouse gases of the kind causing climate change. Advocates say that is a strong argument for continuing research on reactors of the future, which include the small modular kind that appear to be ready to make their debut on the commercial landscape. Fusion power appears to be on the energy horizon as well, a distant but promising hope for the kind of limitless power that fission was supposed to be.

India's prime minister, Narendra Modi, has declared a "national mission" to develop "green hydrogen," a clean fuel made with renewable energy. The U.S. should take note. Hydrogen can be used for transportation, heating, cooking, and industry. It could help decarbonize industries that remain stubborn polluters across the world.

A national energy plan would be useful in spelling out an ideal energy mix for the sake of national security, the economy, the energy infrastructure, updated regulations, and the environment. And what needs to be done to produce that mix so America is not hostage to foreign or domestic energy producers? It's like having a national energy bank account. Because of utility deregulation, free market forces dominate energy trends, making the nation's energy picture an unpredictable free-for-all and unexplainable. Government should explain what is desirable in the national interest, in a way that is understandable to the public. What energy mix appeals to the public, so utility leaders can get behind it? The best we have now is a Department of Energy plan to "increase supply chain resilience" and energy independence. The U.S. EPA has a national action plan for energy efficiency.

U.S. energy planners in the past often expected growing demands for coal, oil, and natural gas, a more-of-the-same outlook. A new outlook needs to include energy conservation that reduces demand, while a growing tide of renewable energy like wind and solar produce reliable, affordable, and environmentally sound energy for America's future.

An energy glutton, the United States has one of the highest living standards in the world and consumes more energy per capita than any other country in the world. That should not be a matter of pride, but a reason to be a model of energy efficiency and conservation as climate change provides another reason for doing that. America should strive to be the world's smartest energy producer and consumer.

By the 1980s, nuclear people came to resemble Vietnam veterans who returned home feeling disillusioned and embittered. They were hostile and defensive, much like my experience with ZionSolutions and the Zion Nuclear Power Station. Their 1980s mindset was described this way:

"They have a sense of resentment toward anybody who they think may portray the industry in a negative way," observed James Keppler, a former Midwest regional administrator for the Nuclear Regulatory Commission. "They feel their job is made more difficult by the perceptions of the public and their regulators— and the media."

The nuclear industry should understand that part of their job is to engage with the public in friendly discourse. Especially important is the public's perception of nuclear power in a democracy. The public has a voice in what goes on in this country. The nuclear industry needs to explain itself better to gain the public's confidence.

"I think we could have done a better job from our side in making our case," admitted Rossin. "The technical people and even executives in the utility business felt they were serving a public need and that people understood that. . . . We didn't have automatic acceptance. We thought we earned it, but didn't get it. We probably misjudged how much we had to work to get our message across, and we never really succeeded. . . . We thought two and two make four and you didn't have to prove that. We thought we had the scientific facts and would go forward, but it didn't work that way."

For everyone involved, nuclear power often seems like a tangle of unanswered questions or disappointments, a puzzle that defies solutions to everyone's satisfaction.

Health effects of radiation exposures while working in a nuclear power station is one of those issues. People working in the power plants usually say they're not worried about it; they tend to shrug it off. It's no surprise that radiation is a health threat, considering the effort put into radiation protection and ALARA (as low as reasonably achievable), the nuclear industry standard.

But where are the statistics? Sixty years into the commercial nuclear power age, some scientific clarification should be expected. I asked the Nuclear Energy Institute if someone there could direct me to a reliable study on cancer incidence among nuclear power plant workers, and I was ignored. Some in the nuclear industry believe that bringing up cancer is anti-nuclear. But the American Nuclear Society did better than that by referring me to the Million Person Study published in 2017.

"The single most important question in radiation epidemiology is determining the level of health risks associated with radiation exposures that occur gradually over time," said the study, which included 150,000 nuclear power plant workers from 1957 to 2011. Nuclear power plant workers comprised about 15 percent of the study, which found 400 deaths from leukemia among 31,200 fatalities. Radiation doses were variable.

Finally, some real figures after years of speculation, some dire and some blindly optimistic. For years, some argued that any level of radiation exposure was potentially hazardous, while others argued that low doses of radiation were harmless.

The U.S. Department of Energy in the early 1940s began studying health effects at DOE and Atomic Energy Commission facilities. The goal was to ensure that workers were adequately protected from radiological hazards at those operations.

In the early 2000s, DOE's Office of Science funded a pilot study into the feasibility of a Million Worker Study, then provided full funding to continue the work started. The study covers DOE workers, atomic bomb test veterans, industrial radiographers, nuclear power plant workers, medical radiation workers, and uranium workers.

"The study of one million early U.S. radiation workers and veterans has been designed to provide information on risk following chronic exposures by focusing on occupational groups with differing radiation exposure patterns, including intakes of radionuclides," it said. "The cost-efficient study builds on the investments made and foundations laid by investigators and government agencies over the past 30–40 years. . . ."

The National Council on Radiation Protection and Measurements in Bethesda, Maryland, was the study's lead agency under the presidency of John D. Boice Jr.

The British Medical Journal in 2005 reported that another study of 400,000 nuclear power plant workers showed they had a slightly higher risk of developing cancer.

"The researchers show that people who are consistently exposed to low doses of radiation—such as those who work in nuclear plants—have about a 10 percent higher risk of death due to all cancers except leukemia. The risk of death due to leukemia was found to be 19 percent higher."

Leukemia is cancer of the body's blood-forming tissues, including bone marrow and the lymphatic system.

Based on those estimates, researchers say about 1 to 2 percent of all deaths among nuclear industry workers may be linked to radiation exposure. It's an occupational hazard.

Some say these reports show a nuclear worker's risk of cancer is lower than previously thought, that the long-term risk of cancer is small but significant. Leukemia in particular stands out as a significant risk.

These findings are important because they help identify the risks more clearly and offer thoughts about safer exposure limits for current and future workers. Now informed by recent health studies, the nuclear power industry should acknowledge the hazards of low-dose radiation exposure over long periods of time, especially leukemia, and discuss these risks openly in nuclear safety training for the sake of worker safety. The industry should recognize potential dangers to the public too, including residents living near nuclear generating facilities.

Let's look at the Three Mile Island accident one more time, in the context of radiation exposure. The average dose from natural background radiation to the U.S. population ranges from 100 millirems to 200 millirems. We live in a radioactive world, with radiation coming from the ground and from the skies. Add to that another 100 millirems each year from medical diagnostics, like X-rays.

An estimated two million people were exposed to small amounts of radiation during the Three Mile Island accident. The report of the president's commission on the accident calculated that the average dose to a person living within 5 miles of the nuclear plant was about 10 percent of annual background radiation or less. It also calculated that the maximum dose to any individual in the general population during the accident was 70 millirems.

In 1979, Joseph A. Califano Jr., the secretary of health, education, and welfare, estimated that at least one to 10 cancer deaths caused by radiation could be expected among the two million people living within 50 miles of the TMI power plant.

I returned to Three Mile Island in 1989, 10 years after the accident, to ask state health officials about public health effects.

The Pennsylvania Health Department was tracking the health of 37,000 people who live, or did live, within five miles of the plant at the time of the accident. "We have found nothing," said Dr. George Tokuhata, the department's director of epidemiology. "We have seen nothing unusual in terms of cancer or the 10 major causes of death."

One of the first health consequences expected would be an increase in leukemia, especially among children. These would have been expected within about five years. But leukemia rates in this area appeared normal. The department also found no evidence of birth defects. It noted low birth weights in infants born in the months after the accident, and that many of the mothers used tranquilizers during the crisis.

In fact, the only proven health hazard from the accident was that it caused a temporary increase in the consumption of alcohol, tobacco, sleeping pills, and tranquilizers.

Commercial nuclear power clearly carries risks. The famous mushroom-shaped clouds of the first atomic bombs demonstrated that, but the Atoms for Peace program tried to tame that image and brush it aside, making nuclear power appear harmless for generating electricity.

That might have gone too far. In 1945, a Pocket Book, *The Atomic Age*, depicted a future where fossil fuel would go unused, with atomic power in everyday objects. One science writer, David Dietz, wrote that instead of filling your car's gas tank two or three times a week with gasoline, you'd travel a year on a pellet of atomic energy the size of a vitamin pill.

Glenn Seaborg, chairman of the Atomic Energy Commission, wrote, "there will be nuclear powered earth-to-moon shuttles, nuclear powered artificial hearts, plutonium heated swimming pools for SCUBA divers, and much more."

Nobody mentioned that accidents involving nuclear-powered cars would leave radioactive wreckage strewn across the highways, spilled radioactive fuel, and contaminated drivers and passengers. The wreckage would have to be handled as radioactive waste, and casualties taken to hospitals prepared to treat them without contaminating the facilities and medical staff.

Nuclear energy was seen almost as a plaything, which played well to Americans who believed technology was the answer to everything.

A clearer vision to the benefits and hazards of commercial nuclear energy was a long time coming. As with all technology, it's a matter of choice and learning from our mistakes.

Nuclear scientists spoke of "defense in depth," meaning redundant safety features would prevent serious accidents. Now we know better. We invest great confidence in our technology, but Fukushima, Chernobyl, and the space shuttle *Challenger* showed it can fail spectacularly. Today, defense in depth should mean planning with the expectation that accidents *will* happen.

We make mistakes and the human consequences are extremely high. We should treat the victims with mercy. Despite our best efforts, some things we believe are probably wrong. Murphy's Law is immutable.

The United States should kick-start efforts to find a final disposal site for highly radioactive used reactor fuel piling up at power plant sites across the nation. The deadline Congress imposed on the Department of Energy for doing that is long past. Or lift the ban on recycling that radioactive waste into something useful and safe, as originally intended. And do it safely or not at all. The U.S. history of handling radioactive material involves spills, leaks, and airborne hazards that harm nearby residents and pollute the environment.

More recent developments have brought new concerns. Critics say the U.S. is in the age of nuclear reactor decommissioning, the biggest environmental cleanup in the nation's history. The nuclear power industry and the Nuclear Regulatory Commission should work together to regulate that enterprise carefully.

All nuclear reactors eventually reach the end of their lifespans. Utilities that own and operate nuclear stations are responsible for demolishing them, a task that demands technical skills and disposing of radioactive waste safely and legally.

Electric power utilities are not demolition companies. This led to a demand for reactor decommissioning services. Companies that seek to demolish nuclear reactors should be carefully scrutinized to ensure they are qualified and have the necessary skills and trained work forces. Multibillion-dollar decommissioning trust funds invite waste and corruption. Decommissioning should include deep expenditure audits to guard against nuclear adventurers and criminal activity.

The demolition phase of nuclear power should get the same amount of attention that was given to the building phase, perhaps more so. That should include socioeconomic impact statements, like environmental impact statements, that look at likely future economic impacts on local communities when a nuclear power station goes out of business. Such closures should not take a community by surprise, throwing them into economic chaos and despair through loss of property taxes. Funds set aside to cushion the loss of revenue from a power plant closure seem like a good idea.

When nuclear plants are demolished, communities should be aware that parts of the plant are likely to remain underground. This helps to decide what the property is suited for in the future. It's a more modern version of property left tainted by the steel and chemical manufacturing industries of the past.

But this is not all a matter of gloom and doom. Much of technology rises and falls to the drumbeat of human need and financial success, blazing a bright trail, as did Zion, that colorful city in northern Illinois still seeking a bountiful future through hope and powerful ideas.

Sources

Cook, Philip L. *Zion City, Illinois: Twentieth-Century Utopia*. Syracuse, NY: Syracuse University Press, 1996.

Denton, Harold R., and Chuck Metz Jr. *Three Mile Island and Beyond: Memories of a Life in Nuclear Safety*. La Grange Park, IL: American Nuclear Society, 2021.

Gardiner, Gordon P. *Out of Zion, Into All the World*. Companion Press, 1990.

Lindsay, Gordon. *John Alexander Dowie: A Life Story of Trials, Tragedies and Triumphs*. Dallas: Christ for the Nations, 1980.

Zion Historical Society. *Images of America: Zion*. Charleston, SC: Arcadia, 2007.

Acknowledgments

A BOOK IS A WINDOW TO A PERSON'S THOUGHTS AND EXPERIENCES, AND THE TIMES in which they happened.

By definition, a window is an opening for air and light. Books do that too. It's an act of faith on the part of many people to make them happen. A lot of people made this book happen.

It took a lot of courage and trust for Commonwealth Edison Co. to allow a *Chicago Tribune* reporter to prowl through the Zion Nuclear Power Station, including the radioactive parts, to see the human and technical forces that drive nuclear reactors. A special thanks to Ken Graesser, the station superintendent, who invited me in. From the Edison ranks came A. David Rossin, Peter LeBlond, David Smith, Ed Fuerst, Dick Principe, Corrine Simon, and James Toscas, who stand out in my mind for their boldness in explaining their beliefs and insights.

From the U.S. Nuclear Regulatory Commission, I will always be grateful to Jan Strasma, who knew how to explain highly technical things in a way that was understandable, and to his public affairs heir, Viktoria Mitlyng. To James Keppler for trying to make nuclear power safer.

And to the *Chicago Tribune*, the exploring instrument of public service that it was when it gave me the time to investigate nuclear safety because it was of public interest. Ellen Soeteber, F. Richard Ciccone, Ronald Yates, and John Crewdson were among the editors who personally played a role in that.

Roger Blomquist of Argonne National Laboratory gave his version of what future nuclear power reactors might look like. David Kraft gave a different version of the future of nuclear power.

The people of Zion, Illinois, were very gracious, including mayors Al Hill and Billy McKinney. Lorna Yates opened the Zion Historical Society archives to me. Rev. Kenneth Langley explained how the religious history that started in Zion lives on powerfully in other parts of the world.

Writing a book is like having an unruly and quarrelsome relative living in my house. My publisher, Doug Seibold of Agate Publishing, and Amanda Gibson, editorial manager, made the process more bearable. I'm grateful they agreed that unlikely elements like a famous, conniving faith-healer and nuclear power could add up to a book, and recognized that history does not spool out in a straight line, as some might prefer. It twists and turns, dwelling in odd but colorful corners that give birth to powerful ideas that influence our lives today. That's our story.

Index